PESTICIDE CONTAMINATION IN FRESHWATER AND SOIL ENVIRONS

Impacts, Threats, and Sustainable Remediation

PESTICIDE CONTAMINATION IN FRESHWATER AND SOIL ENVIRONS

Impacts, Threats, and Sustainable Remediation

Edited by

Mohammad Aneesul Mehmood, PhD

Khalid Rehman Hakeem, PhD

Rouf Ahmad Bhat, PhD

Gowhar Hamid Dar, PhD

First edition published 2022

Apple Academic Press Inc.
1265 Goldenrod Circle, NE,
Palm Bay, FL 32905 USA

4164 Lakeshore Road, Burlington,
ON, L7L 1A4 Canada

CRC Press
6000 Broken Sound Parkway NW,
Suite 300, Boca Raton, FL 33487-2742 USA

2 Park Square, Milton Park,
Abingdon, Oxon, OX14 4RN UK

Library and Archives Canada Cataloguing in Publication

Title: Pesticide contamination in freshwater and soil environs : impacts, threats, and sustainable remediation / edited by Mohammad Aneesul Mehmood, PhD, Khalid Rehman Hakeem, PhD, Rouf Ahmad Bhat, PhD, Gowhar Hamid Dar, PhD.

Names: Mehmood, Mohammad Aneesul, editor. | Hakeem, Khalid Rehman, editor. | Bhat, Rouf Ahmad, 1981- editor. | Dar, Gowhar Hamid, editor.

Description: First edition. | Includes bibliographical references and index.

Identifiers: Canadiana (print) 20200403710 | Canadiana (ebook) 20200403753 | ISBN 9781771889537 (hardcover) | ISBN 9781003104957 (ebook)

Subjects: LCSH: LCSH: Pesticides—Environmental aspects. | LCSH: Agricultural pollution. | LCSH: Water—Pollution. | LCSH: Soil pollution. | LCSH: Bioremediation.

Classification: LCC TD427.P35 P47 2021 | DDC 363.738/498—dc23

Library of Congress Cataloging-in-Publication Data

Names: Mehmood, Mohammad Aneesul, editor. | Hakeem, Khalid Rehman, editor. | Bhat, Rouf Ahmad, 1981- editor. | Dar, Gowhar Hamid, editor.

Title: Pesticide contamination in freshwater and soil environs : impacts, threats, and sustainable remediation / edited by Mohammad Aneesul Mehmood, PhD, Khalid Rehman Hakeem, PhD, Rouf Ahmad Bhat, PhD, Gowhar Hamid Dar, PhD.

Description: First edition. | Palm Bay, FL, USA : Apple Academic Press, 2021. | Includes bibliographical references and index. | Summary: "Taking into consideration that the agricultural industry is greatly dependent on pesticide chemicals to deal with the damage caused due to pests, this new volume details the challenges along with the bioremediation and remediation measures, such as the use of beneficial microorganisms, polymeric nanocomposites for nanoremediation, phytoremediation, and more. It looks at pesticide contamination from agricultural activities in a variety of different environs and a selection of sustainable and eco-friendly remediation approaches. It provides a spectrum of concepts, ideas, and knowledge related to the detrimental actions of pesticides on the environment directly and on human beings indirectly and provides insight into sustainable and advanced pesticide remediation technology. Pesticide Contamination in Freshwater and Soil Environs: Impacts, Threats, and Sustainable Remediation fills a gap in the available literature in this field and will provide valuable for academicians, researchers, agriculturists, and students"-- Provided by publisher.

Identifiers: LCCN 2020054060 (print) | LCCN 2020054061 (ebook) | ISBN 9781771889537 (hbk) | ISBN 9781003104957 (ebk)

Subjects: LCSH: Pesticides--Environmental aspects.

Classification: LCC TD196.P38 P38 2021 (print) | LCC TD196.P38 (ebook) | DDC 628.1/6842--dc23

LC record available at https://lccn.loc.gov/2020054060

LC ebook record available at https://lccn.loc.gov/2020054061

ISBN: 978-1-77188-953-7 (hbk)
ISBN: 978-1-77463-809-5 (pbk)
ISBN: 978-1-00310-495-7 (ebk)

Dedication

Dedicated to our beloved Parents

About the Editors

Mohammad Aneesul Mehmood, PhD, specializes in limnology and environmental toxicology. He has been teaching graduate and postgraduate students for the past two years in the Department of Environmental Science, School of Sciences, Sri Pratap College Campus, Cluster University Srinagar, Jammu and Kashmir, India. He has been supervising many students for their MSc projects. In addition, he has published a number of papers in international journals of repute and a number of books with international publishers. He completed his doctorate from the Division of Environmental Science, Sher-e-Kashmir University of Agricultural Sciences & Technology of Kashmir, with a meritorious certificate from the university. He received the Dr. Mumtaz Ahmad Khan Gold Medal and the Shri Bhushan Memorial Gold Medal during his master's program. He has qualified various national competitive tests in the discipline of environmental science. He was also awarded with an INSPIRE Merit Fellowship (JRF & SRF) by the Department of Science and Technology, GoI, during his doctoral program.

Khalid Rehman Hakeem, PhD, is Professor at King Abdulaziz University, Jeddah, Saudi Arabia. After completing his doctorate (Botany; specialization in Plant Eco-physiology and Molecular Biology) from Jamia Hamdard, New Delhi, India, he worked as a lecturer at the University of Kashmir, Srinagar, India. At Universiti Putra Malaysia, Selangor, Malaysia, he was a Postdoctorate Fellow and Fellow Researcher (Associate Professor) for several years. Dr. Hakeem has more than 10 years of teaching and research experience in plant eco-physiology, biotechnology and molecular biology, medicinal plant research, plant-microbe-soil interactions, as well as in environmental studies. He is the recipient of several fellowships at both national and international levels. He has served as a visiting scientist at Jinan University, Guangzhou, China. Currently, he is involved with a number of international research projects with different government organizations. To date, Dr. Hakeem has authored and edited more than 35 books with international publishers. He also has to his credit more than 80 research publications in peer-reviewed international journals and 55 book chapters in edited volumes with international publishers. At present, Dr. Hakeem serves as an editorial board member and reviewer for several high-impact international scientific journals. He is included in the advisory board of Cambridge Scholars Publishing, UK. He is also a fellow of the Plantae group of the American Society of Plant Biologists, member of the World Academy of Sciences, member of the International Society for Development and Sustainability, Japan, and member of the Asian Federation of Biotechnology, Korea.

 **Rouf Ahmad Bhat, PhD,** is Assistant Professor at Cluster University Srinagar (Jammu and Kashmir), India, and specializes in limnology, toxicology, phytochemistry, and phytoremediation. Dr. Bhat has been teaching graduate and postgraduate students of environmental sciences for the past two years. He is an author of more than 54 scientific articles and 15 book chapters, and has published more than10 books with international publishers. He has presented and participated in numerous state, national and international conferences, seminars, workshops, and symposiums. Dr. Bhat has worked as an associate Environmental Expert in the World Bank-funded Flood Recovery Project and also as Environmental Support Staff on Asian Development Bank-funded development projects. He has received many awards, appreciations, and recognitions for his services to the science of water testing and air and noise analysis. He has served as an editorial board member and reviewer of reputed international journals. Dr. Bhat is writing and experimenting with diverse capacities of plants for use in aquatic pollution.

Gowhar Hamid Dar, PhD, is an Assistant Professor in Environmental Science, Sri Pratap College, Cluster University Srinagar, Department of Higher Education (Jammu and Kashmir), India, where he has been teaching for many years. He has a PhD in Environmental Science with a specialization in Environmental Microbiology (fish microbiology, fish pathology, industrial microbiology, taxonomy and limnology). He has published more than 40 papers in international journals of repute and a number of books with international publishers. He is guiding a number of students for their master's theses. He has been working on the isolation, identification and characterization of microbes, their pathogenic behavior, and impact of pollution on development of diseases in fish fauna for the last several years. He has received many awards and appreciations for his services toward science and development. In addition, he also acts as a member of various research and academic committees.

Contents

Contributors

Tanveer Abbas
Institute of Agricultural Resources and Regional Planning, Chinese Academy of Agricultural Sciences, Beijing 100081, China

Bilawal Abbasi
Agricultural Clean Watershed Research Group, Institute of Environment and Sustainable Development in Agriculture, Chinese Academy of Agricultural Sciences, Beijing 100081, PR China

Charles Oluwaseun Adetunji
Applied Microbiology, Biotechnology and Nanotechnology Laboratory, Department of Microbiology, Edo University Iyamho, PMB 04, Auchi, Edo State, Nigeria

Osikemekha Anthony Anani
Laboratory of Ecotoxicology and Forensic Biology, Department of Biological Science, Faculty of Science, Edo State University Uzairue, Edo State, Nigeria

Mohammad Ashfaq
Multidisciplinary Research Institute for Science and Technology, IIM, University of La Serena, Benavente, La Serena, Chile
School of Life Sciences, BS Abdur Rahman Crescent Institute of Science and Technology, Chennai, India

Ifra Ashraf
College of Agricultural Engineering and Technology, Sher-e-Kashmir University of Agricultural Sciences and Technology of Kashmir Shalimar Campus, Srinagar 190025, Jammu and Kashmir, India

Rezwana Assad
Department of Botany, University of Kashmir, Srinagar 190006, Jammu and Kashmir, India

Muhammad Ashar Ayub
Institute of Soil and Environmental Sciences, University of Agriculture, Faisalabad 38040, Pakistan

Adriana Mera B
Multidisciplinary Research Institute for Science and Technology, IIM, University of La Serena, Benavente, La Serena, Chile

Iqra Bashir
Department of Botany, University of Kashmir, Srinagar 190006, Jammu and Kashmir, India

Divya Chauhan
Department of Chemical and Biomedical Engineering, University of South Florida, Tampa, Florida, USA

Amir Hussain Dar
Department of Food Technology, Islamic University of Science and Technology, Awantipora, Pulwama, India

Pratap Divekar
ICAR-Indian Institute of Vegetable Research, Regional Research Station, Sargatia, Kushinagar 274406, UP, India

Ajinath Dukare
Horticultural Crop Processing Division, ICAR-Central Institute of Post-Harvest Engineering and
Technology, Abohar, Punjab 152116, India

Chukwuebuka Egbuna
Africa Centre of Excellence in Public Health and Toxicological Research (ACE-PUTOR),
University of Port-Harcourt, Rivers State, Nigeria.
Department of Biochemistry, Faculty of Natural Sciences, Chukwuemeka Odumegwu Ojukwu University,
Anambra State 431124, Nigeria

Zia Ur Rahman Farooqi
Institute of Soil and Environmental Sciences, University of Agriculture, Faisalabad 38040, Pakistan

Hilal Ahmad Ganaie
Department of Zoology, Government Degree College Pulwama, J&K 192301, India

Mahendra K. Gupta
Microbiology Research Lab, School of Studies in Botany, Jiwaji University, Gwalior 474011,
Madhya Pradesh, India

Saima Hamid
Centre of Research for Development/P.G. Department of Environmental Sciences,
University of Kashmir 190006, Jammu and Kashmir, India

Allen H. Hu
Institute of Environmental Engineering and Management, National Taipei University of Technology,
Taiwan, ROC

Muhmmad Mahroz Hussain
Institute of Soil and Environmental Sciences, University of Agriculture, Faisalabad 38040, Pakistan

Azra N. Kamili
Centre of Research for Development/P.G. Department of Environmental Sciences,
University of Kashmir 190006, Jammu and Kashmir, India

Abdul Kareem
Institute of Soil and Environmental Sciences, University of Agriculture, Faisalabad 38040, Pakistan

Mohamed S. Khalil
Central Agricultural Pesticides Laboratory, Agricultural Research Center, Al-Sabhia, Alexandria, Egypt

Wael M. Khamis
Plant Protection Research Institute, Agricultural Research Center, Al-Sabhia, Alexandria, Egypt

Khuram Shehzad Khan
Department of Ecological Science and Engineering, College of Resource and Environmental Sciences,
China Agriculture University, China

Sunil Kumar
CSIR-National Environmental Engineering Research Institute, Nagpur, India

Priyank Mhatre
Division of Plant Protection, ICAR-Central Potato Research Station, Ooty, Tamil Nadu 643004, India

Mohammad Yaseen Mir
Centre of Research for Development/P.G. Department of Environmental Sciences,
University of Kashmir 190006, Jammu and Kashmir, India

Muhammad Naveed
Institute of Soil and Environmental Sciences, University of Agriculture, Faisalabad, Pakistan

Suraj Negi
Institute of Environmental Engineering and Management, National Taipei University of Technology, Taiwan, ROC

Charles Oluwaseunadetunji
Applied Microbiology, Biotechnology and Nanotechnology Laboratory, Department of Microbiology, Edo University Iyamho, PMB 04, Auchi, Edo State, Nigeria

Sadhna Pandey
Department of Botany, Government Kamla Raja Girls Post Graduation Autonomous College, Gwalior 474001, Madhya Pradesh, India

Anjali Pathak
Microbiology Research Lab, School of Studies in Botany, Jiwaji University, Gwalior 474011, Madhya Pradesh, India

Sangeeta Paul
Division of Microbiology, ICAR-Indian Agricultural Research Institute, New Delhi 110012, India

Mir Sajad Rabani
Microbiology Research Lab, School of Studies in Botany, Jiwaji University, Gwalior 474011, Madhya Pradesh, India

Iflah Rafiq
Department of Botany, University of Kashmir, Srinagar 190006, Jammu and Kashmir, India

Aishwarya Rani
Chaudhary Brahm Prakash Government Engineering College, Delhi, India

Irfan Rashid
Department of Botany, University of Kashmir, Srinagar 190006, Jammu and Kashmir, India

Nowsheeba Rashid
Amity Institute of Food Technology, Amity University Noida, Uttar Pradesh, India

Gulzar A. Rather
CSIR-Indian Institute of Integrative Medicine, Jammu 180001, India

Carlos A. Rodríguez
Multidisciplinary Research Institute for Science and Technology, IIM, University of La Serena, Benavente, La Serena, Chile

Zafar Ahmad Reshi
Department of Botany, University of Kashmir, Srinagar 190006, Jammu and Kashmir, India

Muhammad Fahad Sardar
Agricultural Clean Watershed Research Group, Institute of Environment and Sustainable Development in Agriculture, Chinese Academy of Agricultural Sciences, Beijing 100081, PR China

Muhammad Tahir Shehzad
Global Centre for Environmental Remediation (GCER), University of Newcastle, Callaghan 2308, Australia

Sulman Siddique
Institute of Soil and Environmental Sciences, University of Agriculture, Faisalabad, Pakistan

Rachna Singh
Microbiology Research Lab, School of Studies in Microbiology, Jiwaji University, Gwalior-474011, Madhya Pradesh, India

Irshad Ahmad Sofi
Department of Botany, University of Kashmir, Srinagar 190006, Jammu and Kashmir, India

Neetu Talreja
Multidisciplinary Research Institute for Science and Technology, IIM, University of La Serena, Benavente, La Serena, Chile

Shivani Tripathi
Microbiology Research Lab, School of Studies in Botany, Jiwaji University, Gwalior 474011, Madhya Pradesh, India

Ali Mohd Yatoo
Centre of Research for Development/P.G. Department of Environmental Sciences, University of Kashmir 190006, Jammu and Kashmir, India

Nukshab Zeeshan
Institute of Soil and Environmental Sciences, University of Agriculture, Faisalabad 8040, Pakistan

Abbreviations

ACF	activated carbon fiber
ALL	acute lymphoblastic leukemia
BCAs	biological control agent
BE	bioemulsifier
BHC	benzene hexachloride
BPH	brown plant hopper
BPs	biopesticides
BS	biosurfactant
Bt	*Bacillus thuringiensis*
CE	carboxylic ester
CI	confidence interval
CN	cyst nematodes
CNF	carbon nanofiber
CNT	carbon nanotube
CPF	chlorpyrifos
DDT	dichlorodiphenyltrichloroethane
EEC	estimate of the environment concentration
EPA	Environmental Protection Agency
ESCAP	Economic and Social Commission for Asia and the Pacific
ET	ethylene
FAO	Food and Agricultural Organization
GAP	good agricultural practices implementation
HCN	hydro cyanide
ICM	Integrated Cultural Management
IPM	integrated pest management
ISR	induced systemic resistance
JA	jasmonic acid
LGG	strain GG
LGR-1	strain GR-1
LPR	liquid-phase polymer-assisted retention
MB	methylene blue
MCL	Maximum Contaminant Level

MF	microfiltration
MIC	methylisocynate
MRLs	maximum residue limits
MV	methyl violet
NF	nanofiltration
NFS	nematode feeding site
NP	nanoparticle
OC	organochlorine
OM	organic matter
OP	organophosphate
OR	odds ratio
PAH	polycyclic aromatic hydrocarbon
PCB	polychlorinated biphenyls
PCDD/F	polychlorine dibenzodioxins and dibenzofuran
PD	Parkinson's disease
PEC	predicted ambient concentration
PECs	predicted environmental levels
PEI	poly-ethylene-imine
PGPR	plant growth promoting rhizobacteria
PPN	plant parasitic nematodes
Ps	pesticides
PUF	polyurethane foam
RKN	root-knot nematodes
RO	reverse osmosis
SA	salicylic acid
SAR	systemic acquired resistance
SDWA	Safe Drinking Water Act
SIT	sterile insect technique
SQIs	soil quality indicators
SVOCs	semivolatile organic compounds
UF	ultrafiltration
USGS	US Geological Survey
VAM	vascular arbuscular mycorrhiza
WHO	World Health Organization

Preface

Today, exposure to pesticides is one of the major concerns related to the safety of the environment worldwide. Huge numbers of people are engaged in agricultural practices, and most use pesticides to protect the food and commercial products. Most people make use of pesticides occupationally for public health programs and in commercial applications, while many others use pesticides in lawn and garden applications and for domestic protection. Although there have been attempts to decrease pesticide use through organic agricultural practices and the use of other alternative technologies to control pests, additional efforts have been made to find alternatives to chemical pesticides.

At present, continued exposure to pesticides from a number of different sources, including occupational exposure, home and garden use, spray drifts, residues in house hold dust, food, soil, and drinking water, remains a serious health problem in both developing and developed countries. Risk assessment plays a crucial role in the process of decision making about the use of pesticides, both new and existing. Accumulating experience suggests that post-market epidemiological surveillance of pesticide safety represents an essential method to ensure public health and the quality of our environment. Epidemiological studies have suggested that pesticides on the market currently may cause deleterious effects, for example, neoplasias and other diseases in nontarget species, including humans. Furthermore, many occupational and agricultural workers experience unintentional pesticide poisoning each year worldwide. In addition to causing environmental damage, wild nontarget species are frequently affected by pesticide exposure because they possess physiological or biochemical similarity to the target organisms.

This book, *Pesticide Contamination in Freshwater and Soil Environs: Impacts, Threats, and Sustainable Remediation*, is intended to provide an overview of toxicology that examines the hazardous effects of common chemical pesticide agents employed everyday in our agricultural practices. We aimed to compress information from diverse sources into a single volume. The chapters include a large variety of pesticide-related topics about the effects of several methods of control on undesired weeds and pests that grow and reproduce aggressively in crops, as well as their management and several empirical methodologies for study. The topics considered include details of the pesticides pollution on target and nontarget organisms; concerns of pesticide pollution in aquatic environs; effects of pesticide pollution on plants and human health; management and challenges of pesticide pollution; biopesticides; bioremediation of pesticide pollution; microbial aspects of pesticide remediation; biocontrol agents; biotechnological aspect of pesticide pollution remediation; biosurfactant for pesticide pollution remediation and polymeric nanocomposite for nanoremediation.

Many researchers have contributed to the publication of this book. The combination of experimental and theoretical pesticide investigations of current interest will make this book significance to researchers, scientists, engineers, and graduate students who make use of those different investigations to understand the toxic aspects of pesticides.

The chief objective of this book is hence to deliver state-of-the-art information for comprehending the toxicity of several pesticides in target and nontarget organisms. We hope that this book will continue to meet the expectations and needs of all interested in the different aspects of pesticide toxicity.

— Editors

CHAPTER 1

Environmental and Pesticide Pollution

HILAL AHMAD GANAIE

Department of Zoology, Government Degree College Pulwama, J&K 192301, India, Email: hilalganie@hotmail.com

ABSTRACT

In the past few years, due to the emergence of technology, human beings have been exposed to a few kinds of substances with wide range. Pesticides are one of these groups of chemical substance. In order to protect the livestock from the pest infections and crops, the pesticides have been a fundamental part of horticulture and agriculture in order to have much yield. Though the use of pesticides is beneficial, their uncontrolled use may pose some risks not only to our food safety but also to the environment and the living things. Their nonjudicious use has driven the attention of the researcher's toward the fate of these substances as these can be emigrated from areas where treated toward air, surrounding land and water bodies like rivers and streams. For the developing countries, the use of pesticides for agriculture is very important. Nevertheless, the risks and hazardous effects which are associated with the use of these pesticides have risen as a key issue for various studies in the developing countries. For the last five decades, the judicious use of these pesticides has increased the quantity as well as quality of the food. However, the worry about the hazardous effects of pesticides on nontarget organisms including humans has also grown due to their nonjudicious usage. The reason for the selection of this chapter is to depict the idea of pesticides and their history, characterization, dangers, and impacts on humans and their environment.

1.1 INTRODUCTION

In the last few decades, because of the environmental pollution, there has been expanding worldwide stress over the general wellbeing of humans.

The industrial revolution has given birth to environmental pollution. In the developing countries due to industrial processes, the people living there are particularly susceptible to toxic pollution. Pollution can be characterized as the presence of poisons into the earth as vitality, for example, warmth, light, or commotion that reason mischief or distress to other living forms or harm the earth. Though, pollutants are normally happening substances yet are viewed as contaminants when found in overabundance of the edge levels. Natural contaminations can be separated into two general classes: biodegradable and nonbiodegradable. Biodegradable pollutants can be degraded into pieces and processed by microorganisms. These include natural waste items, phosphates, and inorganic salts. On the other hand, nonbiodegradable pollutants cannot be degraded by living life forms and subsequently continues in the biological system for significant lots of time. Nonbiodegradable pollutants contain metal pieces, plastics, glass, pesticides, and radioactive isotopes (Santos, 1990).

The pesticides are used worldwide to kill the pests, but their consumption varies from country to country. It had been estimated in 2000 that the consumption of these pesticides was ~5.35 billion pounds (USEPA, 2001). In our country, India, the consumption of these pesticides has been increased approximately four times after Green Revolution era (1966–1999). Nevertheless, the consumption of these pesticides has reduced during the last decade (Kumar, 2011). In terms of pesticide consumption, India is at the lowest rank with 0.5 kg/ha while Taiwan is at the highest rank with 17 kg/ha; still the substantial quantities of pesticide residues are found in food and agricultural products (Assocham, 2007). The major causes are their haphazard and nonjudicious use of chemical pesticides as well as nonobservance of prescribed waiting periods.

The build-up of pesticides utilized ashore eventually discovers its way into oceanic condition, where it presents basic toxicological perils to a crowd of nontarget living things, and finds its way to the developed lifestyle, undermining the ecological balance and the biodiversity of the nature. The contaminators of oceanic biological system have been all around recorded worldwide and establish a noteworthy issue that offers ascent to worries at nearby, provincial, national, and worldwide scales (Dar et al., 2015). Delaplane (2000) opined that there is a need to minimize the damage caused by the pests and hence people make use of these pesticides. One can remember the epidemic, The Black Plague, in which millions of people died because of pests in Europe in the 14th century. That time the people believed that it was God's punishment and now a number of reports are available in literature, art, and public statues which certify the fear and destruction of these epidemics. Later it was discovered that the reason for the plague which destroyed the entire Europe was a bacterial infection that was spread by the rodent bugs. Henceforth, it was the need of hour to control rodents including rats just as bugs to diminish the general recurrence of the event of illnesses (Fishel, 2013).

Pesticides are defined as the compound substances utilized on agrarian land, private nurseries, along railroads, and in other open territories or gardens (Grube et al., 2011). As the world population is growing at an alarming rate, it has thus become essential to protect the crops by using pesticides so that the increased demand for food can be met out. Despite the fact that agricultural produce has increased manyfold due to pesticides, their accumulation in the food chain poses a risk to many animals including the mammals because the pesticides are also having negative effects (Lozowicka et al., 2014). The pesticide applied to the crops enters the surrounding environment through different fates. After applying the pesticide, some portion remains in the farmland while other portions will find their way in surrounding soil, air, and water (Malone et al., 2004). These pesticides have not only the ability to remain in the surrounding environment but also can travel to long distances like the organic compounds. The deposits of pesticides in soil and water are dangers to the earth and these have been named cancer-causing agents. In this manner, the nonreasonable utilization of these pesticides has presented genuine dangers to human wellbeing over the past 50 years (Ouyang et al., 2016).

Despite the fact that the pesticides are having advantages to the humankind, and yet man and other living beings are presented to these dangerous mixes in nature. It has been accounted for by a few epidemiological investigations about the evil impacts of these pesticides on the human wellbeing which incorporates the connection between the malignant growth and the utilization of pesticides, for example, non-Hodgkin's lymphoma, leukemia, and different kinds of strong tumor (Merhi et al., 2007). The concerns of human health regarding the use of these pesticides have increased in the past years. Maximum residue limits for foodstuffs have been established by various organizations and certain regions and countries. So as to keep up the wellbeing of people, show signs of improvement agrarian assets and anticipate ceaseless misfortunes, national sustenance observing projects for pesticides have been instituted around the world (Liu et al., 2016).

1.2   HISTORICAL PERSPECTIVE

Humans are using pesticides for a number of decades in order to fight with pests. Present day pesticides have been created from the synthetic tests during the late nineteenth and mid-twentieth centuries. The new mixtures of pesticides with a correct extent made it conceivable to control the pests. The United States in 1867 produced the first chemical pesticide Paris green; thus marking the beginning of chemical insecticide (Kogan, 1998). Arsenate, nicotine sulfate, and sulfur were used in United States in late 19th century to control the insect pests in crops. These synthetic compounds have been broadly used to control pests since the middle of the 20th century (Chauvel et al., 2012). The weeds were controlled with salt and insect pests were killed by using burning sulfur by ancient Romans. Before the birth of Christ, sulfur (brimstone) was used by pagan priests. In addition, people at that time used sulfur to clean the air and purify a sick room. The people also control ants with mixtures of honey and arsenic in the 16th century. The insecticides derived from the plants like nicotine are used to control aphids, hellebore are used to control body lice and pyrethrins to control a broad diversity of insects.

By the accessibility of dichlorodiphenyltrichloroethane (DDT) for horticultural use has opened the new period of nuisance control. Its broad use has additionally leads to the advancement of other manufactured natural bug sprays. The later was favored for its wide spectrum of activities to control the insect pests of agriculture. The property of persistence of DDT along with its wide spectrum of activities to kill the insect pests has made it a poor choice for its use as pesticide after World War II. With its continuous use, some of the pests have developed resistance against DDT which results in the damages to the nontarget plants and animals as pesticide deposits were seen to be available in surprising spots.

During the first half of 20th century, various problems were found because of the use of these chemical substances to control the pests in agriculture. These chemical methods to control the pests had resulted not only in the resistance of pesticides by the target pests but also harms the nontarget species, contamination of food, water, degradation of ecosystem, and human health problems.

1.3 CLASSIFICATION OF PESTICIDES

"Pesticide" is expansive term that incorporates all synthetic compounds used to execute the various types of vermin. These incorporate bug sprays, fungicides, herbicides, rodenticides, garden synthetic compounds, wood additives, and family unit disinfectants. Pesticides are having various personalities and distinctive physical and substance properties. All the manufactured pesticides can be arranged in different ways. By and large, there are three fundamental approaches to characterize the pesticides:

The classification of pesticides is based on the

(1) mode of action of pesticides
(2) the species which are targeted, and
(3) the chemical composition of pesticides.

1.4 CLASSIFICATION OF PESTICIDES BASED ON THEIR MODE OF ACTION

As per the method of activity, pesticides are arranged depending on the manner in which they act to achieve the ideal impact. The pesticides are grouped into two kinds:

(1) nonsystemic
(2) systemic pesticides.

Nonsystemic pesticides are those pesticides that are not transported within the plant vascular system and thus do not penetrate the plant tissues.

In contrast, systemic pesticides are those that are transported within the plant vascular system and thus effectively penetrate plant tissues and cause their effects.

1.5 CLASSIFICATION OF PESTICIDES BASED ON THE TARGETED PEST SPECIES

This is the most familiar classification of pesticides. For example, insecticides are those pesticides that kill the insects while as herbicides are those pesticides that target unwanted plants in our farm lands. The other types of pesticides in this category include rodenticides which kill rodents, fungicides used to kill fungi, acaricides and miticides, molluscicides, bactericides, avicides, and virucides.

1.6   CLASSIFICATION OF PESTICIDES BASED ON THE CHEMICAL COMPOSITION

This class of pesticides is based on the chemical composition and the active ingredients which are present in these pesticides. This classification is beneficial for those which are pursuing research in the field of pesticides and their effects on the surrounding environment because this kind of grouping gives us data about the adequacy, physical, and substance properties of these pesticides. This also provides us the information about the precautions to be taken while applying these pesticides.

Based on the synthetic properties of pesticides, we can group them into seven sorts:

1. Organochlorines
2. Organophosphorus
3. Carbamates
4. Pyrethroids
5. Amides
6. Anilins, and
7. Azotic heterocyclic compounds.

Organochlorine pesticides are those natural mixes which have at least five chlorine particles in their structure. These were the principal manufactured natural pesticides to be utilized in horticulture and human wellbeing. For the most part, organochlorine pesticides have a steady compound structure and they amass and endure in nature. Organochlorine pesticides act as insecticides and thus used to control the different kinds of insects. These pesticides disrupt the nervous system which results in convulsions and even paralysis of the pest insect and eventually the death of the pest insect. Thus most of these pesticides have been banned for agricultural use worldwide as they cause serious endocrine disorders in all groups of animals like mammals, fish, and birds.

Organophosphates are the second class pesticides that contain a phosphate group. These are highly toxic pesticides and occupied up to 48.6% of all pesticides in 1997 (Zhang, 2007). Their significance has expanded significantly during World War II with their utilization as fighting materials. After

World War II, organophosphates had been utilized in agribusiness, industry, beauty care products, medication, and numerous different territories. The organophosphate pesticides restrain the compound, acetylcholinesterase, in numerous species which hydrolyses acetylcholine in the sensory system of different species including man. They are corrupted effectively than organochlorines. The build-ups of organophosphates represent a more prominent risk to the biological system and nourishment industry on the grounds that their intense toxicities are irreversible. According to Eddleston and Phillips (2007), pesticide self-poisoning is a major public health problem and numerous people are exposed to pesticides because of their occupation. According to Gunnell et al. (2007), every year 250 to 370,000 people die because of exposure to pesticides and more than three million cases of severe poisoning have been observed by the exposure of these toxic pesticides. Accordingly, the utilization of organophosphates has been prohibited everywhere throughout the world.

Carbamates are the organic pesticides which have been derived from carbamic acid. They inactivate the enzyme, acetylcholinesterase, and reversibly. The inhibition of cholinesterase of these pesticides differs from that of organophosphates in that it is reversible and specific to a particular species. The above-mentioned three categories of pesticides, namely, carbamates, organophosphates, and organochlorines are traditionally high toxic whereas pyrethroids, anilines, amides and azotic heterocyclic compounds are normally very less toxic.

The pyrethroids is another group of pesticides which are the synthetic analogs of pyrethrins, are obtained from the flowers of pyrethrum plant (*Chrysanthemum cinerariaefolium*). This class of pesticides has the insecticidal activity from the natural pyrethrum. These pesticides are famous for their fast activity against the pests. These are also having low mammalian toxicity which are easily biodegradable in nature. These pesticides are less toxic to mammals than carbamates and organophosphates as they are nonpersistent sodium channel modulators. Therefore, their use has greatly increased during the last 30 years of usage to control the pests. Koureas et al. (2012) opined that the pyrethroids pose high toxicity to aquatic life such as mollusks arthropods and fish.

Amide herbicides, widely used in recent years, are a group of pesticides used to control herbs. They include acetochlor, butachlor, and metolachlor. The butachlor persists in the environment for up to 10 weeks and it has been identified as mutagens. Aniline and dinitroaniline are another type of pesticides used for controlling the pests. In this group, trifluralin and pendimethalin are widely used. The amide herbicides are extremely toxic to the water animals. Trifluralin and pendimethalin damages the liver and thyroid glands of the aquatic organisms, thus, these have been banned in many European countries. The imidazole and triazole heterocyclic chemicals containing a nitrogen atom are becoming the references for the development of new pesticides. Over the most recent 10 years, these pesticides have involved over 70% of all the recently created synthetic pesticides.

In addition to the above-mentioned classification, we can classify pesticides on the basis of mode of formulation, activity spectrum, and level of toxicity of pesticides. On the basis of mode of formulation, we can classify pesticides into five categories:

1. wettable powders
2. emulsifiable concentrates
3. baits
4. granules, and
5. dusts and fumigants.

On the basis of activity spectrum, they are classified as broad spectrum and selective pesticides. The broad-spectrum pesticides are the pesticides which are intended to kill a wide range of pests and other nontarget organisms. In contrast, selective pesticides are the pesticides which are intended to kill only some specific pests.

The World Health Organization (WHO) has classified all the pesticides into five different classes (Table 1.1) based on the level of toxicity and their effects on human health:

TABLE 1.1   Level of Toxicity and Their Effects on Human Health

Class	Level of Hazardous
Ia	Extreme
Ib	High
II	Moderate
III	Slight
IV	Acute

1.7   PESTICIDE POLLUTION

Due to the use of pesticides for controlling the pests since the middle of 19th century, there has been an extensive release of these toxic compounds into the environment. Their serious use has led to the pollution of the earth in this way presenting unsafe impacts on nourishment security, water assets, and even on the living organisms.

The different kinds of pests which destroy the crops of the farmers pose a challenge to these growers. The pests include the weeds, plant diseases, and insects. It has been estimated that about 30% of the crop yield gets lost because of these pests (Oerke, 2006). It is expected that the population will increase to 30% by 2050; thus the demand for the food will increase to about 70% at that time. Even if other measures are taken to control the pests are of vital importance, this tool, that is, the pesticides to control the pests and food security in future will be a continuing need.

Both pesticides and herbicides are used widely to control the pests and herbs are chemicals used in agriculture. As their definition states that they are toxic in nature which has been validated by finding their bioactive molecules in different animals, vegetal, and fungal species. The use of these pesticides is thus regulated in European countries because of their toxicity. Since the pesticides are having different water solubility and polarity, they can thus have various ways to enter the surrounding water bodies after their application in the agricultural fields. One of the most common pathways for these pesticides in case of surface water from agricultural lands is after irrigation and after rainfall. According to Abrantes et al. (2006), the pesticides influence the biotic architecture in aquatic ecosystems; thus forcing them to change from a less-turbid stage to a high turbid stage because of their harmful effects on macrophytes or zooplanktons.

The humans are responsible for both point and diffuse contamination by means of the pesticides, herbicides, and polycyclic sweet-smelling hydrocarbons. The surface water contaminated with these toxic pesticide chemicals is related to the activities of the humans that take place in the surroundings. Since the pesticides and herbicides are having close relation with the agriculture, therefore it is certain that cultivated lands near or around the lakes will have high intensity of these toxic compounds. Actually, it has been observed that the nutrient concentration in surface waters in a particular area is related to the land use of that area while the concentration of polycyclic aromatic hydrocarbon is with wetlands. In contrast, the areas which are very distant from the water body such as industries, thermal power plants, and far urban areas, the diffuse pollution sources are also related to these toxic compounds as these have been detected in aquatic ecosystems.

The fate of these toxic pesticides in environment is different and can be characterized by various complex processes. Their fate is different in different components of the ecosystems such as soil, plant, air, surface water, and ground water. While one is focusing on the sustainable development, the nonjudicious use of these pesticides is one of the most alarming challenges to the ecosystem. The small amount of the pesticide sprayed to the crops reached the target (only 1%). The persistence of the pesticides in the surroundings for extended duration takes place only because of their accidental release which may be due to leaking of the pipes, waste dumps, spills, and belowground soakage tanks. Therefore, it is the need to correctly monitor and check the status of soil, water, and air for the contamination for better management of the pesticides.

The utilization of pesticide has an imperative role in the modern agricultural practices so as to save the agricultural production from the destruction of the pests. Modern pesticides are not only reliable but also have high capability to protect the crops from the pests. The worldwide consumption of the pesticide has been estimated to be approximately about 5 billion kilograms per year. This huge consumption may have serious threats to the biodiversity, nontarget organisms, and the food chain, therefore, posing threats to human health and environment (Verger and Boobis, 2013). The country zones of the creating nations are at high hazard. It has been evaluated that 3 million ranchers experience the ill effects of genuine pesticide harming and 25 million farmers experience the ill effects of gentle harming, bringing about around 180,000 fatalities among rural laborers every year (Zhang et al., 2011). These immense passings happen in light of off base observations, guidelines, absence of learning, and training among farmers.

1.8 ASSOCIATED RISKS BY USING PESTICIDES

The valuable effects of the pesticides are very low as compared to their risks. These are having extreme consequences for nontarget species and they affect both plant as well as animal diversity whether aquatic or terrestrial. It has been estimated that 80%–90% of these pesticides change into volatile forms within a few days after they are applied to the agricultural land. By using sprayers, it is common that these vaporize in the atmosphere and thus may cause harmful effects to nontarget species. For example, when herbicides are applied, these evaporate from the treated plants and cause harsh damage to other plants and animals in the vicinity (Straathoff, 1986). Several aquatic and terrestrial animals have been reduced in number due to the excessive use of these pesticides. Their uncontrolled use has additionally undermined the survival of some uncommon species, for example, the bald eagle, peregrine bird of prey, and osprey (Helfrich et al., 2009). In addition to this air, we breathe, water we drink and soil in which we grow our vegetables have also been polluted with these pesticides to toxic levels.

In light of the dimension of danger of pesticides, bug sprays are viewed as the most poisonous pursued by fungicides while herbicides are viewed as least dangerous as the two. These pesticides enter the characteristic framework by two unique methods relying on their solubility. The pesticides which are soluble in water get dissolved in water easily and enter the below groundwater, streams, lakes, and rivers by agricultural runoff and thus causing harm to the nontarget species. Rather than this, fat dissolvable pesticides gather in the collections of creatures by a procedure known as "bioamplification." The procedure of bioamplification can be depicted as:

1. The limited quantity of pesticide enters the assemblages of primary consumers which are at low level in the food chain, for example, grasshopper.
2. Secondary consumers, for example, shrews feed on primary consumers (grasshoppers) and thus the absorption of pesticides will rise in their bodies.
3. The pesticide concentration is further increased in the bodies of animals which are still higher trophic levels like the owl.

Therefore, we can say that the pesticide concentration is increased with the increase in the trophic level which is called bioamplification. The process of bioamplification can disrupt the whole ecosystem because the species which are at higher trophic levels die due to the greater toxicity of the pesticides in their bodies. This will at last increment the number of secondary consumers (shrews) and decline the number of primary consumers (grasshoppers).

1.9   THREATS TO BIODIVERSITY

Due to the uncontrolled use of the pesticides on the agricultural land, one cannot oversight the threats that are associated with these toxins. So, it is the need of the hour to evaluate the impacts of the pesticides on the population of both water and land plants, animals, and birds. The predators and raptors get influenced legitimately because of the accumulation of the pesticides in the trophic levels which is a greater concern. The pesticides are also applied to control the weeds and other insects upon which higher organisms feed. Thus, the spraying of fungicide, herbicides, and insecticides is responsible for the decline in population of some rare bird and animal species.

1.10 THREATS TO AQUATIC BIODIVERSITY

Pesticides can find their way to the surface water by different means; they can enter via drift, by runoff, leakage through the soil, or directly into the surface water after their application to control mosquitoes. The water contaminated with these toxic pesticides is a great threat to the aquatic life. The contaminated water affects the aquatic plants thereby decreases the dissolved oxygen content of water which can cause the biochemical and behavioral variations in fish abundance. Some reports revealed the presence of lawn care pesticides in the surface water and in various water bodies like streams, lakes, and ponds. The pesticides reach the aquatic ecosystems after their application to agricultural land and there harm the aquatic fauna including fishes and other nontarget animals. In aquatic ecosystems, these toxic chemicals not only harm the fishes but also interact with harmful algal blooms which act as stressors. By the excessive use of the pesticides, the population of different species of fishes has declined to a noticeable range (Scholz et al., 2012). The animals of the aquatic ecosystem are exposed in three different ways (Helfrich et al., 2009):

- *Dermally:* the pesticide is absorbed directly via skin
- *Breathing:* the pesticide is taken during breathing via gills
- *Orally:* the pesticide enters the body via drinking the contaminated water.

The aquatic plants provide about 80% of dissolved oxygen for the survival of the aquatic life. Thus, if the aquatic plants are killed by applying the herbicides, this will lead to the drastic decrease in dissolved O_2 level and will lead to the hypoxic condition of these fishes and thus their yield will be reduced (Helfrich et al., 2009). The surface waters are having high levels of the pesticides than underground waters because of runoff from the surface agricultural land and contamination by sprays (Anon, 1993). The pesticides come to the groundwater by drainage of debased surface water, inappropriate transfer of pesticides, and inadvertent spills and spillages.

The aquatic life is at significant damage because of the washing of these toxic chemicals from the agricultural lands into these aquatic ecosystems (lakes, ponds, and rivers). It has been found that the pesticide, atrazine, is not only toxic to some fishes but also damages the immune system of the amphibians indirectly (Forson and Storfer, 2006; Rohr et al., 2008). The contaminated surface water with pesticides poses harmful effects to the amphibians because of their overexploitation and habitat loss. The carbaryl class of pesticides has been discovered lethal for a few creatures of land and water, while, herbicide Glyphosate is a causative agent for significant deaths of amphibians including their larvae (Relyea, 2005). The population structure of the planktons and the algal populations has been found changed due to little amounts of pesticide, malathion. The changed population structure of these plankters has affected the population density of amphibians (Relyea and Hoverman, 2008). Last but not least the pesticides, endosulfan, and chlorpyrifos also pose damage to amphibian animals.

1.11   TERRESTRIAL BIODIVERSITY UNDER THREAT

The pesticides impact the terrestrial biodiversity by killing the nontarget plants. The phenoxy herbicides are volatilized; thus, causing injuries to the trees and shrubs in the vicinity (Dreistadt et al., 1994). The vulnerability of plants to diseases is increased by herbicide glyphosate and also reduces the seed quality. The devastating effects on the productivity of nontarget crops, naturally occurring plant communities, and wildlife are found even if minimum doses of herbicides, sulfonylureas, sulphonamides, and imidazolinones are present.

The terrestrial animal life is also not safe from these pesticides. With the continued spray of wide-spectrum insecticides, for example, carbamates, organophosphates, and pyrethroids, the abundance of many of the beneficial insects has declined drastically. The population of insects has been found greater on organic farms than nonorganic farms. The honey bees are affected when aforementioned pesticides are applied to the crops. The neonicotinoids insecticides such as clothianidin and imidacloprid are toxic to bees. The foraging behavior and learning capacity of the bees are reduced even at the low doses of imidacloprid. The unexpected vanishing of the bumble bees during the beginning of 21st century was the biggest destruction that was unleashed by utilization of neonicotinoids. This vanishing of honey bees was a notable worry to the sustenance business as one-third of the nourishment generation is based upon honey bees for pollination. A significant portion of neonicotinoids pesticide was reported from the honey and was which was obtained from the commercial hives.

The pesticide use has also declined the bird populations to about 20%–25% since the preagricultural times. When these pesticides get accumulated in the tissues of these birds, they die. One of the famous examples of bird decline is Bald eagle in USA, which was due to their long exposure time to DDT and its constituents. The fungicides kill the earthworms upon which birds and mammal populations feed, thus declining their populations indirectly. The pesticides that are in granular form are eaten by the birds as food grains. The raptors in the fields get poisoned by organophosphate insecticides as these are highly toxic to the birds. The nervous system of the birds gets affected even by sublethal quantities of the pesticides thus changing the behavior of the target organism.

We can apply the pesticides by different means to control the pests. They can be fused or infused into the dirt or can be connected as granules or can be applied to seeds and can also be applied directly to the crop plant as liquid sprays. Soon after the application of pesticides to the target area, they get disappear by various means. They may get degraded, they may disperse, they may volatile or leach into the surface and groundwater, thus they may be taken by the plants or soil animals or they can remain in the soil as such. The main concern is the leaching of these pesticides into the soil as they affect the microorganisms which are living there. These microbes which live in the soil help the plant in various ways. They not only increase the soil fertility but also take up the nutrients from the soil and help in breakdown of the organic matter thus increasing the fertility of the soil. Further, these microbes play a significant role in human welfare and relay on plants for major needs. Hence, the overuse of these pesticides can have drastic consequences on these organisms and a day may come when our soil will not be fertile and thus may degrade.

Among these soil-dwelling microbes, some fix the nitrogen from air to nitrates. By using chlorothalonil and dinitrophenyl fungicides, it has been found that the nitrification and de-nitrification processes by these bacteria get disrupted (Lang and Cai, 2009). The changing state of ammonia to nitrite by nitrifying micro-organisms gets inhibited when the herbicide triclopyr is used. The population of these microbes in terrestrial environment gets reduced by the use of a nonselective herbicide, Glyphosate, and the transformation of ammonia to nitrite which is carried out by soil bacteria gets inhibited by the use of 2,4-D pesticide. Herbicides also cause considerable damage to fungal diversity in soil due to the use of trifluralin and oryzalin pesticides as both are reported to inhibit the growth of symbiotic mycorrhizal fungi that help in nutrient uptake. Oxadiazon has been known to reduce the number of fungal spores whereas triclopyr is toxic to certain species of mycorrhizal fungi.

The earthworms that act as bioindicators of soil pollution have a major role in terrestrial environs and these are used as model organisms for testing the toxicity of soil. These also help in making the soil fertile. The pesticides also have their toxic effects on these minute organisms and the effect occurs through pore water after contaminating the soil. The DNA of the earthworm gets damaged after their cellular damage by using the pesticides, glyphosate, and chlorpyrifos. The feeding activity and viability of these organisms are affected by the pesticide, Glyphosates.

1.12 PESTICIDE IMPACT ON HUMAN HEALTH

Though the use of pesticides has proved beneficial to the health of human beings by controlling the diseases which are vector-borne its overuse has also resulted in various dreadful diseases. Among the human population, the infants and young children are at high risk of these pesticides as the later are nonspecific in nature. The chances of exposure to these pesticides have increased during the past few decades because of their increased use.

It has been reported that 3,000,000 cases have been reported due to pesticide poisoning mostly in developing countries as per World Health Organization and about 220,000 deaths occur annually (Lah, 2011). According to Hicks (2013), about 2.2 million people who belong to under-developed nations are at high risk of pesticide exposure. In addition to this, some people are at high risk to the toxic effects of these pesticides than others. The newborn infants including the young ones are at high risk than the agricultural workers and the persons who spray the pesticides on the fields.

The pesticides can enter the human body through various processes after their application in the agricultural farmlands. They can enter through the food we eat by ingestion; they can enter by inhaling the contaminated air or can enter through the skin. Among the above-mentioned processes, the main pathway of entrance to the body is by ingestion of the contaminated food. After their entry into our body, they reach to our tissues and storage compartments. Though humans are having a mechanism to excrete these toxins still some of the toxins retained in the body during uptake in the circulatory system. The toxic effects of these chemicals are observed only if their concentration increases than the threshold concentration in the environment. The poisonous impacts of pesticides on strength of people are exceptional factors. Some of the effects show their symptoms within days and thus are instant in nature while others can take long periods (months) or even years to show their symptoms. These long term effects are called chronic effects that are described in the following sections.

1.13   ACUTE EFFECTS OF PESTICIDES

At the point when the people are presented to these pesticides, the people experience different manifestations which incorporate migraine, solidifying of the eyes and skin, skin tingling, aggravation of the nose and throat, appearance of the rash and rankles on the skin, stomach torment, tipsiness, sickness, the runs and regurgitating, obscured vision, visual deficiency and all around seldom passing. There is no need to consult a doctor because these immediate effects are severe.

1.14   CHRONIC EFFECTS OF PESTICIDES

The chronic effects of pesticides are frequently lethal and these effects are not experienced even for years. These chronic effects of the pesticides are long term effects and they cause damage to various body organs. When the pesticides are exposed for long periods of time they may result in the following consequences:

- The exposure of the pesticide can cause damage to nervous system by losing the coordination and memory, decreased visual capacity, and diminished motor signaling (Lah, 2011).
- The long-term pesticide exposure causes damage to the immune system and causing hypersensitive skin, asthma, and other related allergies.
- The pesticide also causes various types of cancers.
- The long-term pesticide accumulation in the body affects the reproductive system by altering the levels of hormones. The changed hormonal level results in various birth defects, abortion, and infertility.
- The long-term exposure to pesticides causes damage to various internal organs like liver, lungs, and kidneys and may cause blood diseases.

The different types of pesticides have different effects on the human body upon their exposure. According to Lah (2011), organochlorines cause several effects to human beings when they are ingested. The effects include hyper-sensitivity to light, touch and sound, vomiting, nausea, seizures, tremors, confusion, and nervousness. Similarly, organophosphates and carbamates interfere with the nerve signal transduction. The symptoms include dizziness, nausea, headache, confusion, and vomiting. Their exposure also causes pain in the muscles and chest. In severe cases, their exposure may also cause complexity in breathing, convulsions, coma and even death may occur.

The pyrethroids are also having their toxic effects on human beings. They can cause allergy in the skin, hyper-excitation, and aggressiveness. They not only cause reproductive and developmental effects but also cause tremors and seizures in addition to (Lah, 2011). Casida and Durkin (2013) observed a relationship between pesticides and Alzheimer's disease and Parkinson's disease.

1.15 CONCLUSION AND FUTURE PROSPECTS

Although the pesticides are very beneficial not only for the producers but also for people around the globe as they helped in increasing the yield and thereby providing benefit to the society indirectly. The harmful effects that are caused by the spray of toxic chemicals have raised the concerns about the environment and human well-being. The hazards associated with the use of pesticides cannot be eliminated but we can avoid the use of these pesticides by one or the other way. The hazardous effects of these toxic chemicals may be minimized by various ways. We can have alternate methods of cropping and the spraying equipment should be well-maintained. The harmful effects of pesticides can also be reduced if safe and environment friendly pesticides can be produced. The judicious use of these pesticides, that is, their appropriate amount and at appropriate time can also help in minimizing their risks. If the pesticides are formulated in such a way that they will be less toxic and then a low dose of such pesticides cannot create any havoc. One can remember the famous proverb of Paracelsus "The right dose differentiates a poison from a remedy."

Though, the organochlorine pesticides have been banned in many countries because of their nonbiodegradable nature still they are used in many countries. Thus their use results in serious health problems. These pesticides pollute the water bodies, thus pose a serious threat to the aquatic life at very low concentration (Agrawal et al., 2010). The ordinary persons like the farmers are not well aware of the hazardous effects of pesticides because they are having no information about the various classes of pesticides and their toxicity level. They do not know the precautionary measures to be taken before the application of these pesticides. It is because of this reason that these pesticides remain continuously in the surroundings so as to kill the pests hence the farmers are exposed to them. These chemicals that persist in the environment have long-term effects on human health. There must be mass awareness camps for the growers to minimize the use of these toxic chemicals (Sharma et al., 2012).

The need of the hour is that these pesticides should be formulated in such a way so that there is combination of chemical pesticides with usual treatments which can result in the abolition of the pests in a sustainable way. These new combinations can have varied applications in controlling the pests and invasive species and will promise environmental sustainability (Gentz et al., 2010). The biological integrity of marine as well as aquatic ecosystems is under threat because of the nonjudicious use of pesticides. Last but not least, it is the need of time to see immediate and aberrant impacts of pesticides on the environment by putting together various disciplines like toxicology, natural science, population biology, community ecology, conservation biology, and landscape ecology (Macneale et al., 2010).

KEYWORDS

- **pesticides**
- **environment**
- **pollution**
- **health**
- **systemic pesticides**
- **carbamates**
- **biodiversity**

REFERENCES

Abrantes, N.; Pereira, R.; Gonçalve, F. First step for an ecological risk assessment to evaluate the impact of diffuse pollution in Lake Vela (Portugal). *Environ Monit Assess* **2006**, *117*, 411–431.

Agrawal, A.; Pandey, R.S.; Sharma, B. Water pollution with special reference to pesticide contamination in India. *J Water Res Prot* **2010**, *2*(5), 432–448.

Anon. The environmental effects of pesticide drift, *English Nature, Peterborough*. **1993**, pp. 9–17. Benefits of pesticides and crop protection chemicals. In: Crop life America. Available from .

Casida, J.E.; Durkin, K.A. Neuroactive insecticides: targets, selectivity, resistance, and secondary effects. *Annu Rev Entomol* **2013**, *58*, 99–117.

Chauvel, B.; Guillemin, J.P.; Gazquez, J.; Gauvrit, C. History of chemical weeding from 1944 to 2011 in France: Changes and evolution of herbicide molecule. *Crop Prot* **2012**, *42*, 320–326.

Dar, S.A.; Yousuf, A.R.; Balkhi, M.H.; Ganai, F.A.; Bhat, F.A. Assessment of endosulfan induced genotoxicity and mutagenicity manifested by oxidative stress pathways in freshwater cyprinid fish crucian carp (*Carassius carassius* L.). *Chemosphere* **2015**, *120*, 273–283.

Delaplane, K.S. Pesticide Usage in the United States: History, Benefits, Risks, and Trends [Internet]. **2000**. Available at: .

Dreistadt, S.H.; Clark, J.K.; Flint, M.L. Pests of landscape trees and shrubs. An integrated pest management guide. *University of California Division of Agriculture and Natural Resources*. **1994**, Publication No. 3359.

Eddleston, M.; Phillips M.R. Self poisoning with pesticides. *BMJ* **2004**, *328*, 42–44.

Fishel, F.M. Pest Management and Pesticides: A Historical Perspective. **2013**, Available at: http://edis.ifas.ufl.edu

Forson, D.D.; Storfer, A. Atrazine increases Ranavirus susceptibility in the tiger salamander (*Ambystoma tigrinum*). *Ecol Appl* **2006**, *16*, 2325–2332.

Gentz, M.C.; Murdoch, G.; King, G.F. Tandem use of selective insecticides and natural enemies for effective, reduced-risk pest management. *Biol Control* **2010**, *52*(3), 208–215.

Grube, A.; Donaldson, D.; Kiely, T.; Wu, L. Pesticides industry sales and usage: 2006 and 2007 market estimates. U.S. Environmental Protection Agency, Washington, DC. **2011**.

Gunnell, M.; Eddleston, M.; Phillips, M.R.; Konradsen, F. The global distribution of fatal pesticide self-poisoning: Systematic review. *BMC Public Health* **2007**, *7*, 357.

Helfrich, L.A.; Weigmann, D.L.; Hipkins, P.; Stinson, E.R. Pesticides and aquatic animals: a guide to reducing impacts on aquatic systems. In: Virginia Polytechnic Institute and State University. **2009**, Available from https://pubs.ext.vt.edu/420/420-013/420-013.html.

Hicks, B. Agricultural pesticides and human health. In: National Association of Geoscience Teachers. **2013**, Available from .

Kogan, M. Integrated pest management: Historical perspectives and contemporary developments. *Annu Rev Entomol* **1998**, *43*, 243–270.

Koureas, M.; Tsakalof, A.; Tsatsakis, A.; Hadjichristodoulou, C. Systematic review of biomonitoring studies to determine the association between exposure to organophosphorus and pyrethroid insecticides and human health outcomes. *Toxicol Lett* **2012**, *210*, 155–168.

Kumar, N.; Antony, J.P.P.; Pal, A.K.; Remya, S.; Aklakur, M.; Rana, R.S.; Gupta, S.; Raman, R.P.; Jadhao, S.B. Anti-oxidative and immuno-hematological status of Tilapia (*Oreochromis mossambicus*) during acute toxicity test of endosulfan. *Pestic Biochem Phys* **2011**, *99*, 45–52.

Lah, K. Effects of pesticides on human health. In: Toxipedia. **2011**, Available from https://www.pesticidereform.org/pesticides-human-health/#:~:text=Pesticides%20can%20cause%20short%2Dterm,%2C%20dizziness%2C%20diarrhea%20and%20death.

Lang, M.; Cai, Z. Effects of chlorothalonil and carbendazim on nitrification and denitrification in soils. *J Environ Sci* **2009**, *21*, 458–467.

Liu, Y.; Li, S.; Ni, Z.; Qu, M.; Zhong, D.; Ye, C. Fubin Tang pesticides in persimmons, jujubes and soil from China: Residue levels, risk assessment and relationship between fruits and soils. *Sci Total Environ* **2016**, *542*, 620–628.

Lozowicka, B.; Kaczynski, P.; Paritova, A.E.; Kuzembekova, G.B.; Abzhalieva, A.B.; Sarsembayeva, N.B.; Alihan, K. Pesticide residues in grain from Kazakhstan and potential health risks associated with exposure to detected pesticides. *Food Chem Toxicol* **2014**, *64*, 238–248.

Macneale, K.H.; Kiffney, P.M.; Scholz, N.L. Pesticides, aquatic food webs, and the conservation of Pacific salmon. *Front Ecol Environ* **2010**, *8*, 475–482.

Majewski, M.; Capel, P. Pesticides in the atmosphere: distribution, trends, and governing factors. Pesticides in the hydrologic system, vol. 1. Ann Arbor Press Inc., Boca Raton, FL, USA. **1995**, p. 118.

Malone, R.W.; Ahuja, L.R.; Ma, L.; DonWauchope, R.; Ma, Q.; Rojas, K.W. Application of the root zone water quality model (RZWQM) to pesticide fate and transport: An overview. *Pest Manag Sci* **2004**, *60*(3), 205–221.

Merhi, M.; Raynal, H.; Cahuzac, E.; Vinson, F.; Cravedi, J.P.; Gamet-Payrastre, L. Occupational exposure to pesticides and risk of hematopoietic cancers: Meta-analysis of case-control studies. *Cancer Cause Control* **2007**, *18*, 1209–1226.

Oerke, E.C. Crop losses to pests. *J Agric Sci* **2006**, *144*, 31–43.

Ouyang, W.; Cai, G.; Huang, W.; Hao, F. Temporal-spatial loss of diffuse pesticide and potential risks for water quality in China. *Sci Total Environ* **2016**, *541*, 551–558.

Relyea, R.A. The lethal impact of roundup on aquatic and terrestrial amphibians. *Ecol Appl* **2005**, *15*, 1118–1124.

Relyea, R.A.; Hoverman, J.T. Interactive effects of predators and a pesticide on aquatic communities. *Oikos* **2008**, *117*, 1647–1658.

Rohr, J.R.; Schotthoefer, A.M.; Raffel, T.R.; Carrick, H.J.; Halstead, N.; Hoverman, J.T.; Johnson, C.M.; Johnson, L.B.; Lieske, C.; Piwoni, M.D.; Schoff, P.K.; Beasley, V.R. Agrochemicals increase trematode infections in a declining amphibian species. *Nature* **2008**, *455*, 1235–1239.

Santos, M.A. Managing Planet Earth: Perspectives on Population, Ecology, and the Law. Bergin & Garvey, Westport, CT, USA. **1990**, p. 44.

Scholz, N.L.; Fleishman, E.; Brown, L.; Werner, I.; Johnson, M.L.; Brooks, M.L.; Mitchelmore, C.L. A perspective on modern pesticides, pelagic fish declines, and unknown ecological resilience in highly managed ecosystems. *Bioscience* **2012**, *62*(4), 428–434.

Sharma, D.R.; Thapa, R.B.; Manandhar, H.K.; Shrestha, S.M.; Pradhan, S.B. Use of pesticides in Nepal and impacts on human health and environment. *J Agric Environ* **2012**, *13*, 67–72.

Straathoff, H. Investigations on the phytotoxic relevance of volatilization of herbicides. *Mededelingen* **1986**, *51*(2A), 433–438.

Verger, P.J.P.; Boobis, A.R. Re-evaluate pesticides for food security and safety. *Science* **2013**, *341*, 717–718.

Zhang, Y. New Progress in Pesticides in the World. Chemical Industry Press, Beijing. **2007**.

Zhang, X.J.; Zhao, W.Y.; Jing, R.W.; Wheeler, K.; Smith, G.A.; Stallones, L.; Xiang, H.Y. Work related pesticide poisoning among farmers in two villages of southern China: A cross sectional survey. *BMC Public Health* **2011**, *11*, 429.

CHAPTER 2

Pesticide Contamination in Water: Perspectives and Concerns

NOWSHEEBA RASHID[1*], AMIR HUSSAIN DAR[2], and IFRA ASHRAF[3]

[1]*Amity Institute of Food Technology, Amity University Noida, Uttar Pradesh, India*

[2]*Department of Food Technology, Islamic University of Science and Technology, Awantipora, Pulwama, India*

[3]*College of Agricultural Engineering and Technology, Sher-e-Kashmir University of Agricultural Sciences and Technology of Kashmir Shalimar Campus, Srinagar 190025, Jammu and Kashmir, India*

Corresponding author. E-mail: nowsheebaft@rediffmail.com

ABSTRACT

The most significant aspect pertaining to the excellence of environment is water pollution. Because of various reasons, these days the eminence of surface as well as groundwater is waning. In various urbanized countries, the presence of pesticide in water has become a vital and concerning problem. As a known fact, pesticides were used and are still being used in order to elevate production of crops. Therefore, the important reason behind surface and groundwater pollution is intensive use of these pesticides. The most important aspects distressing pollution of water because of pesticides and their remainders comprise soil temperature, microbial activity, drainage, treatment surface, rainfall, submission speed, and also the mobility, solubility as well as half-life of the pesticides. In general, the image is not as miserable as a person might be thinking as the methodologies have been developed to make these vital water resources accessible currently and also in the near future. Due to the intimidation to these water systems and mechanisms, ones causing water pollution are now fully recognized and safety measures have been implemented to save the water from pollution. The possession of pesticide pollution on drinking water in India has been talked about in this chapter.

2.1 INTRODUCTION

Water, a natural endowment, is indispensable for the life to sustain on the planet Earth. The maximum portion of the water present on earth is saline existing in seas and oceans which cannot be tapped by humans without processing. The water present below the earth in the soil pores referred to as groundwater is the only source of freshwater which could meet the standards of palatable water, hence used for drinking purposes. Despite the groundwater volume present on earth is limited, yet it has potential to meet the needs of all the living creatures provided that its quality would have been lofty. Water quality is of great significance for animates owing to its essentiality in supporting the fundamental physiological functioning of any biological cell. In the current epoch, there has been a growing concern with regards to the quality of both surface as well as subsurface water reserves. The reason for these budding concerns is owed to the entrainment of the pesticides in both ground as well as surface water throughout the world.

Globally, water pollution is a great cradle of trepidation because it causes inception of a number of lethal ailments (Agrawal et al., 2010); causing death of human beings in multitude. The predicament is more distressing in developing countries as compared to the developed countries. Together with the pesticides, natural phenomena, for instance, algal blooms, quakes, volcanic eruptions, and storms, too instigate major vicissitudes in ecological state of water and its quality thereof. Water pollution has numerous sources and traits. In case of entrainment of toxic chemicals in water, the quality of water degrades and makes it life staking instead of life sustaining.

The introduction of pesticides is not only frequent in surface water reserves but they find their way also in the groundwater reserves (Ongley, 1996) since the last several decades. Entrée of pesticides in surface water and groundwater in particular needs to be prevented. The contamination of groundwater is more gruesome issue because it takes a longer span of time to degrade the pesticides and to dilute the contaminant concentration over there as compared to other ambiances; and this underground water is mostly used for irrigation and drinking purposes for humans and animals.

Contrarily, pollution of surface water is less portentous in contrast with the groundwater. The reason behind this can be avowed to the swift turnover rate of most surface water sources excluding deep lakes, which implies the faster dilution of contaminants in the freshwater. Moreover, surface water reserves largely embody the higher concentration of free oxygen which increases the breakdown of the pesticides by microbial organisms. However, the pollution of surface water should not be regarded as nonchalantly. A highly noxious pesticide even at low concentrations is potent enough to kill fishes and other aquatic creatures. The turnover rate for the water present underground may last for a few months, but generally years or decades are taken by water to get replaced below in aquifers. The microbial organisms present in groundwater are less effectual in degrading the pesticides on account of absence of oxygen in groundwater. Tremendously protracted dilution and disintegration of contaminants imply that the contaminants will have greater residence time in groundwater (Tschirley, 1990).

Many researchers conducted to assess the cause of water pollution have arrived upon that the main cause of water contamination is most likely due to the entrainment of most commonly used pesticides. The most rational method to tackle with the quandary of water contamination possibly will be endeavoring to prevent introduction of any perilous materials in the water bodies without purpose, because the outcome may be decline of water quality. The apparent scenario is not as austere as it appears. As the potent risks to water environments and the mechanisms that govern the contamination of water systems are well fathomed, the measures are looked-for to safeguard the quality of water (Agrawal et al., 2010). Considering the gravity of pesticide pollution of water systems and its bearing on animate things besides the milieu, an attempt has been made in this chapter to compile and disseminate the existing information available with regards to this issue.

2.2 ANTAGONISTIC SWAYS OF PESTICIDES ON WATER

Pesticides can pollute water, soil, grass, and other flora. Besides exterminating the target weeds or insects, pesticides can be noxious to a horde of other nontarget living things comprising fishes, birds, beneficial insects, and even desired plants/crops. Insecticides are usually the most acutely toxic class of pesticides, but herbicides can also pose risks to nontarget organisms.

2.3   SURFACE WATER POLLUTION VIA PESTICIDES

Pesticides can find their way in the surface water bodies via runoff from treated floras and soil. Contamination of water by means of pesticides is rampant. The disquieting results surfaced on ground out of wide-ranging suite of studies carried out by the US Geological Survey (USGS) in the early to mid-1990s on major river basins throughout the country. The fish and water samples containing one or more pesticides exceeded 90% of all the samples analyzed (Kole et al., 2001). The pesticides were realized in all the samples of the major rivers having the influences of mixed urban and agrarian land use, while about 99% of samples from rivers having urban influence only (Bortleson and Davis, 1997). USGS also observed that insecticide concentrations in urban river courses outstripped the recommendations for safeguarding the aquatic life (Aktar et al., 2009). According to USGS, urban watercourses outweighed the agrarian watercourses in the number of pesticides (Mills et al., 2005). Commonly used herbicides for instance 2,4-D, prometon, and diuron, and insecticides like diazinon and chlorpyrifos, by urban landowners and school localities, were among the most identified 21 pesticides in ground and surface water throughout the country (Aktar et al., 2009). Trifluralin and 2,4-D were found in water samples taken abstracted from 19 out of 20 analyzed river basins (Wall et al., 1998). The most frequently recovered pesticide was herbicide 2,4-D, spotted in 12 out of 13 rivulets. In Puget Sound Basin, 23 pesticides including 17 herbicides were spotted in the watercourses. The diazinon insecticide and weedicides including diuron, dichlobenil, glyphosate, and triclopyr were also identified in Puget Sound basin streams. The intensities of diuron and diazinon pesticides were perceived to exceed the guidelines by the National Academy of Sciences for the protection of aquatic life (Bortleson and Davis, 1997).

2.4   GROUND WATER CONTAMINATION BY PESTICIDES

Groundwater contamination owing to pesticides is a global nuisance. As reported by USGS, no less than 143 diverse pesticides and 21 transformation products have been claimed in groundwater, comprising of pesticides from each major chemical class. In the recent few decades, pesticides have been encountered in the groundwater of more than 43 states. In Bhopal (India), 58% of drinking water excerpts tapped from several wells and hand pumps were found to be polluted with organochlorine pesticides in excess of Environmental Protection Agency (EPA) recommendations (Islam et al., 2011).

The toxic contamination of groundwater with poisonous chemicals is very problematic as it may entail multitude of years for pollutants to fritter away or get cleansed. Cleaning if not possible, may incur huge investment and is very complex. Agricultural growth is directly associated with the exercise of pesticides. The exploitation of pesticides has abetted in averting the damages brought about by pest assault and has enhanced the production competence of crops; but in turn their excessive application has caused the leaching down of pesticides to groundwater, thereby its contamination. The contamination of groundwater due to pesticides is meticulously associated with the perseverance of pesticides in soil. The possibility of getting leached down to groundwater is determined by the capability of a pesticide to get absorbed/adsorbed in soil horizon. The pesticides having low adsorption/absorption capability will leach down to groundwater and thereby cause its pollution. Numerous records have been reported concerning the groundwater contamination owing to the pesticide residues. Due to rapid solubility of pesticides in fat and their accretion in target beings, they pose severe health jeopardies (Agrawal et al., 2010).

Khanna and Gupta (2018) gathered the 56 groundwater abstracts at shallow depth from agricultural plot in Taibu basin of China, and detected 13 types of organochlorine pesticides in samples of groundwater. In another study carried out in Shandong and Hebei provinces of China, the groundwater samples showed the presence of pharate, aldicarb, and terbulos treasured from agricultural land with sweet potato cultivation (Kong et al., 2004). Likewise, Khanna and Gupta (2018) confirmed the presence of a highly noxious insecticide namely aldicarb in the groundwater.

Further, the presence of aldicarb in groundwater was also reported by Rothschild et al. (1982) in Central sand plain of Wisconsin. They confirmed the presence of aldicarb in high intensities in shallow wells located just over the water table, while the absence of aldicarb was perceived in deeper wells located 60 feet beneath the water table. High intensities of aldicarb, that is, greater than 10 mg/L, including sulfoxides and sulfones of aldicarb were found in eight territories of United States (Moye and Miles, 1988). In Luzon (Philippines), the average pesticide content was detected to the extent of one to two orders less than single (0.1 µg/L) and multiple (0.5 µg/L) pesticide thresholds set by World Health Organization (WHO), yet transitory high levels of 1.14–4.17 µg/L were also substantiated. In Portugal, the sway of practice of intensive horticulture on groundwater pollution was studied by Gonçalves et al. (2007). The maximum spotted pesticides were Pendimethaline (49%), Lindane (53%), Endosulfan (38%), and Endosulfan sulfate (44%). Khanna and Gupta (2018) revealed the existence of atrazine in groundwater with low concentration of about 0.1 mg/L in Anglian Water, UK. The groundwater in Hawai islands was also found to be polluted with ethylene dibromide and dibromochloropropane (Khanna and Gupta, 2018).

Similarly in India (Howrah, West Bengal), the raised pesticide levels were perceived in groundwater and in this way deemed to be inapt for consumption/drinking (Chaudhary et al., 2002). The samples of groundwater arranged from another region of India (Kanpur) also revealed the existence of high levels of both organophosphorus as well as organochlorine pesticides. Even the groundwater excerpts arranged from many hand pumps situated in industrial and agricultural extents also indicated the existence of dieldrin, malathion, and γ-HCN in concentration of 16.227, 29.835, and 0.9000 µg/L, respectively. Kumari et al. (2008) gathered 12 testers of groundwater from different fields of sugarcane, wheat-paddy, cotton-paddy, and tube wells from the places around Hissar, Haryana. They detected the residues of DDT, hexachlorocyclohexane (HCH), cypermethrin, and endosulfan in the groundwater samples recurrently. Out of 12 samples, 10 samples showed the presence of only one at the concentrations above safety guidelines, hence deemed to be unsafe for human consumption. Another instance of groundwater contamination in India was reported in Thirvallur district of Tamil Nadu, which showed the presence of HCH, DDT, endosulfan, and their by-products (Jayashree and Vasudevan, 2007). The residues of γ-HCH among the by-products of HCH were ascertained to present in high level (9.8 µg/L) in open wells of Arumbakkam. Also, op-DDT and pp-DDT with concentration of 0.8 and 14.3 µg/L, respectively, were reported. A borewell located in Kandigai village showed the maximum remnants of endosulfan sulfate (15.9 µg/L).

In rural areas in vicinity of Farrukhabad, India, many groundwater samples were ascertained for pesticide contamination and almost all the groundwater excerpts were found to be contaminated with remnants of DDT and HCH pesticides. The remnants of heptachlor, aldrin, and endosulfan were also seen in multitude of samples. The groundwater contamination in these areas was attributed to the seepage from the polluted Ganga River and the drift of pesticides with the downward movement of surface runoff (Mohapatra et al., 1995).

Tariq et al. (2004) also testified the presence of six types of pesticides in the groundwater samples arranged from four cotton cultivating districts

(Muzafargarh, Bahwalanagar, Rajanpur, and D.G. Khan). These six types of pesticides were generally used to eradicate the pests on cotton which included λ-cyhalothrin, bifenthrin, endosulfan, carbofuran, monocrotophos, and methyl parathion; and were found in 5.4%, 13.5%, 8.0%, 59.4%, 35.1%, and 5.4% concentration, respectively in July, whereas their concentration was 13.55%, 16.2%, 8.0%, 43.2%, 24.3%, and not detectable, respectively, in October. In another study carried out at Jaipur city of Rajasthan, India, the pesticides were found not only in the groundwater samples but also the wheat flour samples. The contamination was caused due to organochlorine pesticide DDT residues and its by-products, isomers of HCH and HCH itself, epoxide of heptachlor, aldrin, and heptachlor (Bakore et al., 2004). In North Eastern Regions of India, pesticide remnants were not only seen in ground-water but in surface water as well. HCH, α-HCH, and aldrin were seen in all the testers. Total DDT was observed in 82% surface water testers and 90.9% groundwater testers. The groundwater testers indicated the presence of β-HCH, α-HCH, γ-HCH, pp-DDT, op-DDT, pp-dichlorodiphenyldichloroethane (DDD), pp- dichlorodiphenyldichloroethylene (DDE), aldrin, and endosulfan in elevated concentrations of 14.06%, 4.6%, 2.4%, 15.6%, 24.2%, 174%, 101.2%, 24.73%, and 57.85%, respectively, in comparison with the surface water (Kumar et al., 1995). The organochlorine pesticides were also reported in groundwater samples procured from Kolkata (Ghose et al., 2009). The groundwater in Gurgaon and Ambala districts of northern state, Haryana, India, was also found to be polluted due to endosulfan, isomers of HCH, and by-products of DDT (Kaushik et al., 2012). The southern state of India, that is, Kerala is also not sheltered from this menace because the samples from open wells of Kasargod district also showed the presence of organochlorine pesticides. Among organochlorine pesticides, endosulfan was found in higher concentrations tailed by hexachlorobenzene in open wells (Akhil and Sujatha, 2012). Simazine and Atrazine have been reported in groundwater in Delhi, India by Aslam et al. (2013). The experimental results also illustrated that Atrazine concentration was comparatively lower as compared to Simazine, which was found in highest intensities in northern parts of Delhi Howbeit, Atrazine and its by-products have turned out to be the principal noxious pollutants of the groundwater in Canada/USA (Khanna and Gupta, 2018). Methylisocynate (MIC), an intermediate compound of carbamate class of pesticide, is highly dangerous to human health. MIC has been reported in 5 out of 12 surface and groundwater samples procured from Mysore, India (Somashekar et al., 2015).

2.5 MAXIMUM CONTAMINANT LEVEL (MCL)

This phrase signifies the noxious chemicals deliberated as contaminants in accordance with the Safe Drinking Water Act (SDWA). The use of MCLs is not limited to the pesticides in particular, but they are also generally used. Pesticides under SDWA are flocked with the grander group of toxic chemicals that often influence human health when found at particular concentrations in potable water in excess of customary MCLs. An attempt is being made with the usage of SDWA along with its connected guidelines for stopping pollution of drinking water from attaining MCLs via ceaseless monitoring of water supplies. Protocols of SDWA ascertain MCLs much in the similar manner as Federal Food, Drug, and Cosmetic Act, Federal Insecticide, Fungicide, and Rodenticide Act, and the Food Quality Protection Act of 1996 make it possible to know tolerances of pesticides with ignorable residues (Rai and Pandey, 2003).

2.6 CHARGE OF WATER SYSTEMS WITH PESTICIDES

Sankhla et al. (2018) revealed that pesticides can make the way into the water by means of surface runoff or via leaching. These two basic processes are associated with the hydrological cycle of Earth. When we take account of civic water usage in surface runoff, the remnants of pesticides in municipal wastewater do fit hydrological model. When water makes its way into a specific water body or moves backward behind a barricade, it transports dissolved substances with it which it lifted up in the middle of its flow. But it is an uphill task to find out as to how substances that turn out to be pollutants practically go into the sources of water. Generally, the action of water itself triggers the contaminants to flow into water bodies.

The entry of pollutants into the water bodies may be natural or human source of water; natural by means of rainfall or humanoid by means of irrigation or diversion of water. It is by way of wind or by its passive movement that the pollutants may also make their way into water bodies. Passage of pollutants is very intricate system. Pesticides may emanate either from nonpoint sources (NPS) or point sources. Point sources are minute, clearly recognized entities or locales of elevated pesticide intensity for instance containers, tanks, or spills. NPS include wide and undefined expanses wherein the pesticide remains are extant (Agrawal et al., 2010).

2.7   SURFACE RUNOFF

Water that flows through the surface, the origin of which be either from rain or irrigation or from the method of releasing onto the surface, finally flows downstream down the hill till it meets with a blockade, a water body, or starts to infiltrate into the soil.

2.8   AGRICULTURE AND WATER QUALITY DISCORDS

Water quality nuisances (Table 2.1), considered to be instigated to some extent by NPS pollution or by cropland runoff, muddle the country's lakes and drinking water, watercourses, and estuaries. Action plan being adopted by civic officials to save our water reserves may result in vicissitudes to the diversity, amount, and quality of farm stuffs, production techniques, and finally the rates consumers pay. The distorted water quality can cause the loss of huge amount of money of the order of billions of dollars not only to agriculture but to commercial fishing, recreation, and river routing and civic water management also.

TABLE 2.1   Impacts of Agriculture on Quality of Water

Agricultural Activity	Impacts	
	Ground Water	**Surface Water**
Irrigation	Enriches the groundwater with nutrients especially nitrates and salts.	Salinization of surface waters because of runoff of salts; ecological damage via runoff of pesticides and fertilizers to surface waters and bioaccumulation inedible fish species, etc.; serious environmental damage together potential human health impacts could result due to high levels of trace elements such as selenium.
Pesticides	Few of the pesticides seep into groundwater leading to various human health problems via impure wells.	Public health emergencies because of consumption of polluted fish. These pesticides move as dust by wind to very long distances thus contaminating marine systems 1000s of miles away (e.g., humid and subtropical pesticides were present in Arctic mammals)

TABLE 2.1   *(Continued)*

Agricultural Activity	Impacts	
	Ground Water	**Surface Water**
Tillage/plowing		Overspill of pesticides is responsible for infectivity of surface water and biota; dysfunction of environment in surface waters by failure of top predators because of inhibition of growth and failure of reproductive system. *Deposits/turbidity*: the deposits transmit phosphorus and pesticides adsorbed to them; *siltation* of river beds as well as loss of habitat, etc.
Manure distribution	Infectivity of groundwater, particularly by nitrogen	Accepted as a fertilizer action; distribution on ice-covered land outcomes in towering levels of infectivity of receiving waters by metals, pathogens, nitrogen, and phosphorus causing potential infectivity and eutrophication.
Fertilizing	Leakage of nitrate to groundwater; extreme concentrations area hazard to public health.	Overflow of nutrients, particularly phosphorus, causes eutrophication leading to pungent taste and odor in municipal water supply, surplus algal expansion causes deoxygenation of water thus killing fish.
Aquaculture		Liberation of pesticides, e.g., TBT1 and also elevated levels of nutrients to surface and groundwater via feed and feces, cause severe eutrophication.
Comprehensible cutting	Interruption of hydrologic system, frequently with bigger surface runoff and smaller groundwater revive; affects surface water by diminishing flowing arid times and intent nutrients and pollutants in surface water.	Wearing of land, causing elevated levels of turbidity in rivers, and siltation of bottom habitation, etc., interruption and alteration of hydrologic regime, frequently with loss of recurrent streams; leading to public health troubles because of loss of drinkable water.
Feedlots/animal corrals	Probable leakage of metals, nitrogen, etc., to groundwater.	Infectivity of surface water with number of pathogens such as bacteria, viruses, etc. causing severe human health issues. In addition infectivity by metals enclosed in urine and feces.
Silviculture		Wide variety of effects: pesticide overflow and pollution of surface water and fish; wearing of soil and sedimentation issues.

The sources of groundwater are amenable to pollution from various paths. When fertilizers, insecticides, fungicides, herbicides, and manures from animal wastes are applied on agricultural fields, some remnants do persist in the soil after plant uptake and percolate down into groundwaters, or remnants may make their way into surface water due to dissolution in runoff or absorption/adsorption to sediment. During pesticide application, spray drift can also cause the transport of the pesticides to the surface water. The products formed due to physical or chemical transformation of the pesticide residues can also cause contamination of water. For instance, nitrogen from animal waste or nitrogenous fertilizer may be changed firstly into ammonium and thereafter into nitrates. Nitrates can also change into nitrites and both are harmful to human wellbeing.

Nutrients, especially phosphorus and nitrogen from fertilizers when accumulate in water bodies cause eutrophication, that is, increase algal growth and hasten the aging of lakes, estuaries, and streams. The sediments in suspension diminish the sunlight, harm the spawning bases, and may prove noxious to aquatic creatures, thereby causing damage to aquatic life. The residues of pesticides which extend to surface water structures may also cause adverse effect on the vigor and healthiness of aquatic beings (Agrawal et al., 2010).

2.9 IMPACT OF IRRIGATION ON QUALITY OF SURFACE WATER

United Nations' projections of worldwide populace upsurge upto the year 2025 necessitate a surge in food production by about 40%–45%. Although irrigated agriculture encompasses about 17% of total cultivated land, yet it contributes to 36% of world food production. Focusing on irrigated agriculture will be a vital constituent of any policy to enhance the food supply at global level. Besides the predicaments of erosion, desertification, waterlogging, salinization, etc., that impact the irrigated areas; the nuisance of downstream squalor of water quality owing to noxious leachates, agro-chemicals and salts is a grievous environmental issue. Recently, it has been established that a major and global phenomenon of salinization of water bodies is of greater trepidation than the salinization of soil itself. In fact, it has surfaced on the ground that the trace elements like Mo, Se, and As in waters drained from the agricultural lands may stem the pollution issues threatening the continuance of irrigation in various schemes/projects (Azab, 2012).

2.10 CIVIC HEALTH BRUNTS

Contaminated water is a big source of human ailments, suffering, and death. As stated by WHO, diarrhea at the hands of water pollution causes the death of about four million children annually. The bacteria coliforms are mainly present in the soiled water which are excreted by humans. NPS pollution driven by surface runoff plays a significant part in the pooling of maximum pathogens in water bodies. Indecorously devised rural sanitary amenities also introduce the contaminants in the groundwater. Agricultural pollution serves as both a direct as well as an indirect source of human vigor sways. In line with the intensification of the agrarian practices, the nitrogen concentrations in the groundwater have swelled in different regions of the world (World Health Organization, 2004). This incident is quite prevalent in the extents of Europe. The surge in the nitrate levels in some nations has reached to such an echelon that even more than 10% of populace is under the sway of nitrate levels in excess of 10 mg/L guideline in potable water. Even though WHO perceives no substantial relations between nitrate/nitrite concentration in human body and cancers, yet the potable water guideline has been framed to preclude methemoglobinemia which mainly affects the infants owing to their large intake of water containing nitrate (World Health Organization, 2004). Even if the issue has been recorded very less, the developing countries emerge to be affected by the nitrate nuisance in groundwater. King (2003) registered approximately 40–45 mg N/L nitrate strengths in irrigation wells that subsist proximate to the intensively irrigated rice fields.

2.11 APPREHENSIONS REGARDING WATER QUALITY

Exposure of wells to pesticide and fertilizer residues contaminates them and hence the water present in them which is used by humans for consumption (Table 2.2). Infant methemoglobinemia is a state in which nitrates are transformed into nitrites within digestive system thus, damaging the capability of infants' blood to transmit oxygen; it is the common and known health hazard of nitrate contamination. Several analysts declared nitrates as tumor causing (carcinogenic). In drinking water, the amount of these nitrates and/or other pesticides may possibly be lower than levels at which severe health effects have been observed. On the other hand, continuous contact possibly will result in persistent effects such as impairments of reproductive system, cancer, etc., to human beings or other organisms. Besides the extent of health risk related to drinking water holding minute quantities of pesticides or nitrates at, or lower than, levels where health of humans could be scarce is weakly implicit.

TABLE 2.2 Polluted Stretches in Rivers and Lakes in States of India

Name of State	River	Number of Water Bodies	Lake/Tank/Drain, etc.
Delhi	1	1	–
U.P.	8	8	–
Tamil Nadu	7	7	–
Assam	2	2	–
Jharkhand	1	1	–
Andhra Pradesh	3	8	5
Gujarat	9	10	1
West Bengal	1	1	–
Sikkim	1	1	–
Punjab	3	3	–
Haryana	2	3	1
Karnataka	4	6	2
Madhya Pradesh	4	5	1
Rajasthan	3	3	–
Orissa	5	5	–
Meghalaya	1	5	4
Maharashtra	15	15	–
Total	71	86	15

Maximum permissible contagion levels for several pesticides have been issued as health standard by EPA because several pesticides are considered carcinogenic if administered in large doses. Birds or aquatic organisms, a number of which are in danger of extinction in the environment and nontargeted plants, are also affected as the contaminated groundwater resurfaces.

The share of pollution via point sources, such as emancipation from sewage treatment plants or manufacturing sources, emerges to be shrinking because of numerous years of control efforts. The major reason for the contamination of surface water these days as per EPA is the NPS pollution ensuing from pesticide application, agricultural tillage, urban development sites. The most extensive single source of surface water pollution is the agricultural runoff, which includes nearly about 55% of damaged river miles and 58% of damaged lake acres evaluated by the states in 1986 and 1987 (Agrawal et al., 2010). On the other hand, the degree to which the nation's groundwater resources are exaggerated by these compounds is less well recognized. Sighting of these chemical residues in groundwater all through the late 1970s and near the beginning of 1980s dispersed the normally apprehended observation that groundwater was secluded from farming chemicals by impermeable sheets of soil, rock, and clay. Contamination of groundwater may also be because of other sources, including nonagricultural utilization of fertilizers, and pesticides as well as leak subversive storage tanks.

2.12   QUALITY OF WATER AS A WORLD WIDE CONCERN

On a global scale, the one and only single largest user of freshwater is farming sector. Thus, making this sector the main reason for contamination of surface and groundwater resources through erosion and chemical runoff. Hence, a major cause to be worried about the global repercussions of water quality (Table 2.3). The allied agro-food dispensation industry is besides an important resource of organic contamination in the majority of countries. Aquaculture is nowadays acknowledged as a chief crisis in coastal environments, freshwater, estuarine thus, causing eutrophication and severe damage to ecosystem. The major ecological and public health proportions of the universal freshwater eminence crisis are enumerated below:

1.   Due to waterborne disease about five million people expire every 12 months
2.   Environmental dysfunction as well as loss of biodiversity.
3.   Pollution of aquatic environment from soil and other agricultural activities.
4.   Degradation of groundwater reserves.
5.   Universal pollution by unrelenting crude pollutants.

Specialists expect that, since contamination can no longer be cured by dilution in various countries, quality of freshwater will turn out to be the chief constraint for sustainable advancement in these realms timely in the subsequent century. This "calamity" is calculated to have the subsequent comprehensive magnitude:

1.   Depletion in sustainable food resources such as freshwater and coastal fisheries because of effluence.

2. Growing outcome of deprived supervision of water resource assessment due to insufficient water quality records in several countries.
3. Various countries can no longer administer contamination by dilution thus, causing elevated levels of marine contamination.
4. Probable loss of "creditworthiness" and increasing expenditure of remediation

TABLE 2.3   Principal Causes of Water Eminence Mutilation

Rank	Lakes	Rivers	Estuaries
1	Agriculture	Agriculture	Municipal point source
2	Urban runoff/storm sewers	Municipal point source	Urban runoff/storm sewers
3	Hydrologic/habitat modification	Urban runoff/storm sewers	Agriculture
4	Municipal point source	Resource extraction	Industrial point sources
5	On-site wastewaters	Industrial point sources	Resource extraction

Various countries distinguished the actual and impending loss of growth opportunities because of utilization of lots of funds for remediation of water pollution. An Expert Meeting which was held in 1994 on Water Quantity and Quality Management and was organized by the Economic and Social Commission for Asia and the Pacific (ESCAP). In this meeting, the representatives of Asia agreed over an affirmation which entitled to national and international action to review failure of financially viable opportunities because of water pollution. Also, it determined the prospective of financially viable brunt's of the "alarming water calamity." The major concern of Asian delegates interestingly to attend the meeting of ESCAP was specifically to exhibit the financial rather than only the ecological brunts of water pollution on sustainable advancement. Praise worthiness (Mackay and Paterson, 1991) is an apprehension so far as lending associations nowadays come across at the expenditure of remediation comparative to the money-making gains. Presently, the apprehension is that if the expenditure for remediation surpasses financial reimbursements, improvement projects possibly will no longer be praiseworthy. Agriculture sustainability without doubt will be obligatory to feature into its water reserve scheduling the superior concern of sustainable financially viable growth transversely with profitable segments. These inclusive advances to organization of water reserves have been tinted in the (Grigg, 2008) guiding principle on water resource advancements. In various human health concerns, older chlorinated agricultural pesticides have been occupied as sources of noteworthy and prevalent environmental dysfunction all the way through their poisonous consequence on individuals. Normally prohibited in the urbanized countries, there is at present an intensive global attempt to forbid these universally as piece of a code of behavior for Persistent Organic Pollutants. The most common case of such an attempt was the Intergovernmental Convention on the safety of the Aquatic Environment from land-based actions; summoned in Washington DC in 1995 in cooperation with UNEP.

2.13 ASPECTS DISTRESSING PESTICIDE TOXICITY IN WATER SYSTEMS

The environmental brunts of pesticides in water are indomitable by the subsequent criteria:

1. *Toxicity*: Toxicity of mammalian and nonmammalian species is generally articulated as LD50 ("Lethal Dose": Is defined as deliberation of the pesticide which will exterminate partially the test organisms over a particular assessment epoch). *Values of 0–10 are very toxic* as lower the LD50 greater the toxicity (Adedeji and Okocha, 2012). Risk-based review techniques are used to determine drinking water and food guidelines. Normally, risk is defined as contact, that is, quantity and/or period toxicity. Toxic reaction (consequence) can be *severe* (fatality) or *persistent* (an outcome that does not cause fatality through the test interlude however sources apparent results in the trial organism like tumors and cancers, reproductive breakdown, development reticence, teratogenic possessions, etc.).

2. *Perseverance*: Calculated as half-life, that is, the time requisite for the ambient absorption to reduce by 50%. Perseverance is resolute by biotic and abiotic dilapidation procedure. Biotic procedures are metabolism and biodegradation and abiotic procedures are chiefly photolysis, hydrolysis, and oxidation (Bozoglu, 2011). Present pesticides have a propensity to have small half-lives that reproduce the era over which the pest necessitate to be restricted.

3. *Depredates*: The squalor practice may direct to development of "depredates" which may have superior, equivalent, or minor toxicity

than the major composite. The example is the degradation of DDT to DDD and DDE.

4. *Providence (Ecological)*: The ecological providence, that is, performance of a pesticide is exaggerated by the innate similarity of the substance to one among the four ecological sections (Bozoglu, 2011): solid substance such as mineral matter and particulate crude carbon, liquid matter such as solubility in surface and soil water, gaseous form, that is, volatilization, and biota. These activities are often referred to as "separation" and engross, correspondingly, the purpose of solubility; the soil sorption coefficient (Koc); *n*-octanol/water partition coefficient (Kow), and Henry's constant (*H*). These constraints are well identified for pesticides and are utilized to envisage the ecological providence of the pesticide.

The existence of impurities in the pesticide formulation can be an auxiliary factor but does not act as a fraction of the vigorous components. The latest illustration is of 3-trifluoromethyl-4-nitrophenol (TFM), a lampricide utilized in branches of Great Lakes for several years for the management of the sea lamprey. Even though for many years the ecological fortune of TFM has been well known, the latest investigation by Munkittrick et al. (1994) has established a fact regarding the formulation of TFM that it comprises one or more extremely intoxicating contaminations that damage hormonal structure of fish and cause liver damage.

2.14 ENVIRONMENTAL CONSEQUENCE OF PESTICIDES

In a wide range of natural micropollutants, pesticides have been included which have environmental brunt. Diverse groups of pesticides have dissimilar varieties of consequences on existing organisms, hence simplification is complicated. Even though global brunts by pesticides do transpire, the primary conduit that causes environmental impacts is that of water polluted by pesticide overflow. The two main prime mechanisms are bio-magnification and bio-concentration.

1. *Bio-magnification*: The word portrays the rising amounts of chemicals as food energy is changed inside the food chain. The amounts of pesticides and supplementary substances are progressively more exaggerated in tissue and additional organs as minor organisms are

consumed by bigger organisms. Large amounts are experienced in top predators, together with man.

2. *Bio-concentration*: In this mechanism, there occurs the movement of a harmful substance via the immediate medium into an organism. The chief "reservoir" for a number of pesticides is "lipids," that is, fatty tissues. A number of pesticides, like DDT, are "lipophilic" (fat loving), that is, they accumulate and are soluble in fatty tissues like human fatty tissue and edible fish tissue. Besides another pesticide called glyphosate is metabolized and excreted.

The environmental brunts of pesticides and additional natural pollutants are diverse and are frequently interconnected. Ecological or organism level effects are typically measured to be early on admonition signs of impending human healthiness impacts. Main forms of brunts related to pesticides are enumerated below besides these effects will contrast depending on the type of pesticide and the organism under investigation. The simplification is extremely complicated as different pesticides have noticeably diverse effects on aquatic life. The central view is that various effects among these are chronic, that is, nonlethal thus, are often unnoticed by informal viewers; however, have consequences for the whole food chain.

1. Organism's death.
2. Lesions, tumors, and cancers on aquatic mammals and animals.
3. Failure or inhibition of reproductive system.
4. Hormonal system disorder.
5.   DNA as well as cellular damage.
6. Teratogenic effects, that is, bodily malformation like curved beaks of birds.
7. Deprived health of fish noticeable by stumpy ratio of red to white blood cell as well as too much slime on scales and gills of fish, etc.
8. Intergenerational possessions: these possessions are not visible until succeeding generations of that organism.
9. Various additional physiological possessions like the thinning of eggshell.

The above effects are not wholly and solely because of contact with pesticides and other natural contaminants but may also be linked to a mixture of various ecological stresses like pathogens and eutrophication. These connected anxieties do not require being huge for synergistic possessions with natural microcontaminants. Environmental possessions of pesticides enlarge ahead of individual organisms and are capable of extension to ecological unit (Husna, 2012)

2.15 CONCLUSION

The significant remedies for the management agrochemical wastes on farm are helpful and harmless to the surroundings; on the other hand, it is uniformly essential that they are realistic, cost-effective, and do not over-trouble the grower. As a well-known fact that water is the most vital resource of environment, however, it is being overstimulated. By the passing time, this precious resource is getting polluted and much less exertion is remunerated toward its proficient and prudent consumption. Contamination of the water is an extremely severe trouble as it distresses all the living creatures, both directly and indirectly. Therefore, it has turned out to be crucial to administer ground and hence protect it. Agricultural activities serve up as a prospective pollutant of water which is undoubtedly apparent from the preceding facts. Hence, approaches are required to check the flow of disproportionate agrochemicals out of arming fields to groundwater as well as surface water.

KEYWORDS

- pesticides
- drinking water
- water pollution
- environment
- methodologies
- drainage
- rainfall
- soil temperature

REFERENCES

Adedeji, O. B.; Okocha, R. O. Overview of pesticide toxicity in fish. *Advances in Environmental Biology* **2012**, 2344–2352.

Agrawal, A.; Pandey, R. S.; Sharma, B. Water pollution with special reference to pesticide contamination in India. *Journal of Water Resource and Protection* **2010**, *2*(5), 432.

Akhil, P. S.; Sujatha, C. H. Prevalence of organochlorine pesticide residues in groundwaters of Kasargod District, India. *Toxicological & Environmental Chemistry* **2012**, *94*(9), 1718–1725.

Aktar, W.; Sengupta, D.; Chowdhury, A. Impact of pesticides use in agriculture: Their benefits and hazards. *Interdisciplinary Toxicology* **2009**, *2*(1), 1–12.

Aslam, M.; Alam, M.; Rais, S. Detection of atrazine and simazine in ground water of Delhi using high performance liquid chromatography with ultraviolet detector. *Current World Environment* **2013**, *8*(2), 323.

Azab, A. M. *Integrating GIS, Remote Sensing, and Mathematical Modelling for Surface Water Quality Management in Irrigated Watersheds.* **2012**. CRC Press, Balkema.

Bakore, N.; John, P. J.; Bhatnagar, P. Organochlorine pesticide residues in wheat and drinking water samples from Jaipur, Rajasthan, India. *Environmental Monitoring and Assessment* **2004**, *98*(1–3), 381–389.

Bortleson, G. C.; Davis, D. *Pesticides in Selected Small Streams in the Puget Sound Basin 1987–1995*, US Geological Survey **1997**, No. 67–97.

Bozoglu, F. Impact of pesticides as organic micro-pollutants on the environment and risks for mankind. In *Environmental Security and Ecoterrorism*, Hami Alpas, Simon M. Berkowicz and Irina Ermakova (Eds.), **2011**, 73–82.

Burgan, B. *National Water Quality Inventory: 1994 Report to Congress* (No. PB-96-185699/XAB; EPA-841/R-95/005). Environmental Protection Agency, Washington, DC, USA. Office of Water **1995**.

Chaudhary, V.; Jacks, G.; Gustafsson, J. E. An analysis of groundwater vulnerability and water policy reform in India. *Environmental Management and Health* **2002**, *13*(2), 175–193.

Debnath, M. Insecticide toxicity on indigenous cyanobacteria from alluvial rice fields. In *The Role of Microalgae in Wastewater Treatment,* Sukla, Enketeswara Subudhi and Debabrata Pradhan (Eds.), **2019**, 137–151, Springer.

Ghose, N. C.; Dipankar, S. A. H. A.; Gupta, A. Synthetic detergents (surfactants) and organochlorine pesticide signatures in surface water and groundwater of greater Kolkata, India. *Journal of Water Resource and Protection* **2009**, *1*(4), 290.

Gilliom, R. J.; Barbash, J. E.; Crawford, C. G.; Hamilton, P. A.; Martin, J. D.; Nakagaki, N.; Wolock, D. M. Pesticides in the nation's streams and ground water. *US Geological Survey* **2006**, *No. 1291*, 1992–2001.

Goldar, B.; Banerjee, N. Impact of informal regulation of pollution on water quality in rivers in India. *Journal of Environmental Management* **2004**, *73*(2), 117–130.

Gonçalves, C. M.; Esteves da Silva, J. C.; Alpendurada, M. F. Evaluation of the pesticide contamination of groundwater sampled over two years from a vulnerable zone in Portugal. *Journal of Agricultural and Food Chemistry* **2007**, *55*(15), 6227–6235.

Grigg, N. S. Integrated water resources management: Balancing views and improving practice. *Water International* **2008**, *33*(3), 279–292.

Husna, A. H. *Determination of Pesticides in Water Samples Using Dispersive Liquid–Liquid Microextraction (DLLME) and Gas Chromatography-Micro Electron Capture Detector (GC-μECD)/Husna Binti A Hamid* (Doctoral dissertation, University of Malaya). **2012**.

Islam, N.; Sadiq, R.; Rodriguez, M. J.; Francisque, A. Reviewing source water protection strategies: A conceptual model for water quality assessment. *Environmental Reviews* **2011**, *19*, 68–105.

Jayashree, R.; Vasudevan, N. Organochlorine pesticide residues in ground water of Thiruvallur district, India. *Environmental Monitoring and Assessment* **2007**, *128*(1–3), 209–215.

Kaushik, C. P.; Sharma, H. R.; Kaushik, A. Organochlorine pesticide residues in drinking water in the rural areas of Haryana, India. *Environmental Monitoring and Assessment* **2012**, *184*(1), 103–112.

Khanna, R.; Gupta, S. Agrochemicals as a potential cause of ground water pollution: A review. *IJCS* **2018**, *6*(3), 985–990.

King, C. *UNU/IAS Working Paper No. 101*, **2003**.

Kole, R. K.; Banerjee, H.; Bhattacharyya, A. Monitoring of market fish samples for endosulfan and hexachlorocyclohexane residues in and around Calcutta. *Bulletin of Environmental Contamination and Toxicology* **2001**, *67*(4), 554–559.

Kong, D. Y.; Zhu, Z. L.; Shi, L. L.; Shan, Z. J.; Cai, D. J. Effect of pesticides on groundwater under sweet-potato-based cropping systems in northern China. *Journal of Agro-Environment Science* **2004**, *23*(5), 1017–1020.

Kumar, S.; Singh, K. P.; Gopal, K. Organochlorine residues in rural drinking water sources of Northern and North Eastern India. *Journal of Environmental Science & Health Part A* **1995**, *30*(6), 1211–1222.

Kumari, B.; Madan, V. K.; Kathpal, T. S. Status of insecticide contamination of soil and water in Haryana, India. *Environmental Monitoring and Assessment* **2008**, *136*(1–3), 239–244.

Mackay, D.; Paterson, S. Evaluating the multimedia fate of organic chemicals: A level III fugacity model. *Environmental Science & Technology* **1991**, *25*(3), 427–436.

Mills, P. C.; Kolpin, D. W.; Scribner, E. A.; Thurman, E. M. Herbicides and degradates in shallow aquifers of Illinois: Spatial and temporal trends 1. *Journal of the American Water Resources Association* **2005**, *41*(3), 537–547.

Mohapatra, S. P.; Kumar, M.; Gajbhiye, V. T.; Agnihotri, N. P. Ground water contamination by organochlorine insecticide residues in a rural area in the Indo-Gangetic plain. *Environmental Monitoring and Assessment* **1995**, *35*(2), 155–164.

Moye, H. A.; Miles, C. J. Aldicarb contamination of groundwater. In *Reviews of Environmental Contamination and Toxicology* **1988**, 99–146. Springer, New York, NY, USA.

Munkittrick, K. R.; Servos, M. R.; Parrott, J. L.; Martin, V.; Carey, J. H.; Flett, P. A.; Van Der Kraak, G. J. Identification of lampricide formulations as a potent inducer of MFO activity in fish. *Journal of Great Lakes Research* **1994**, *20*(2), 355–365.

Ongley, E.D. Control of Water Pollution from Agriculture. FAO Irrigation and Drainage, Paper 55, FAO, Rome **1996**.

Rai, P.; Pandey, N. D. Impact of Pesticide Usage on Water Resources: Indian Scenario **2003**.

Rothschild, E. R.; Manser, R. J.; Anderson, M. P. Investigation of aldicarb in ground water in selected areas of the Central Sand Plain of Wisconsin. *Groundwater* **1982**, *20*(4), 437–445.

Sankhla, M. S.; Kumari, M.; Sharma, K.; Kushwah, R. S.; Kumar, R. Water contamination through pesticide & their toxic effect on human health. *International Journal for Research in Applied Science and Engineering Technology* **2018**, *6*, 967–970.

Somashekar, K. M.; Mahima, M. R.; Manjunath, K. C. Contamination of water sources in Mysore city by pesticide residues and plasticizer—A cause of health concern. *Aquatic Procedia* **2015**, *4*, 1181–1188.

Tariq, M. I.; Afzal, S.; Hussain, I. Pesticides in shallow groundwater of Bahawalnagar, Muzafargarh, DG Khan and Rajan Pur districts of Punjab, Pakistan. *Environment International* **2004**, *30*(4), 471–479.

Tschirley, F. *Managing Pesticides to Avoid Surface and Groundwater Contamination.* Cooperative Extension Service, Michigan State University, **1990**.

Wall, G. R.; Riva-Murray, K.; Phillips, P. J. *Water Quality in the Hudson River Basin, New York and Adjacent States* **1998**, *1992–95* (Vol. 1165). US Geological Survey.

World Health Organization. **2004**. *Guidelines for Drinking-Water Quality* (Vol. 1). World Health Organization.

CHAPTER 3

Use of Pesticides in Agriculture: Impacts on Soil, Plant and Human Health

ZIA UR RAHMAN FAROOQI[1*], ABDUL KAREEM[1],
MUHAMMAD ASHAR AYUB[1], MUHMMAD MAHROZ HUSSAIN[1],
NUKSHAB ZEESHAN[1], and MUHAMMAD TAHIR SHEHZAD[2]

[1]*Institute of Soil and Environmental Sciences, University of Agriculture, Faisalabad 38040, Pakistan*

[2]*Global Centre for Environmental Remediation (GCER), University of Newcastle, Callaghan 2308, Australia*

Corresponding author. E-mail: ziaa2600@gmail.com

ABSTRACT

Pesticides in agriculture have an important place. A lot of insects, pests, and rodents are controlled by them to save the food and cash crops. Farmers get different benefits by using these pesticides like more crop yield, decreased pest infestation to their crops and increased economic returns. They are also playing an important role in food security right from green revolution till now. Discovery of the pesticides occurred when lots of food crop was destroyed on the field due to the insect and pest attack. With the scientific advancement, many specific and multipurpose pesticides were developed which were toxic to both the pests and environment. They caused serious soil, plant, and human health effects by bioaccumulating in them. Overexploitation of these pesticides was also a problem that aggravated the serious effects. To overcome this situation, different methods and techniques were developed to control the toxic effects of pesticides on soil, plant, and humans. Along with the remediation techniques, different alternatives to these pesticides were also developed to avoid progress and advancements in agriculture sector. This chapter briefly discusses the history, effects, impacts, and remediation techniques for sustainable agriculture.

3.1 INTRODUCTION

Pesticides (Ps) are synthetic complexes which are made for the eradication of any yield limiting factor from the field (Khan, 2016; Jacobson, 2018). In general, pesticides have become an integral part of global agricultural systems in the last century, resulting in a significant increase in crop production and food security. Nevertheless, exponential population growth underscores the need to strengthen food production. The conflict between food production and the displacement of millions of refugees has aggravated this need, along with the impact of climate change on agricultural activities, exacerbating food shortages in many areas of the world, and calling for renewed efforts in food production. Meanwhile, we have noticed in recent decades that pesticides are distributed in the environment, causing serious soil and food pollution. In addition, the global pollution of aquatic systems by Ps residues—a case study of tropical coastal ecosystems—is repeatedly damaging food resources and fisheries (Carvalho, 2017). The use of Ps was limited in ancient times. The oldest testimony is that Homer mentioned (around 1000 BC), Odysseus burned "sulfur" by vacating the halls, houses, and courts (Odyssey XXII, 492–494). Pliny the Elder (23–79 AD) collected in his natural history many anecdotes about the use of Ps in the last three or four centuries. Dioscorides, a Greek physician (40–90 AD), knew the toxicity of sulfur and arsenic. History showed that around 900 AD, Chinese used arsenic sulfide to stop invading attacks in gardens. *Veratrurn album* and *V. nigrum*, two fake sorghum roots were used by the Romans to kill rodents. In 1669, arsenic was first time known as Ps in the world, referring to the use of honey as an ant bait. Tobacco use was also implemented in the same century as contact pesticide for mites. It has been known since 1807 that copper compounds are fungicidal, and in 1883 France first used Bordeaux mixture as Ps. There are also evidences of hydrocyanic acid use as fumigant. Carbon disulfide has also been used since 1854 as a fumigant for insects (Costa, 1987; Kinkela, 2016).

There are many types of Ps. The importance of plant protection products for agriculture in surface waters may be underestimated, as it lacks comprehensive quantitative analysis in comparison to nutrient content and habitat degradation. The increasing contamination with Ps has led to a reduction in regional aquatic biodiversity, that is, pesticide concentrations are in line with the legally recognized limits and the incidence of large invertebrates has fallen by approximately 30% (Stehle and Schulz, 2015; Dev, 2017; Simon-Delso et al., 2015). Today, these Ps contain several elements, with a global share sale in market about 25% (Bass et al., 2015).

Despite the widespread Ps use, only a few examples of resistance have been reported. The complex segregation indicates that several genes are responsible for resistance, although the numbers of offspring were too small to substantiate any genetic model. The number of reports documenting resistance to dinitroanilines is relatively small given the long term and widespread use of these herbicides. Those resistant biotypes that have been selected offer an intriguing view as to what perturbations to the cytoskeletal proteins can occur and their consequences. Evidence of the mode of action of acetolactate synthase herbal herbicides comes from various scientific disciplines. Noncompetitive binding of imazaquin to pyruvate was observed using plant, but noncompetitive binding of imidazolinone has also been reported. The reduced susceptibility of the plants to several herbicides is associated with a lower uptake or relocation rate from the ingested to the active site. The commercial form of the herbicide is an inactive methyl ester, the active form of the herbicide is the free acid imazamethabenz. Sulfonylureas, imidazolinones, triazolopyrimidines, and pyrimidinyl thiobenzoates include current or upcoming commercial inhibitor Ps (Saari et al., 2018; Smeda and Vaughn, 2018). Pesticides are still efficient, cost-effective, and fast controlling way to eradicate the invading pest problem. The strategies are now being developed to clean these substances in an economical and eco-friendly manner (Singh and Singh, 2016; Duke, 2018).

In recent times, a variety of oils extracted from plants are tested as broad-spectrum Ps and they gave promising results. These extracts showed high efficacy, manifold mechanisms against insects, and minimal effects on nontargeted species. However, less number of commercially produced biopesticides (BPs) are available. It is a need of time for their promotion to save soil and human health for sustainable food production (Pavela and Benelli, 2016; Lovett and St. Leger, 2018; Mishra et al., 2015). In this chapter, we have discussed the main strengths and weaknesses arising from the use of Ps, their solutions, and key challenges for future research.

3.2 APPLICATION OF PESTICIDES IN AGRICULTURE AND IMPORTANCE IN SUSTAINABLE AGRICULTURE

In current scenario, world population is increasing @ 70 million persons per year with expected 10 billion people by the end of 21st century (Gerland et al., 2014). We have to increase food production to feed this large population as after green revolution crop yield increased sharply but now have attained maximum potential and now lagging phase has been started due to pest and insect diseases (Grassini et al., 2013). As reported till today, approximately 67,000 species of pests are known affecting crops and without Ps, up to 70% of agriculture production can be lost (Donatelli et al., 2017). These Ps are beneficial in controlling pest-related yield compromise but their intensive use is of main concern nowadays as it can lead to environmental damage and toxicity to living organisms and making pest resistant to their use (Fountain and Wratten, 2013; Kumar, 2012).

A wide range of Ps is being applied in field to control insects, pests, and weeds and diseases (Juraske et al., 2007). Though application rate of these pesticides is very uncontrolled and excessive, only a fraction of applied amount reaches targeted pests and remaining part moves deep into ecosystem and can leave long-lasting impacts on human lives and health (Davidson, 2004; Bouchard et al., 2010). Various controlling agencies and stakeholders (growers, policymakers, and protection agencies) are making great efforts to reduce pesticide risk as usually these pesticides have long-lasting residual impacts (Zhan and Zhang, 2018). A wide range of models is being used to assess Ps risk globally (van Bol et al., 2002; Bockstaller et al., 2009; Trevisan et al., 2009). These models work on estimated exposure level and toxicity effects. Models of risk assessment involve algorithms involving wide range of data as pesticide half-life and residue toxicity to assess pesticide environmental risk (Levitan et al., 2000; Bockstaller et al., 2008; Stenrod et al., 2008; Pierlot et al., 2017). Soil features and environmental factors are key influencers of Ps fate and thus included in more sophisti-cated indicators of more realistic and site-specific pesticide risk evaluation (Zhan and Zhang, 2018). Integrated pest management practices have been reported to be efficient in effective pest management in modern agriculture (USEPA, 2017) which involves, biological, cultural, mechanical, physical, and chemical control of pests (Zhan and Zhang, 2018). World usage of Ps is about two million tons per year which includes; 45% usage in Europe, 25% in USA, and 25% in the rest of world (De et al., 2014).

3.3 FATE OF PESTICIDES IN ENVIRONMENT; PESTICIDES ACTIVITY AND ITS BIOAVAILABILITY

Application of Ps in soil can lead to various fates depending upon its persis-tence and mobility (Kerle et al., 2007). Figures 3.1 and 3.2 explain the entry of pesticides in environment involves its application drift, soil erosion, and dry deposition (Cessna et al., 2005, 2006). Based on environmental persistence, major classes of Ps involve organochlorine insecticides, organophosphate insecticides, triazine herbicides, and acetanilide herbicides (Toth and Buhler, 2009).

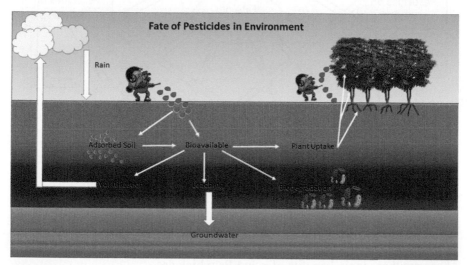

FIGURE 3.1 Fate of pesticides in environment.

Pesticide entry in aquatic, soil, and atmosphere are main sinks of pesticides in environment (Tiryaki and Temur, 2010). Fate of Ps involves its absorption by plants (Juraske et al., 2009; Hwang et al., 2017; Motoki et al., 2015). Second fate of Ps is adsorption on soil constituents which depends upon physico-chemical properties of soil (Zbytniewski et al., 2002; Kah and Brown, 2006; Kodešová et al., 2011). Pesticide leaching and runoff are other fates which involve solubilization in soil solution, wind, and soil erosion (Cohen et al., 1995; Kellogg et al., 2002). Pesticide degradation is the most concerned factor nowadays and its success is dependent upon the physico-chemical properties of media and nature of Ps (Kah et al., 2007). Soil microbial community is a leading factor in pesticide degradation (Kumar et al., 1996; Pal et al., 2006). If these pesticides are not degraded in soil then can leach down and pollute groundwater (Arias-Estévez et al., 2008) or runoff from field causing water pollution (Cohen et al., 1995; Kellogg et al., 2002). Various factors are involved in pesticide activity and bioavailability in soil like soil pH (Kah et al., 2007), growth of plant, and soil adsorption of pesticides (Edwards, 1975).

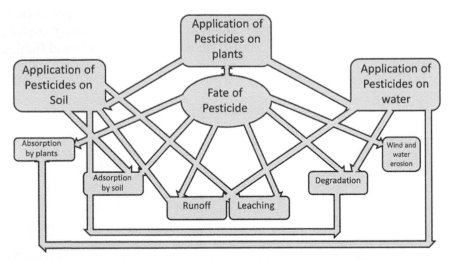

FIGURE 3.2   Pesticide application in agriculture, its fate, and bioavailability.

3.4   BENEFITS OF PESTICIDES USE IN AGRICULTURE

There are three main benefits of using Ps which are

(1) They control agricultural crops insects, pests, and pathogens (including diseases and weeds) and plant vectors.
(2) Control disease carriers and pathogens of human and animals
(3) Prevention of that endangers other human activities and structures.

Over the past six decades, farmers accomplished important development in manufacture of foodstuff by means of Ps. They used this principally to avoid and decrease agricultural losses due to action of pests which cause increase in production and fighting food security, at a sensible price in all seasons. The use of Ps in agriculture has increased the productivity of different crops in different countries, such as wheat production in the United Kingdom and corn production in the United States and many other examples. It has long been believed that foods containing fresh fruits and vegetables cannot offset the potential risk of very low Ps residues in crops. Better nutrition and less work can improve quality of life and longevity. Strengthening health care, drug treatment, and sanitation plays an important role in extending life, but the value of nutrition, safety, and adequate nutrition should not be underestimated as a health promoter of increased life expectancy.

3.5 PRIMARY BENEFITS

The main benefit lies in the importance of the effect of Ps—a direct improvement over the expected use. For example, killing of caterpillars that feed on crops. These three main effects have brought 26 secondary benefits, from the protection of recreational lawns to the rescue of humanity. Secondary benefits are less immediate or incomprehensible benefits of greater profits. They can be indirect, less intuitive, or long term. The use of pesticides in forestry, public health, the family sector, and, of course, agriculture has brought incredible benefits. The increase in productivity is due to various problems, including the use of fertilizers, better grades, and the use of machinery. Pesticides are an important part of this process as they reduce weeds and the loss of pests and diseases and significantly reduce the number of products that can be harvested.

3.5.1 MONETARY

The control of a variety of human and animal disease carriers can reduce the number of diseased units, and the suppression of death by international disease transmission is an important and significant benefit of the widespread use of pesticides. Killing the vector is the most effective method. Pesticides are used to maintain the decorum of the sports grounds. Pesticides enhance the durability of buildings and other wood structures from termites and hibiscus insects.

A common misconception among the public is that only farmers' income got increase by pesticide application. There are three main benefit types: (1) public health, quality of life, and wellbeing, (2) farm income, costs, and profits, (3) environmental issues reduction, including global warming. The benefits of these areas can be in the community, nationally, or globally. For example, the use of herbicides has saved costs and the efforts to control mechanical weeds at local level have led to a medium-term public interest in reducing housework, improving the living environment of public and private institutions or sports facilities at national level. Reducing the environmental benefits of using fossil fuels, soil disturbances, and water losses from agriculture—the global dimension is beneficial to all of us.

3.6   IMPACTS OF PESTICIDES USE IN AGRICULTURE

Combine the use of pesticides with other technologies, such as improved plant varieties, and agricultural equipment to promote the green revolution and the highest capacity to produce food and fiber in history. With the increasing use of Ps, the potential effects on public health and the environment have evolved. The environmental deterioration of using pesticides may also be higher than recommended, accidental spillage, or long-term in situ soil residence. Pesticide application practices have intensified many environmental problems, including human and animal health risks. Foods that are contaminated with toxic pesticides have serious effects on human health as they are a prerequisite for life. Soil, water, and plants contamination by pesticides is well known. In addition to killing insects or weeds, Ps can be toxic to many other organisms, including birds, fish, beneficial, and nontarget plants. Insecticides are usually the most toxic class of pesticides, but herbicides can also pose a threat to nontarget organisms.

3.6.1   SOIL HEALTH

Pesticides are an important component of many agricultural management systems and should consider their effects and their capability to destroy soil health. Limited information is available on standard test protocols for the assessment of lethal or sublethal toxicity in Ps-contaminated soils. The relationship between the impact of Ps of one or more indicator species on the health of soil ecosystems is more complex. Factors such as soil resilience to harsh conditions, persistence of Ps, or indirect effects play an important role in this assessment. Many pesticides quickly disintegrate. Others can accumulate so they are more concentrated in the soil when reused (Brasil et al., 2018).

Soil fauna (e.g., ticks, nematodes, micro-arthropods, protozoa) has certain importance in the transformation of organic matter and soil structure and are useful biological indicators for studying the toxicity in soil. The toxicity of Ps to soil fauna and flora is an integral part of the standardized ecotoxicity test for the registration of Ps. The metabolic capacity of the soil is usually characterized by the quantification of the activity of various hydrolysis and oxidoreductases. Pesticide has considerable impact on the enzyme activity in the soil. The activity of soil enzymes is a useful indicator of soil health and was used to determine whether adverse effects of farming practices affect soil biochemical functions. The effect of pesticides on enzyme activity depends on soil conditions and pesticide application rate (Hung et al., 2019).

3.6.2   PLANT HEALTH

Pesticides not only kill the invading pests but also showed the detrimental effects on the surrounding environment including plants and beneficial microbial communities. Groundwater contamination by Ps is an emerging issue globally. According to the US Geological Survey, at least 143 different pesticides and 21 transformation products have been found in groundwater, including Ps in each of the major chemical categories (Kim et al., 2017). Plants are susceptible to the indirect effects of Ps use when soil microbes and beneficial organisms are harmed. In arctic environmental samples such as Dacthal, Chlorthalonil, Chlorpyrifos, Metolachlor, Terbufos, and Triflu-ralin, including a new generation of pesticides have been detected. Once the groundwater is contaminated with toxic chemicals, it may take many years for the contaminants to dissolve or be eliminated. Cleaning can also be very expensive and complicated, which is not impossible.

3.6.3   HUMAN HEALTH

There is a credit of Ps to increase food as well as other features; they have harmful effect on human life. Different pesticides can cause different diseases in humans. Its toxicity can result from ingestion, inhalation, or dermal absorp-tion (Fantke and Jolliet, 2016). Their continuous toxicity results in various diseases like neurological, hormonal imbalance, immune system dysfunc-tion, cancer, blood disorder, genotoxicity, defects of reproductive system (Li and Jennings, 2017). Farmers and working staff dealing with pesticides and its formulators, sprayers, mixers, loaders, and farm workers are at high risk. The manufacturing and formulation process involves high risks and there is a high probability of danger. In industrial environments, workers are at an increased risk of handling a variety of toxic chemicals, including pesticides, raw materials, toxic solvents, and inert carriers. Early health examinations, including liver function, immune function, nerve damage, and reproductive effects, yielded incredible results. Excessive mortality from heart and respi-ratory diseases may be related to the psychosocial value of the accident and chemical infections. There was also an increase in diabetes cases. Follow-up results of cancer incidence and mortality showed an increase in the number of cancers in gastric and lymphatic, and hematopoietic tissues (Matthews, 2016).

3.7   PESTICIDES CONTAMINATED SOILS AND THEIR REMEDIATION TECHNIQUES

The uncontrolled use of Ps in the agricultural crop production is posing serious threats to the soil which is among alarming environmental issues; this widespread use of Ps worldwide and the lack of large-scale tested remediation technologies have increased this problem. Contaminated soil can be defined as soil with contaminant concentrations above the levels required by relevant regulations. Environmental guidelines on air and water pollution are commonly used in the most developed countries, but few countries have soil pollution regulations.

The purpose of this chapter is to outline the real research and development techniques for the restoration of Ps contaminated soils over the past few years. Since soils contaminated with Ps have complex mixture of various chemical compounds instead of a single toxin, so their remediation is a complex process. Based on relative nature and fate of pesticides to the remediation process, these techniques are divided into three classes: separation, containment fixation, or destruction. This includes some of the current states of emerging techniques and their advantages, limitations, and treatment of pesticides. Emerging technologies still require research and development initiatives to enable them to be fully implemented.

On contingent to technology used, soil remediation techniques can be applied in two ways: (1) In situ application of intermediation methods without excavation of soil, where contaminants are treated where pollution occurs; (2) ex situ, in which soils contaminated with Ps are excavated and transported to another place for its remediation (MoscaAngelucci and Tomei, 2016). To decrease, remove, separate, or stabilize Ps, physical, chemical, thermal, and biological processes are used as remediation techniques. Choosing the right technology depends on many factors, likewise site characteristics and pollution (prompt or diffusion), the concentration and nature of Ps to be isolated, and the dumping of the contaminated medium (Gavrilescu, 2009). For agricultural land, soil properties must be preserved, and aggressive techniques used for industrially contaminated soils cannot be used in agricultural soils. Therefore, most of the collected references are biological processes (bioremediation and phytoremediation).

3.8   FATE OF PESTICIDES BY REMEDIATION

1. Remediation technologies
2. In-situ remediation
3. Ex-situ remediation

3.8.1   EX-SITU VERSUS IN-SITU REMEDIATION

The two main remediation types are ex-situ and in-situ. Ex-situ involves physically removing the contaminated site and moving it to another location, while in-situ repair treats the contaminants on site. If the contaminants outside the site are only present in the soil, the soil is excavated. When the pollution reaches the groundwater, pump it and remove contaminated soil and water. Both ex-situ and in-situ techniques have specific advantages and costs. The main advantage of in-situ technology is that contaminated soil does not have to be removed or transported. A disadvantage of in-situ technology is that they remove contaminants less efficiently than ex-situ repair solutions. In addition, in-situ remediation techniques are preferred over ex-situ remediation techniques because the excavation costs are high, and the individual excavators are exposed to unfavorable health risks that are exposed to contaminants. Despite the high cost, the ex-situ treatment typically requires less time to achieve effective cleaning of the contaminants and is easier to monitor and more even. Decontaminated soil can be used for landscape purposes after an ex-situ treatment.

Processes Involved in Remediation

1. Physical–chemical
2. Thermal
3. Biological

3.9   PHYSICO-CHEMICAL TECHNIQUES

Physico-chemical techniques are based on physical and/or chemical phenomena. This section describes the main techniques for using soil-based methods for soil remediation.

3.9.1   SOIL VAPOR EXTRACTION

Among the physico-chemical techniques the most commonly used is the soil vapor extraction, a nonbiotechnology for the treatment of volatile organic compounds, semivolatile organic compounds (SVOCs) and polychlorinated biphenyls in the polychlorinated biphenyls, and in dioxins present in the unsaturated zone of the soil with a vacuum source applied to the soil matrix to produce a pressure gradient resulting from movement of the air present in the extraction well. Biological ventilation is one soil vapor extraction-like technology. In soil vapor extraction, the main removal mechanism is the volatilization, which promotes biodegradation of biological emissions and, therefore, has a lower rate of volatilization in the wellbore, which typically results in a gas phase that continues to be processed before entering the atmosphere must be (e.g., by adsorption in activated carbon) (Kinney et al., 2017).

3.9.2   AIR SPARGING

This technique also referred to as in-situ air stripping or in-situ volatilization involves injecting nonpolluting air into the subsurface of the saturation zone to convert hydrocarbons from a dissolved state to a gas phase. Then the air is vented through the unsaturated zone (Lok et al., 2017).

3.9.3   DECHLORINATION

Dechlorination is a technique that converts poisonous compounds into less poisonous intermediate substances based on the loss of halogen group (i.e., chlorine, fluorine, bromine, and iodine) in halogenated organic molecules. They are often water soluble and thus promote separation from the soil. This technology applies the nuclear power plant substitution reaction of chlorine atoms (or other halogens) to other less hazardous reactions using dehalogenation agents, sodium hydroxide, and potassium hydroxide, as well as polyethylene glycol. Basically, there are two variants of this technology: base-catalyzed decomposition and glycolic acid/basic polyethylene glycol. The first of these is used as an active reagent for sodium bicarbonate and has been used to treat chlorinated organic compounds, mainly polychlorinated biphenyls, dioxins, and furans, contaminated soils and sediments. In an alternative method, the reaction chemistry used is a basic polyethylenediol which produces residual water and must be further processed (He et al., 2016; Moorman, 2016).

3.9.4   SOIL FLUSHING

Soil flushing may also be accomplished by in-situ cleaning of the soil, inter layer by removing contaminants from the water, suspending in aqueous solution, or by contaminating the soil layer with a chemical reaction with the passing liquid. Soil cleaning can also be done ex-situ, with the following steps: digging, crushing, separating different particle sizes, washing, and treating different fractions. This technology removes both the organic and inorganic Ps from the contaminated soils and materials. The efficacy can be improved by using suitable additives. The sludge produced by this process can be reused for a treatment such as solvent extraction, solidification, or vitrification when mixed with soil or impurities. This technique is generally considered to be a pretreatment which reduces the amount of contaminated soil or material to be treated by any other remediation technique (Saez et al., 2016; Risco et al., 2016).

3.9.5   SOLVENT EXTRACTION

Although water and water additives used as cleaning agent rinse the Ps contaminated soils, the technique uses organic chemicals as the solvent. Commercially operated units are set up for this technique which can be differing in soil type, nature of equipment in it, and their modes of operation. Their operation starts by soil excavation and transportation, then, soil is prepared for adding into the extraction unit together with the solvent stream. Two streams are discharged from the extractor: one with treated soil that normally needs to be processed and one stream is sent to the separator to recover the solvent and concentrate the contaminants which are extracted from contaminated soil. They are then eliminated by chemical or biological treatment. Solvent extraction techniques do not eliminate impurities, but only they are separated from the soil and therefore are considered a pretreatment technique that resembles soil rinsing. The efficacy of this technique depends on the choice of solvent, and the solvent must be selected according to the type of impurity to be extracted (Duodo et al., 2016; Florindo et al., 2017).

3.9.6   SOLIDIFICATION/STABILIZATION

Pesticides contaminated soil remediation is done by hardening/stabilizing techniques in which reactive materials such as cement or concrete are mixed with solids, semi-solids, and sludge to fix contaminants. By adding stabilizers such as leached ash and wastes from the furnace, a block with high physical stability is generated during curing to limit the flow and solubility of the residue components. There are several variants of this technology: cement-based (adding cement directly to the soil), silicate-based curing (whereby the leached ash is added to the cement as material and then to the stabilizer then mix with the soil), and microencapsulation (Ochoa et al., 2018).

3.10   THERMAL TECHNIQUES

3.10.1   THERMAL INCINERATION

It is among the most commonly used techniques for decontamination of organic Ps contaminated soils, which burns Ps at high temperatures in the presence of oxygen so that Ps are converted into carbon dioxide and water. These technologies can efficiently de-contaminate the soils which are contaminated with halogenated and nonhalogenated Ps. This technique is operated on industrial scale. Depending on the type of energy recovery system used, incineration plants have two types: regenerative (shell exchange systems) and regenerative (ceramic exchange systems). The regeneration system can withstand high temperatures and allow recovery of approximately 90% of the combustion energy. However, there are higher capital costs than a regeneration system, which is offset by lower operating costs (Yoon et al., 2017).

3.10.2   THERMAL DESORPTION

It is a process designed to destroy any impurities. It involves heating the soil to volatilize water and contaminants. The gas stream or vacuum system supplies the previously volatilized compound to the gas processing unit. The temperature reached and the residence time are determined so that the volatilization of the selected impurities is promoted, but they cannot oxidize. Depending on the operating temperature, the process can be divided into two categories of high and low temperature thermal desorption. Thermal desorption systems can treat soils that are contaminated with different types of organic Ps but with different efficiencies. The low temperature thermal desorption process is typically used to treat halogen-free and combustible volatile organic carbons. The target group of pollutants in the high temperature thermal desorption process are SVOC, PAH, PCBs, and pesticides. It also removes volatile heavy metals. This process is also applied for separating organic residues in the wood industry as well as for treating soil contaminated with synthetic rubber, ink, creosote, or hydrocarbon (Lee et al., 2016; Shiea et al., 2015).

3.10.3 VITRIFICATION

Another thermal technique for floor cleaning is glazing, which converts contaminated soil into a glassy product and stabilizes it. This technique can be applied in-situ or ex-situ. In-situ glazing, a graphite electrode is inserted into the soil to generate a high current so that the released heat causes fusion of the soil matrix. As the glazed area grows, it combines with inorganic contaminants. The pyrolyzed organic components migrate into the vitrification zone where they are burned in the presence of oxygen. This is necessary to provide a treatment zone for the gas before it is released into the atmosphere. The ex-situ glazing is based on a similar treatment, except that the soil is excavated and placed in the glazing system, the function of which is identical to the described in situ glazing process (Tomašević and Gašić, 2015).

3.11 BIOLOGICAL TECHNIQUES

Biological remediation techniques include:

1. Bioremediation
2. Phytoremediation
3. Bioremediation techniques.

Bioremediation process increases the rate of microbial degradation of pollutants by providing these microorganisms with nutrients and energy sources (Aparicio et al., 2018). This can be achieved by using microbiota or by adding a microbially rich culture with specific enzymes which allow them to more rapidly degrade. Ideally, bioremediation leads to complete mineralization of contaminants to water and carbon dioxide without any other intermediates. Bioremediation process can roughly be divided into two categories: ex-situ and in-situ. Ex-situ restoration techniques include bioreactors, biofilters, tillage, and some composting methods. In-situ bioremediation techniques include biological ventilation, bioinjection, biostimulation, fluid delivery systems, and some composting methods. In-situ processes are often more attractive to suppliers and managers because they require less equipment, are more economic, and low environmental risks (Chen et al., 2015; Diaz et al., 2016). Bioremediation involves different techniques which are as follows:

Land Farming: Solid-phase treatment system for contaminated soils: may be done in-situ or ex-situ (Bicki et al., 2018).

Composting: Aerobic, thermophilic treatment process in which contaminated material is mixed with a bulking agent, which can be done using static piles or aerated piles (Epstein, 2018).

Bioreactors: Biodegradation in a container or reactor; may be used to treat liquids or slurries (Svobodova et al., 2018).

Bioventing: Method of treating contaminated soils by drawing oxygen through the soil to stimulate microbial activity (Parween et al., 2018).

Biofilters: Use of microbial stripping columns to treat air emissions (Wolfand et al., 2018).

Bioaugmentation: The addition of bacterial cultures to a contaminated medium; frequently used in both in situ and ex situ systems (Cycon et al., 2017).

Biostimulation: Stimulation of indigenous microbial populations in soils or groundwater by providing necessary nutrients (Johansen et al., 2015).

3.12   PHYTOREMEDIATION

The process that utilizes the ability of plants to absorb, accumulate and/or degrade constituents present in soil and water environments. All plants extract essential ingredients from these environments, including nutrients and heavy metals. Some plants are called hyperaccumulators because they can store a large amount of these metals, which apparently are not used for their function. It is also known that plants absorb and process various organic substances for physiological processes. There are five basic types of phytoremediation: (1) rhizofiltration, a water treatment technique that absorbs contaminants from the roots of plants; (2)

phytoextraction whereby impurities are absorbed from the soil; (3) phyto-transformation, applicable in soil and water involved in the degradation of pollutants by plant metabolism, (4) plant stimulation or plant-assisted biological remediation involving microbial degradation by plant activity in the root zone, (5) plant stabilization, use of plants to reduce pollutants due to soil migration.

3.13   ALTERNATE APPROACHES TO OVERCOME PEST ISSUES

3.13.1   BIOPESTICIDES

BPs are an environmentally friendly alternative to BPs, including a wide range of microbial Ps, biochemicals from microbes and other natural sources, as well as processes that involve the genetic incorporation of DNA into agricultural commodities to protect against pest damage. The potential benefits of using BPs for agricultural and health programs are considerable. The interest in them is due to deficiencies related to chemical Ps. Global BPs production exceeds 3000 tonnes per year, a rapid growth rate. BPs have enormous economic potentials. The adoption for farmers, however, requires training to maximize yields. The market share of BPs represents only 2.5% of the total pesticide market. The pressure on organic farming and other raw materials will certainly encourage farmers to make more use of BPs.

It is believed that pest-specific BPs are relatively safe for nontarget organisms, including humans. The most commonly used BPs are organisms that are pathogenic to the pest of interest. Some examples of PBs include Bt and Azain pesticides which are being widely used in the world (Pavela and Bannelli, 2016).

3.13.2   CULTURAL CONTROL

The culture control approach reduces the attractiveness of the environment to pests, adversely affects survival, transmission, growth, and reproduction, and promotes natural pest control. The aim is to reduce the number of pests below the level of economic damage or to enable natural or biological control.

3.13.2.1 STRATEGIES ON WHICH CULTURAL PRACTICES ARE BASED

Disruption of spawning preferences, host plant discrimination, or adult and immature sites make crops or pest habitats unacceptable. (1) Knowledge of the life history of pests, their habitat, and hibernation habits, so that pests cannot access crops spatially and temporally. (2) Reduce the survival of pests in crops by strengthening their natural enemies or by changing the susceptibility of crops to pests. In order to implement, it is necessary to understand the biology, ecology, and phenology of crops and pests as well as the vulnerabilities in the interaction between crops and pests.

3.13.2.2 ADVANTAGES

Cultural control is usually the economic approach. Sometimes they do not even need extra work, just plan carefully. Often, they are the only control measures that are conducive to the cultivation of high quality and low-value crops. Cultural control is reliable and usually specific. Above all these do not leave the residues in the food commodities, no killing of other beneficial organisms, and create no resistance.

3.13.2.3 DISADVANTAGES

Cultural control requires large strategies for maximum effectiveness and planning for its maximum efficacy. They are often based on the substitution of knowledge and skills in the purchase of equipment and therefore require greater capacity for farmers. These approaches may eliminate one pest species and may be ineffective for another. The efficacy of this type of approach cannot be determined and also it is not economically accepted way to control invading pests.

3.14 BIOCONTROL AGENTS

In recent decades, people have become more aware of the environmental and health risk of pesticides and are therefore endeavoring to lessen their dependence on chemical control. Many countries have adopted stricter rules for the production, registration, and use of Ps, which increases costs and reduces the availability of these tools. In many cases, it has been shown that the pest itself requires change, and Ps resistance is now a common reality in many weeds, insects, and diseases. The use of natural enemies to reduce pest impact has a long history. The ancient Chinese observed that ants are an effective predator for many citrus pests. They increased their populations by collecting nests from surrounding habitats and planting them in orchards. Today's insects and natural enemies in the air across the country or around the world are just modern adaptations of these original ideas.

3.15 IMPORTATION

When exotic pests are the target of a biocontrol, natural imports, known as classical biological control, are used. The constant import of pests into non-native countries, either by accident or in some cases, is intentional. Many of these introductions do not lead to the establishment or if so, the organism may not become a pest. However, due to the lack of pathogens to inhibit their populations, it is not uncommon for some of these introduced organisms to become pests. In these cases, entering pathogens can be very effective.

3.16 AUGMENTATION

Improvement in environment is to manipulate microbes directly to enhance their efficacy. This can be achieved by one or both two methods: mass production and colonization. The most common of these methods is the first one in which microbes are mass produced and introduced into natural environment. In areas where certain natural enemies cannot survive the winter, the population can detect and control pests with each spring vaccination. Therefore, proliferation usually does not permanently prevent pests, as may be the case with import or protection methods.

3.17 CONSERVATION

The protection of natural enemies is an integral part of any biological control. This includes identifying factors that can limit the effects of natural enemies and use them to increase their efficiency. The import, distribution, and protection of natural enemies are the three basic methods of biological control of insects. Specific technologies for these methods are being continuously developed and adapted to the changing requirements of pest management. Improvements in feeding and release techniques and genetic enhancement of natural enemies have led to more effective improvement programs. The application of new ecological theories changes our perception of the protection of natural enemies. If the full potential of this bio-based pest management strategy is to be exploited, biological control methods and applications need to be further refined and adapted.

3.18   GENETICALLY MODIFIED ORGANISMS

Through the expression of the various *Bacillus thuringiensis* (Bt) δ-endotoxin, insect resistance traits confer the second most common commercial herbicide-resistant transgenic crops. Four types of internal Bt toxin gene (cry1Ab, cry1Ac, cry2Ab, and cry9C) currently in corn and cotton prevent attacks on commercial lepidopteran pests. For one-year crops grown on a large scale worldwide, more Bt transgenic plants are being developed which express delta-endotoxin. Vegetative insecticidal proteins (non-δ-endotoxins) from Bt (VIP3A) also undergo field trials. Protease inhibitors and lectins, their pesticidal activity in the range of usually wider than the Bt toxin, are also used in many crop experiments. Biotin-binding protein, a toxin of bacterial symbionts of an entomopathogenic nematode of chitinase, an enzyme control of aromatic aldehyde synthesis, peptide spider venom, *Aedes aegypti* setting trypsin inhibitor, of enhancing insects, plants, and plant defensins hormones are some of the newer insecticidal transgenes used for agricultural purposes be used. By rapidly adopting the diversity of possible approach new features transgenic can introduce and genetically modified crops and the public debate, it is urgent to complete the regulatory framework for genetically modified pesticide protective products (Mandell et al., 2015).

3.19   CONCLUSIONS

In general, pesticides have become a key constituent of global agricultural sector in the last century, resulting in a significant increase in crop production and food security. Nevertheless, exponential population growth underscores the need to strengthen food production. Entry of pesticides in environment involves its application drift, soil erosion, and dry deposition. The environmental costs of using Ps may also be higher than recommended, accidental spillage, or long term in-situ soil residence. Pesticide application practices have intensified many environmental problems, including human and animal health risks. Foods that are contaminated with toxic Ps have serious effects on human health as they are a prerequisite for life. They can contaminate soil, water, pastures, and other plants. Their pollution is posing a serious threat to not only humans but also the other factors of environment. So, to remediate Ps, different approaches have been used but every approach has its own limitations. Need of time is to discourage the use of BPs and synthetic Ps and to start the use of organic approaches to eradicate the pest attack issue.

ACKNOWLEDGMENT

All the authors appreciate the time given by Mr Muhammad Usman Ghani and Waqas Mohy-Ud-Din for reviewing this chapter.

KEYWORDS

- **pesticides application**
- **biopesticides**
- **bioaccumulation**
- **environmental health**
- **remediation**

REFERENCES

Aparicio, J. D.; Raimondo, E. E.; Gil, R. A.; Benimeli, C. S.; Polti, M. A. Actinobacteria consortium as an efficient biotechnological tool for mixed polluted soil reclamation: experimental factorial design for bioremediation process optimization. *Journal of Hazardous Materials*, **2018**, *342*, 408–417.

Arias-Estévez, M.; López-Periago, E.; Martínez-Carballo, E.; Simal-Gándara, J.; Mejuto, J. C.; García-Río, L. The mobility and degradation of pesticides in soils and the pollution of groundwater resources. *Agriculture, Ecosystems & Environment*, **2008**, *123*(4), 247–260.

Bass, C.; Denholm, I.; Williamson, M. S.; Nauen, R. The global status of insect resistance to neonicotinoid insecticides. *Pesticide Biochemistry and Physiology*, **2015**, *121*, 78–87.

Bicki, T. J.; Felsot, A. S. Remediation of pesticide contaminated soil at agrichemical facilities. In *Mechanisms of Pesticide Movement into Ground Water*, **2018**, pp. 81–100. CRC Press, Boca Raton, FL, USA.

Bockstaller, C.; Guichard, L.; Keichinger, O.; Girardin, P.; Galan, M. B.; Gaillard, G. Comparison of methods to assess the sustainability of agricultural systems. A review. *Agronomy for Sustainable Development*, **2009**, *29*, 223–235.

Bockstaller, C.; Guichard, L.; Makowski, D.; Aveline, A.; Girardin, P.; Plantureux, S. Agri-environmental indicators to assess cropping and farming systems. A review. *Agronomy for Sustainable Development,* **2008**, *28*, 139–149.

Bouchard, M. F.; Bellinger, D. C.; Wright, R. O.; Weisskopf, M. G. Attention deficit/hyperactivity disorder and urinary metabolites of organophosphate pesticides. *Pediatrics,* **2010**, *125*(6), e1270–e1277.

Brasil, V. L. M.; Ramos Pinto, M. B.; Bonan, R. F.; Kowalski, L. P.; da Cruz Perez, D. E. Pesticides as risk factors for head and neck cancer: a review. *Journal of Oral Pathology & Medicine,* **2018**, *47*(7), 641–651.

Carson, R. L. *Silent Spring,* Riverside Press: Cambridge, MA, USA, **1962**.

Carvalho, F. P. Pesticides, environment, and food safety. *Food and Energy Security* **2017**, *6*(2), 48–60.

Cessna, A. J.; Wolf, T. M.; Stephenson, G. R.; Brown, R. B. Pesticide movement to field margins: routes, impacts and mitigation. Field boundary habitats: implications for weed. *Insect and Disease Management,* **2005**, *1*, 69–112.

Cessna, A. J.; Larney, F. J.; Kerr, L. A.; Bullock, M. S. Transport of trifluralin on wind-eroded sediment. *Canadian Journal of Soil Science,* **2006**, *86*, 545–554.

Chen, M.; Xu, P.; Zeng, G.; Yang, C.; Huang, D.; Zhang, J. Bioremediation of soils contaminated with polycyclic aromatic hydrocarbons, petroleum, pesticides, chlorophenols and heavy metals by composting: applications, microbes and future research needs. *Biotechnology Advances,* **2015**, *33*(6), 745–755.

Cohen, S. Z.; Wauchope, R. D.; Klein, A. W.; Eadsforth, C. V.; Graney, R. L. Pesticides report 35. Offsite transport of pesticides in water: mathematical models of pesticide leaching and runoff (Technical Report). *Pure and Applied Chemistry,* **1995**, *67*(12), 2109–2148.

Costa, L. G. *Toxicology of Pesticides: A Brief History* 1987, Berlin, Heidelberg.

Cycoń, M.; Mrozik, A.; Piotrowska-Seget, Z. Bioaugmentation as a strategy for the remediation of pesticide-polluted soil: a review. *Chemosphere,* **2017**, *172*, 52–71.

Davidson, C. Declining downwind: amphibian population declines in California and historical pesticide use. *Ecological Applications,* **2004**, *14*, 1892–1902.

De, A.; Bose, R.; Kumar, A.; Mozumdar, S. Worldwide pesticide use. In *Targeted Delivery of Pesticides Using Biodegradable Polymeric Nanoparticles,* **2014**, (pp. 5–6). Springer, New Delhi.

Dev, S. *Insecticides of Natural Origin,* **2017**. Routledge, Abingdon, United Kingdom.

Diaz, J. M. C.; Delgado-Moreno, L.; Núñez, R.; Nogales, R.; Romero, E. Enhancing pesticide degradation using indigenous microorganisms isolated under high pesticide load in bioremediation systems with vermicomposts. *Bioresource Technology,* **2016**, *214*, 234–241.

Donatelli, M.; Magarey, R. D.; Bregaglio, S.; Willocquet, L.; Whish, J. P.; Savary, S. Modelling the impacts of pests and diseases on agricultural systems. *Agricultural Systems,* **2017**, *155*, 213–224.

Duke, S. O. Effects of herbicides on nonphotosynthetic biosynthetic processes. In *Weed Physiology,* **2018** (pp. 91–112): CRC Press, Boca Raton, FL, USA.

Duodu, G. O.; Goonetilleke, A.; Ayoko, G. A. Optimization of in-cell accelerated solvent extraction technique for the determination of organochlorine pesticides in river sediments. *Talanta,* **2016**, *150*, 278–285.

Edwards, C. A. Factors that affect the persistence of pesticides in plants and soils. In *Pesticide Chemistry–3,* **1975**, (pp. 39–56). Butterworth-Heinemann.

Epstein, E. *The Science of Composting,* **2017**, Routledge, Abingdon, United Kingdom.

Fantke, P.; Jolliet, O. Life cycle human health impacts of 875 pesticides. *The International Journal of Life Cycle Assessment*, **2016**, *21*(5), 722–733.

Florindo, C.; Branco, L. C.; Marrucho, I. M. Development of hydrophobic deep eutectic solvents for extraction of pesticides from aqueous environments. *Fluid Phase Equilibria*, **2017**, *448*, 135–142.

Fountain, E. D.; Wratten, S. D. *Conservation Biological Control and Biopesticides in Agricultural.* **2013.**

Gavrilescu, M.; Demnerová, K.; Aamand, J.; Agathos, S.; Fava, F. Emerging pollutants in the environment: present and future challenges in biomonitoring, ecological risks and bioremediation. *New Biotech*, **2015**, *32*(1), 147–156.

Gavrilescu, M. Behaviour of persistent pollutants and risks associated with their presence in the environment–integrated studies. Environmental Engineering and Management Journal. **2009**, *8*(6), 1517–1531.

Gerakis, P. A.; Sficas, A. G. The presence and cycling of pesticides in the ecosphere. In *Residue Reviews*, **1974**, (pp. 69–87). Springer, New York, NY, USA.

Gerland, P.; Raftery, A. E.; Ševčíková, H.; Li, N.; Gu, D.; Spoorenberg, T.; Bay, G. World population stabilization unlikely this century. *Science*, **2014**, *346*(6206), 234–237.

Grassini, P.; Eskridge, K. M.; Cassman, K. G. Distinguishing between yield advances and yield plateaus in historical crop production trends. *Nature Communications*, **2013**, *4*, 2918.

He, W. Y.; Fontmorin, J. M.; Hapiot, P.; Soutrel, I.; Floner, D.; Fourcade, F.; Geneste, F. A new bipyridyl cobalt complex for reductive dechlorination of pesticides. *Electrochimica Acta*, **2016**, *207*, 313–320.

Huang, A. L.; Meng, L. Y.; Zhang, W.; Liu, J. Y.; Li, G. Y.; Tan, H. H.; Zheng, X. Effects of five pesticides on toxicity, detoxifying and protective enzymes in Phaudaflammans Walker (Lepidoptera: Zygaenidae). *Pakistan Journal of Zoology*, **2019**, *51*(4), 1457–1463.

Hwang, J. I.; Lee, S. E.; Kim, J. E. Comparison of theoretical and experimental values for plant uptake of pesticide from soil. *PLoS One*, **2017**, *12*(2), e0172254.

Jacobson, M. *Focus on Phytochemical Pesticides*, **2018**, CRC Press, Boca Raton, FL, USA.

Johansen, A.; Nielsen, T. K.; Tsitonaki, K.; Roost, S.; Larsen, L.; Tuxen, N.; Gosewinkel, U. B. In situ biostimulation and bioaugmentation to remove phenoxy-acids in subsoil. In *Science for the Environment*, **2015**.

Juraske, R.; Anton, A.; Castells, F.; Huijbregts, M. A. J. Pest Screen: a screening approach for scoring and ranking pesticides by their environmental and toxicological concern. *Environment International*, **2007**, 33, 886–893.

Juraske, R.; Castells, F.; Vijay, A.; Muñoz, P.; Antón, A. Uptake and persistence of pesticides in plants: measurements and model estimates for imidacloprid after foliar and soil application. *Journal of Hazardous Materials*, **2009**, *165*(1–3), 683–689.

Kah, M.; Beulke, S.; Brown, C. D. Factors influencing degradation of pesticides in soil. *Journal of Agricultural and Food Chemistry*, **2007**, *55*(11), 4487–4492.

Kah, M.; Brown, C. D. Adsorption of ionisable pesticides in soils. In *Reviews of Environmental Contamination and Toxicology*, **2006**, (pp. 149–217). Springer, New York, NY, USA.

Kellogg, R. L.; Nehring, R. F.; Grube, A.; Goss, D. W.; Plotkin, S. Environmental indicators of pesticide leaching and runoff from farm fields. In *Agricultural Productivity*, **2002**, (pp. 213–256). Springer, Boston, MA, USA.

Kerle, E. A.; Jenkins, J. J.; Vogue, P. A. *Understanding Pesticide Persistence and Mobility for Groundwater and Surface Water Protection.* Oregon State Univ Extension Service, EM8561-E. **2007.**

Khan, S. U. *Pesticides in the Soil Environment*, **2016**, Elsevier.

Kim, K. H.; Kabir, E.; Jahan, S. A. Exposure to pesticides and the associated human health effects. *Science of the Total Environment* **2017**, *575*, 525–535.

Kinkela, D. *Banned: A History of Pesticides and the Science of Toxicology*. By Frederick Rowe Davis. **2016**, Oxford University Press.

Kinney, K. A.; Wright, W.; Chang, D. P.; Schroeder, E. D. Biodegradation of vapor-phase contaminants. In *Fundamentals and Applications of Bioremediation*, **2017**, (pp. 601–632). Routledge, Abingdon, United Kingdom.

Kodešová, R.; Kočárek, M.; Kodeš, V.; Drábek, O.; Kozák, J.; Hejtmánková, K. Pesticide adsorption in relation to soil properties and soil type distribution in regional scale. *Journal of Hazardous Materials*, **2011**, *186*(1), 540–550.

Kumar, S. Biopesticides: A need for food and environmental safety. *Journal of Biofertilizers & Biopesticides*, **2012**, *3*(4), 1–3.

Kumar, S.; Mukerji, K. G.; Lai, R. Molecular aspects of pesticide degradation by microorganisms. *Critical Reviews in Microbiology*, **1996**, *22*(1), 1–26.

Kumar, S.; Singh, A. Biopesticides for integrated crop management: environmental and regulatory aspects. *Journal of Biofertilizers & Biopesticides*, **2014**, *5*, e121.

Lee, C. W.; Su, H.; Chen, P. Y.; Lin, S. J.; Shiea, J.; Shin, S. J.; Chen, B. H. Rapid identification of pesticides in human oral fluid for emergency management by thermal desorption electrospray ionization/mass spectrometry. *Journal of Mass Spectrometry*, **2016**, *51*(2), 97–104.

Levitan, L. "How to" and "why": assessing the enviro-social impacts of pesticides. *Crop Protection* **2000**, 19, 629–636.

Li, Z.; Jennings, A. Worldwide regulations of standard values of pesticides for human health risk control: a review. *International Journal of Environmental Research and Public Health*, **2017**, *14*(7), 826.

Lok, A.; Wray, H.; Bérubé, P.; Andrews, R. C. Optimization of air sparging and in-line coagulation for ultrafiltration fouling control. *Separation and Purification Technology*, **2017**, *188*, 60–66.

Lovett, B.; St. Leger, R. J. Genetically engineering better fungal biopesticides. *Pest Management Science* 2018, *74*(4), 781–789.

Mandell, D. J.; Lajoie, M. J.; Mee, M. T.; Takeuchi, R.; Kuznetsov, G.; Norville, J. E.; Church, G. M. Biocontainment of genetically modified organisms by synthetic protein design. *Nature*, **2015**, *518*(7537), 55.

Matthews, G. *Pesticides: Health, Safety and the Environment*, **2015**, John Wiley & Sons.

Mishra, J.; Tewari, S.; Singh, S.; Arora, N. K. Biopesticides: where we stand? In N. K. Arora (Ed.), *Plant Microbes Symbiosis: Applied Facets*, **2015**, (pp. 37–75). Springer India, New Delhi.

Moorman, T. B. Pesticide degradation by soil microorganisms: environmental, ecological, and management effects. In *Soil Biology*, **2018**, (pp. 127–172). CRC Press, Boca Raton, FL, USA.

MoscaAngelucci, D.; Tomei, M. C. Ex situ bioremediation of chlorophenol contaminated soil: comparison of slurry and solid-phase bioreactors with the two-step polymer extraction-bioregeneration process. *Journal of Chemical Technology & Biotechnology*, **2016**, *91*(6), 1577–1584.

Motoki, Y.; Iwafune, T.; Seike, N.; Otani, T.; Akiyama, Y. Relationship between plant uptake of pesticides and water-extractable residue in Japanese soils. *Journal of Pesticide Science,* **2015,** D15–017.

Ochoa, V.; Maestroni, B. Pesticides in water, soil, and sediments. In *Integrated Analytical Approaches for Pesticide Management,* **2018,** (pp. 133–147). Academic Press.

Pal, R.; Chakrabarti, K.; Chakraborty, A.; Chowdhury, A. Degradation and effects of pesticides on soil microbiological parameters—A review. *International Journal of Agricultural Research,* **2006,** *1*(33), 240–258.

Parween, T.; Bhandari, P.; Sharma, R.; Jan, S.; Siddiqui, Z. H.; Patanjali, P. K. Bioremediation: a sustainable tool to prevent pesticide pollution. In *Modern Age Environmental Problems and their Remediation,* **2018,** pp. 215–227. Springer, Cham.

Pavela, R.; Benelli, G. Essential oils as ecofriendly biopesticides? Challenges and constraints. *Trends in Plant Science,* **2016,** *21*(12), 1000–1007.

Pavela, R.; Benelli, G. Essential oils as ecofriendly biopesticides? Challenges and constraints. *Trends in Plant Science,* **2016,** *21*(12), 1000–1007.

Pierce Jr, R. H.; Olney, C. E.; Felbeck Jr, G. T. Pesticide adsorption in soils and sediments. *Environmental Letters,* **1971,** *1*(2), 157–172.

Pierlot, F.; Marks-Perreau, J.; Réal, B.; Carluer, N.; Constant, T.; Lioeddine, A.; van Dijk, P.; Villerd, J.; Keichinger, O.; Cherrier, R.; Bockstaller, C. Predictive quality of 26 pesticide risk indicators and one flow model: a multisite assessment for water contamination. *Science of the Total Environment,* **2017,** 605–606, 655–665.

Risco, C.; Rodrigo, S.; Vizcaíno, R. L.; Yustres, A.; Saez, C.; Cañizares, P.; Rodrigo, M. A. Removal of oxyfluorfen from spiked soils using electrokinetic soil flushing with linear rows of electrodes. *Chemical Engineering Journal,* **2016,** *294,* 65–72.

Saari, L.; Cotterman, J.; Thill, D. Resistance to acetolactate synthase inhibiting herbicides. In *Herbicide Resistance in Plants,* **2018,** (pp. 83–140), CRC Press, Boca Raton, FL, USA.

Saez, C.; Vieira dos Santos, E.; Cañizares, P.; Souza, F. L.; Lanza, M. R.; Martínez-Huitle, C. A.; Rodrigo, M. A. Application of electrokinetic soil flushing to four herbicides: a comparison. Chemosphere, **2016,** *153,* 205–211.

Shiea, C.; Huang, Y. L.; Liu, D. L.; Chou, C. C.; Chou, J. H.; Chen, P. Y.; Huang, M. Z. Rapid screening of residual pesticides on fruits and vegetables using thermal desorption electrospray ionization mass spectrometry. *Rapid Communications in Mass Spectrometry,* **2015,** *29*(2), 163–170.

Simon-Delso, N.; Amaral-Rogers, V.; Belzunces, L. P.; Bonmatin, J.-M.; Chagnon, M.; Downs, C.; Girolami, V. Systemic insecticides (neonicotinoids and fipronil): trends, uses, mode of action and metabolites. *Environmental Science and Pollution Research,* **2015,** *22*(1), 5–34.

Singh, B.; Singh, K. Microbial degradation of herbicides. *Critical Reviews in Microbiology,* **2016,** *42*(2), 245–261. doi:10.3109/1040841X.2014.929564.

Smeda, R. J.; Vaughn, K. C. Resistance to dinitroaniline herbicides. In *Herbicide Resistance in Plants,* **2018** (pp. 215–228): CRC Press, Boca Raton, FL, USA.

Stehle, S.; Schulz, R. Agricultural insecticides threaten surface waters at the global scale. *Proceedings of the National Academy of Sciences,* **2015,** *112*(18), 5750–5755.

Stenrod, M.; Heggen, H. E.; Bolli, R. I.; Eklo, O. M. Testing and comparison of three pesticide risk indicator models under Norwegian conditions—A case study in the Skuterud and Heiabekken catchments. *Agriculture, Ecosystems & Environment,* **2008,** 123, 15–29.

Svobodová, K.; Novotný, Č. Bioreactors based on immobilized fungi: Bioremediation under non-sterile conditions. *Applied Microbiology and Biotechnology*, **2018**, *102*(1), 39–46.

Tiryaki, O.; Temur, C. The fate of pesticide in the environment. *Journal of Biological and Environmental Sciences*, **2010**, *4*(10), 29–38.

Tomašević, A.; Gašić, S. Photochemical processes and their use in remediation of water containing pesticides. In *Proceedings of the 7th Congress on Plant Protection" Integrated Plant Protection—A Knowledge-Based Step Towards Sustainable Agriculture, Forestry and Landscape Architecture."* November 24–28, 2014, Zlatibor, Serbia, **2015**, pp. 365–369. Plant Protection Society of Serbia (PPSS).

Toth, S.J,; Buhler, W.G. *Environmental Effects of Pesticides*, **2009**, Department of Entomology and Horticultural Science, North Carolina State University.

Trevisan, M.; Di Guardo, A.; Balderacchi, M. An environmental indicator to drive sustainable pest management practices. *Environmental Modelling & Software*, **2009**, *24*, 994–1002.

USEPA. *Integrated Pest Management (IPM) Principles*. United States Environmental Protection Agency, **2012**. http://www.epa.gov/opp00001/factsheets/ipm.htm (accessed November 2017).

van Bol, V.; Debongnie, P.; Pussemier, L.; Maraite, H.; Steurbaut, W. *Study and Analysis of Existing Pesticide Risk Indicators*, **2002**, Veterinary and Agrochemical Research Center: Teruren, Belgium.

Wolfand, J.; Seller, C.; Bell, C. D.; Cho, Y.; Oetjen, K.; Hogue, T. S.; Luthy, R. G. Managing urban-use pesticides with enhanced green infrastructure on the watershed scale. In *AGU Fall Meeting Abstracts*. **2018**, December

Yoon, Y. S.; Kwon, E. H.; Bae, J. S.; Lee, S. Y.; Jeon, T. W.; Shin, S. K. A *Study on the Thermal Treatment Possibility of Organochlorine Pesticide Containing Wastes*. **2017**, pp. 92–92.

Zbytniewski, R.; Buszewski, B. Sorption of pesticides in soil and compost. *Polish Journal of Environmental Studies*, **2002**, *11*(2), 179–184.

Zhan, Y.; Zhang, M. Pesticide Use Risk Evaluation (PURE), a self-evaluation tool of pesticide use. In *Managing and Analyzing Pesticide Use Data for Pest Management, Environmental Monitoring, Public Health, and Public Policy,* **2018**, (pp. 518–534). American Chemical Society.

CHAPTER 4

Pesticide Pollution: Management and Challenges

SURAJ NEGI,[1] AISHWARYA RANI,[2] ALLEN H. HU,[1] and SUNIL KUMAR[3*]

[1]*Institute of Environmental Engineering and Management, National Taipei University of Technology, Taiwan, ROC*

[2]*Chaudhary Brahm Prakash Government Engineering College, Delhi, India*

[3]*CSIR-National Environmental Engineering Research Institute, Nagpur, India*

[*]Corresponding author. E-mail: s_kumar@neeri.res.in

ABSTRACT

Chemicals have found extensive application in the agricultural system due to its benefits in increasing agricultural productivity and in minimizing the pests' population. It is widely accepted that pesticides are capable enough in the rapid and successful removal of target pests. However, the negative and chronic irreparable impacts it imposes on the environment and on the health of living beings cannot be overruled. Pesticide resistance, pest resurgence, and pest residues are the major challenges caused by the prolonged use of pesticides. Alternate pest management techniques, integrated efforts of man and natural pest agents and the judicious Integrated Pest Management, are needed to safeguard the human, other living organisms, and the environment. In this chapter, the development of pesticide resistance in pests and resurgence has been discussed in detail, along with the solution tactics for effective pesticide pollution management.

4.1   INTRODUCTION

The proliferation of human percentage worldwide is resulting in enlargement of global croplands. Booming food demands require an increase in the production of crops. According to the Food and Agricultural Organization (FAO) of the United Nations, world food production needs to be hiked by 70% in order to meet the food demands of the burgeoning population (Gill and Garg, 2014). There has been a tremendous pressure on the global agricultural system to meet the food requirements with the same available natural resources. Food safety and saving crop from pests injuries is an important factor in order to increase food availability. To provide safety against 8000 species of weeds (13% loss); 9000 species of mites and insects (14% loss); and 50,000 plant pathogens' species (13% loss), pesticides have found an indispensable role in crop production (Zhang et al., 2011). Loss of 78%, 54%, and 32% of fruits, vegetables, and cereals may occur annually without the application of pesticides (Zhang et al., 2011). Herbicides, fungicides, pesticides, fertilizers, nematicides, and soil amendments have found extensive application in the agricultural system in order to increase crop production. The pesticides were introduced first in the year 1940 when organochlorines were employed for pest management. Prior to this, most of the weeds, insects, and pests were controlled and managed by physical, mechanical, and cultural strategies.

Approximately $38 billion USD are used up for pesticides worldwide (Pan-Germany, 2012). To ensure food safety against pests, more than thousands of pesticides are used worldwide. The trend of application is different in India when compared to the world. A total of 76% of the total pesticide used by India is insecticide and 45% of the total pesticides have main application for cotton crops followed by paddy and wheat (Agrawal et al., 2010). Herbicides and fungicides are used in the small and nearly the same proportion. In general, it has been observed that only 0.1% of the applied pesticides reach the target organism and the remaining major portion gets discharged in the environment and contaminates the surroundings (Carriger et al., 2006). It is introduced in the environment followed by its mixing with air, water, soil, and plants. The pesticides are produced for target pests but the disproportionate application makes it poison for nontarget living organisms. Pesticides have also been found to be bioaccumulated in the living organisms by entering into the food chain. Many acute and chronic diseases are detected in humans and animals due to exposure to pesticides. This chapter discusses the challenges and management techniques of pesticidal pollution along with its chronic and irreparable effects on the biodiversity and environment.

4.2 CHALLENGES IN PESTICIDES MANAGEMENT

The pesticides have the ability to remain for longer duration in the environment, and even in far places from the initial applications. The water solubility of most of the pesticides results in its global dispersion. The pesticides are responsible for the toxicity of plankton, which is the source of 80% of the earth's oxygen budget. The most difficult challenge is to make the other methods more effective and to ensure their adoption by pest control operators and crop growers. There is an exigency to reduce environmental contamination and to increase the effectiveness of pesticides against target pests. A hike has been observed in the pesticide development to target a wide range of pests. The targeted pests are facing major challenges either to disperse to a novel environment and/or get adapted to new conditions. When pests adapt to the new environmental conditions, several phenomena can be observed, such as multiplication of generations, fluctuations in population growth rates, and gene mutation. This results in pesticide resistance and pest resurgence, and pesticide residual which are the major challenges to overcome.

4.3 INHERENT TOXICITY

Pesticides have the ability to disturb the normal activity of an organism by poisoning. This property is known as toxicity. The poisoning, acute or chronic, depends on the exposure to pesticides. The quantity of application, ways of the entrance, duration of exposure, the environment, and the state of an organism are the factors on which toxicity of pesticides depends. When a toxic substance is ingested into an organism and it kills 50% of it on a single exposure, that dose of the toxic matter is known as lethal dose 50% (LD_{50}).

4.4 PESTICIDE RESISTANCE

Resistance is when pesticides fail repeatedly in achieving the expected level of control over target pests caused due to heritable change in the sensitivity of pests (IRAC, 2013). With the application of the same chemicals over and over to control the pest, pesticide resistance can become a problem, that is, the pest develops resistance against the applied chemical and the pesticide cannot affect the pest any longer. Resistant pests (weeds, insects, or microbes which manage to survive due to its genetic makeup) are rarely found in normal pest population but the application of indiscriminate chemicals destroy the normal susceptible population. The resistant pests take advantage of this situation and multiply and grow their population (as the genetic trait is inherited by young ones). In such a case, the resistant pests grow in numbers and effectively decrease the efficiency of pesticides. For instance, in Figure 4.1, the green plants represent the normal suscep-tible weeds and the yellow plants represent the resistant weeds. The same herbicide is applied to the field for a period of 4 years. In the following year, the resistant weeds increased in percentage (Figure 4.1b–d). In the fourth year, most of the weeds become resistant. The intensive use of pesticides has resulted in resistance development in most of the target pests. Taking advantage, resistance pests have increased in number, that is, from 600 to 700 by the end of 1990 to 2001, respectively (Gill and Garg, 2014). Different pesticides against which pests have developed resistance are, presented in Table 4.1 (Dhaliwal et al., 2016). The development of pesticide resistance depends on factors, such as frequency of application of pesticides to the croplands, the mechanisms of resistance and the genetics involved in it (single or multiple resistance mechanisms), the size of the gene pool, and the reproduction period of pests.

TABLE 4.1   Different Pesticides Against Which Pests Have Developed Resistance

Sl. No.	Pesticide	No. of Species Developed Resistance
1.	Cyclodiene	291
2.	Dichloro diphenyl trichloroethane	263
3.	Organophosphates	260
4.	Carbamates	85
5.	Pyrethroids	48
6.	Fumigants	12
7.	Others	40

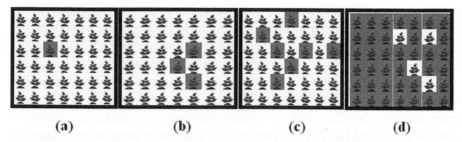

(a)   **(b)**   **(c)**   **(d)**

FIGURE 4.1   A resistant weed dominating the cropland applied with same herbicide in 4 years; (a) the first year: the beginning of resistant population; (b) second year, (c) third year, (d) fourth year: the resistant weed multiplied and grown the population.

Resistance developed by parasites, disease vectors, and crop and urban pests are exceedingly challenging. Due to the pesticide resistance, the farmers apply pesticides to the croplands more frequently which results in increased resistance and escalated crop losses. This continued cycle of spraying of pesticides is often termed as pesticide treadmill. In the United States, 383 million pounds of herbicides were applied to the genetically engineered crops in the first 13 years of their introduction to the markets (Harriott, 2010). During the 1940s in the United States, dichloro diphenyl trichloroethane was employed to control the bed bugs in the initial years, soon synthetic pyrethroids replaced it. Due to the intensive application, bed bugs developed resistance (by multiple mechanisms) against most of the commercially available insecticides. As a result, instead of seconds or minutes, insecticides started taking 150 h to kill bed bugs. According to the past studies, the insecticides were unable to penetrate the cuticle (exoskeleton), and if they are able to penetrate, the bed bugs eliminated it from their bodies because of the heightened enzymatic action.

4.5 PESTICIDE RESURGENCE

Pest resurgence is another major challenge and is a dose-dependent process. It is the escalated abnormal reappearance of a pest population after the application of pesticides. For example, brown plant hopper (BPH) is kept under control by mirid bugs, spiders, and ladybird beetles. But, due to the pesticide application, these natural enemies get destroyed and it also affected the fecundity of BPH females. It accelerates the rate of resurgence. Similar cases of resurgence are reported for cotton bollworm *(Helicoverpa armigera)* and bed bug *(Cimex lectularius)*. It arises due to the elimination of natural enemies by applying pesticides of the widespectrum and persisting nature. This problem is mainly raised due to economic constraints, that forced the farmers to apply low-dose insecticides, which leads to ineffective management of pests. The other reasons for pest resurgence are reduced competition with weeds; hiked feeding and rate of reproduction of pests; employment of pesticides' sublethal doses; direct stimulation of pests, reduced biological control; killing of primary pests; and improved crop growth. There are many cases of pesticide resurgence reported in past in walnut (*Juglans regia*), cotton (*Gossypium hirsutum*), soybeans (*Glycine max*), and hemlock (*Conium macaulatam*).

4.6   PEST REPLACEMENT

When the pesticides successfully kill the targeted pests but a different species (formerly a minor pest) replace the target, this phenomenon is termed as pest replacement or upsurge. This becomes severe when replacing pests causes more crop loss comparatively.

4.7   PESTICIDES RESIDUES

4.7.1   PESTICIDES IN WATER

Water requirements have been increased for agricultural purposes. Sixty percent of the withdrawn groundwater is applied to the agricultural system only. Rainwater can be held by soil to some extent; afterward, it gets infiltrated and percolated deep into the soil. The pesticides and agricultural residues (including nitrates and phosphates) quickly get transported to a nearby surface, subsurface, and groundwater resources. Leaching or surface runoff introduce pesticides into water systems. Both the processes are linked to the hydrologic cycle of the earth. Different pesticides have different mobility properties including both horizontally and vertically. Pesticides are introduced in the water by accidental spillage, surface runoff, industrial effluent, runoff from pesticide-treated soils, drift into water streams, ponds and lakes, spray equipment washing, postspray operation, and aerial spray for controlling water-inhibiting pests. They get into the water system either from a point or nonpoint sources. Tanks, containers, or spills are some of the point sources. Nonpoint sources can wide undefinable areas where residues of pesticides are present. Pesticides are made soluble in water so that they can be dissolved in water and can be absorbed by the target.

The runoff from croplands deteriorates the quality of water in the river, lake, estuaries, and streams. Billions of dollars would be required to make up for the loss on the part of water quality in addition to the river navigation, municipal water treatment, recreation, and commercial fishing. Pesticides in aquatic system adversely affect aquatic life. For instance, in Europe, 27 freshwater species of fish are observed to be affected by plant production products (Ibrahim et al., 2013). It not only affects the aquatic organism, but also the food webs they are part of. According to a study conducted by Hargrave et al. (1992), majorly in the Arctic Ocean, food webs were observed to be influenced by persistent pesticides (polychlorinated biphenyls and organochlorine pesticides). When pesticides are employed to the agricultural fields, after plant uptake, some fraction remains in the soil and may end up to surface streams by being dissolved in runoff or may leach into subsurface waters. Physico-chemical transformations in pesticide residues make it more capable to contaminate water. For instance, the transformation of nitrogen fertilizer or nitrogen from animal waste into ammonium followed by nitrates which further change into nitrites. Both the nitrates and nitrites leave negative impacts on human and animal health. Nitrate causes methemoglobinemia (blue baby syndrome) in infants by impairing the ability of its blood to carry oxygen. Due to the carcinogenic nature of some pesticides when applied in large doses; for 26 pesticides, the maximum allowable contamination levels are issued by the United States Environmental Protection Agency (2018 Edition of the Drinking Water Standards and Health Advisories Tables).

When pesticides reach the stagnant water bodies, such as lakes and ponds, the high nitrogen and phosphorus concentrations of pesticides cause the algal blooms and premature aging of the water bodies. This phenomenon is known as eutrophication. As per the FAO of the United Nations (2017), the chief causes of deterioration of surface water are agricultural tillage, urban development sites, and pesticide discharge. As per the study conducted by the United States Department of Agriculture Economic Research Service; 99 watersheds were examined and nutrients/sediments in excessive concentrations were found in 48 watersheds (Agrawal et al., 2010). Agricultural source of sediment was reported in 34 watersheds. Agricultural runoff was found to be contributing source of nutrients in 22 out of 78 estuaries (Agrawal et al., 2010).

4.7.2   PESTICIDES IN SOIL

Soil is the primary ecosystem to encounter immediate and rebound impacts of agrochemicals and their adverse effects depend on the intensity and perseverance of parent chemicals and their metabolites. The pesticides are one of the leading polluter cropland soil, thus, shifting the equilibrium of the soil process. It defiles the soil strata by removing nontarget organisms along with the pest. With a myriad of transport pathways, prediction of transport and fate of pesticides in the soil is a complex task. However, the behavior and fate of pesticides are mainly affected by some factors including adsorption, movement, volatilization, and decomposition. Adsorption, directly or indirectly, affects the behavior of other factors in the soil by affecting the interaction of solid phase and pesticides in the soil environment. Different processes, such as runoff, volatilization, and leaching are responsible for the movement of the pesticides. The aforementioned factors are discussed.

4.7.3   ADSORPTION

Organic characteristics of pesticides undergo various processes in soil and thus affecting their potential activity. Characteristics of soil and pesticides, such as structure and physiochemical properties play an important role in evaluating the adsorption behavior. Clay and organic matter are the foremost components in the soil which is significant to adsorption. Some of the factors like shape, pH, molecular weight, charge distribution on the cations, chemical character, water solubility, the polarity of molecule, and polarizability of pesticides affect the sorption process in the soil colloids. Generally, organic pesticides possess low molecular weight with low water solubility. The mobility reduces with the sorption of pesticides on the soil. Due to the inaccessibility of microorganisms, UV-light and dissolve oxidants chemicals, the degradation of adsorbed pesticides is low. The mechanisms of adsorption of pesticides are van der Waals interaction, hydrophobic bonding, hydrogen bonding, charge transfer, ion exchange, and ligand exchange.

4.7.4   MOVEMENT

4.7.4.1   DIFFUSION

Distribution pattern of pesticides in the soil is significantly influenced by diffusion. The diffusion of pesticides can occur in vapor as well as in the nonvapor phase. Factor like adsorption, porosity, bulk density, diffusion coefficient, temperature, solubility, soil water content, and vapor density affect the diffusion of pesticides in the soil. Graham-Bryce (1969) derived an equation providing a relationship between adsorption and diffusion coefficient as shown in Equation (4.1)

$$D = D_L V_L f_L / b + V_L \qquad (4.1)$$

where D_L is the diffusion coefficient in the free solution, V_L is the fraction of soil occupied by the liquid phase, f_L is the tortuosity factor for a soil, b is the slope of adsorption isotherm, and is the bulk density. It can be observed from Equation (4.1) that diffusion reduces with the increase in adsorption. Shearer et al. (1973) observed that no diffusion occurs in dry soil and diffusion increases with an increase in water content up to 4% and decreases afterward. Diffusion coefficient increases with the rise of temperature and decrease of bulk density.

4.7.4.2 MASS FLOW

Water, soil, and air act as a carrier for the mass flow of pesticides. The sorption of pesticide on soil, rate, and direction of flow of water are the factors which influence the mass flow of pesticide with water as a carrier. Several studies reported the inverse relationship between the adsorption and movement of pesticides by water through soil (Ashton, 1961; Guenzi and Beard, 1967). The soil particles need external force by water or air to act as a carrier for the mass flow of pesticide. Amount of pesticide adsorbed by transporting soil is directly proportional to the amount of pesticide transported by soil erosion. The wind has great potential to transfer the soil to large distances which resulting in the presence of pesticides within dust particles in the atmosphere.

4.7.5 VOLATILIZATION

Volatilization of pesticides can be defined as the process in which dissolved pesticide within the soil is vaporized and eventually lost by surface evaporation. Temperature and adsorption of pesticide on the soil are some factors which influence the vapor pressure of pesticide. The organic matter present in the soil is inversely proportional to the rate of pesticide volatilization. Fine sandy loam soil exhibits a maximum rate of pesticide volatilization. Volatilization is directly proportional to the water content present in the soil, hence, pesticides present in the wet soil volatilize faster than dry soil. A direct and indirect effect of airflow on volatilization of pesticides has been observed in past studies. Loss of pesticide from the target area is mainly due to the volatilization process. Volatilization often exceeds runoff and chemical degradation.

4.7.6 DECOMPOSITION

4.7.6.1 CHEMICAL DEGRADATION

Dissipation of pesticides from the soil is mainly due to the chemical degradation process. In most of the reactions, water acts as a reactor, reaction medium, or both. Major compounds are chemically degraded by two processes, mainly, oxidation and hydrolysis. For degradation of certain specific compounds, isomerization or chemical reduction is used. Common pesticide, such as organophosphorus compounds undergoes alkaline hydrolysis which results in detoxification of these compounds.

4.7.6.2 BIOLOGICAL DEGRADATION

Microbes present in the soil mainly help in the biodegradation of pesticides. An initial lag in the degradation of pesticides by microbes can be observed due to the adaption of microbes in the new environment. Due to the high resistance of fungi to the unfavorable environment, it plays an important role in the biological decomposition of pesticides. The enzyme produces in the fungi is responsible for the degradation of pesticides. Soil microbe activity mainly depends on the chemical structure and selected substances. Karpouzas and Walker (2014) found that the organophosphate pesticide can be effectively degraded by *Pseudomonas putida* epI strain. Some microbes used pesticides as a source of energy, carbon, and other nutrients which results in the increase of their number in pesticides contaminated soil.

4.7.7 PESTICIDES IN THE SURROUNDING AIR

Pesticides reach in the air from spray drifts, aerial pesticide application, and volatilization from the treated surfaces. The speed of wind and size of the droplet of pesticide sprayed decide the extent or drift. The humidity, ambient temperature, wind speed, the vapor pressure of ingredients, the surface on which settling of pesticides occurred, and the time duration after the application of pesticides drive the rate of volatilization. The complete and partial volatility of pesticide ingredients contributes to the atmospheric pollution of large cities. For instance, in California and Washington, traces of organophosphorus pesticides were found in the air sampling after the pesticidal spray.

4.7.8 BHOPAL GAS TRAGEDY: A CASE STUDY

In 1969, a pesticide plant was built in Bhopal for the manufacturing of Sevin (a pesticide used for the elimination of weevils, beetles, and worms in Asia). The operation control was under Union Carbide India, Limited (major stock held by An American Company: Union Carbide Corporation). On December 2, 1984, leak began due to the introduction of water in the tank of methyl isocyanate (key ingredient for the production of Sevin and toxic gas). The tank exploded due to the extreme pressure and heat created in the tank caused by the formation of hydrogen cyanide formed by the reaction between water and methyl isocyanate. It released 40 tons of poisonous gas in Bhopal's air. The cloud formed by methyl isocyanate occupied the skies of Bhopal and caused deadlier respiratory problems among humans. The health effects were chronic and still can be observed in gas survivors in the form of respiratory problems, blindness, immune and neurological disorders, and many more. Rain caused the wet deposition of the pollutants in the air and the same leached into the groundwater. A series of water qualitative tests were carried out on the water samples collected from the wells of Bhopal in 1996 by the State Pollution Control Board. Traces of pesticides were reported in most of the local wells. In 2004, considering all the reports and the health conditions of people, the federal government ordered the state to equipped the community with clean drinking water.

4.8 THE SOLUTION STRATEGIES

Pesticides can be managed by a few management tactics that can prove to be successful under appropriate circumstances. Monitoring of pest population in the agricultural environment is imperial prior to the application of pesticides in the field. Pesticides can be altered by employing various modes of physical, mechanical, and cultural actions. The number of applications must be restricted over time and space. Creating or exploiting refugia, avoiding unnecessary persistence, targeting pesticide applications against the most vulnerable stages of the pest life cycle, using synergists which can enhance the toxicity of given pesticides by inhibiting the detoxification mechanisms. The consideration of the rational use of pesticides along with their physiological and ecological selectivity plays a vital role in the management of pesticides. For a given pesticide, differential toxicity between taxa helps in the characterization of physiological activity. On the other hand, in order to reduce unwanted harm to nontarget organisms, the modifications of the operational procedure are referred to as ecological selectivity. There are four main ways to improve safety against pesticide use:

(1) Aim to minimize the toxicity in addition to lowering persistence and increasing selectivity.
(2) Practicing optimal ways of applying pesticides such as strip and band treatments, treatment of seeds prior to planting, and the use of granulated formulations.
(3) Application of pesticides as per the requirements considering the economic aspects.
(4) Strict norms and regulations for pesticide applications and other agricultural practices that can risk human health.

4.8.1 INTEGRATING NATURAL PEST CONTROL AGENTS AND MAN-MADE EFFORTS

Integrated efforts of both man-made technologies and natural pest agents are important in elimination and killing of pests. The integrated system must be utilized to a great extent to limit or cease the vital activity of harmful organisms. The main aim is to keep the number of pests to that level where they cannot be able to cause any perceptible harm. Collective measures, as following, should be taken actions and upon:

(1) Prior knowledge of ecological situations of croplands, the strength of harmful pests, and most importantly, the population of beneficial natural enemies for limiting the development of pests.
(2) Highly selective pesticides must be applied to safeguard the beneficial organisms.

(3) Prior treatment should be encouraged, such as seed treatment and application of pesticides along with the fertilizers at the time of plantation, which preserves beneficial insects and hikes the crop yield.

(4) Spraying in ultra-low volume or low volume must be performed as it completes the elimination process of pests in a shorter span of time.

(5) Spraying during nights is more productive as low or no winds do not allow the pesticide to drift away and let it settle on the plant parts uniformly. The pesticides (mosquitoes, cutworms, etc.) which are active during nights are also get eliminated by the same.

(6) Early spring and autumn days are best for pesticide application, as the beneficial natural enemies, that is, endomorphs are in early and inactive stages (eggs, pupae) and the pesticides will not cause any harm to them.

(7) Persistent pesticides should be applied in the case of multiple crop rotation; it prevents the accumulation of toxic substances in the soil.

4.8.2 RESISTANCE MANAGEMENT STRATEGIES

Resistance can be managed by introducing gaps between pesticide applications. It allows pesticide-susceptible individuals to dilute the resistant population. However, it is quite difficult to predict the time duration by the resistant population would be diluted. The best management tactics are to reduce resistance development at the very first stage. Some of the management tactics are as follows:

(1) Plantation of varieties of pest-resistant crops;

(2) The healthier the crop, the more competitive it is with weeds and less prone to be attacked by diseases and insects; therefore, maintaining crops through timely irrigation and fertilization.

(3) Crop rotation (crops with different pest problems); and

(4) tillage of croplands for controlling weed (except where soil erosion is a problem).

Integrated pest management practices are also helpful in preventing resistance, some of them are discussed below:

(1) Regular scouting of fields as a quick response for the changes in pest populations. Pest monitoring helps in determining the necessary pesticides and their quantities.

(2) The best application time for the pesticides is when pests are most susceptible. It reduces the frequency of applications.

(3) Special care and assistance should be taken in the case of large weeds and late instar insect larvae or diseases. Pesticide applications in these cases can accelerate resistance development.

(4) Avoid the use of persistent pesticides, instead apply the selective pesticides which can break down quickly.

(5) Identify the zones where pest control is required. Use barrier or banded treatments and spot treatments for targeting the pests.

(6) Controlling alternate diseases and hosts of insects. For instance, removal of junipers for minimizing cedar-apple rust.

(7) Label directions should be followed properly, including crop, pest species, location, conditions, volume, carrier type, use of an adjuvant, etc.

(8) In case of the same crop or location, pesticide rotation with a different mode of action can be a better option to prevent resistance development.

(9) When the pesticides are applied to the maximum label rate, the combination of many resistance management strategies is required to eliminate the high chances of resistance development.

The tank mix or prepacks are emergent management tactics to minimize the risk of resistance development. It is a combination of two or more pesticides. Tank mixes allow the applier to adjust the ratios of the different pesticides as a single mixture. While the prepacks are already mixed on the manufacturer's end. It is designed to increase the potential of pesticides with different modes of action for the better prevention/management of pesticide resistance. The resistance developed for one of the pesticides in a mix is balanced by the other partner pesticide. There are some other methods based on biochemical and molecular techniques, such as allele-specific polymerase chain reactions, enzyme electrophoresis, enzyme, immuno-assays, etc.

4.8.3 RESURGENCE MANAGEMENT STRATEGIES

The management of pest resurgence can be performed by the following solution strategies:

(1) It can be managed by the correct and recommended dosage of effective pesticides.

(2) Spraying of recommended insecticides at proper intervals using high volume sprayers.

(3) Cultural practices should be owned such as manipulation of an optimum dose of fertilizers.

(4) Pest monitoring should be performed through traps to determine the right time for the application of pesticides.

(5) Predators, parasitoids, and other bioagents should be preserved and utilized.

(6) To prevent the favorable environmental conditions for pest multiplication, recommended spacing should be provided.

(7) Crop rotation and insecticide rotation with a different mode of action

(8) New insecticides and plant extracts should be introduced for alternative and effective control of pests.

(9) Inoculative release of natural enemies can be performed to decrease the pest population.

4.8.4   INTEGRATED PEST MANAGEMENT (IPM)

IPM is a well-defined program that focuses on prevention, monitoring, and control to decrease or limit the quantity of pesticide being used. It is an environmentally sensitive and effective program. The life cycles of pests and their interaction with the environment are taken into consideration. It inherits cultural structural and biological strategies to control the pest population. The six essentials of IPM are briefly discussed as follows:

(1) *Monitoring*. It includes site inspections regularly and the trapping for the determination of types of pests present in agricultural lands and their infestation levels.

(2) *Record-Keeping*. Keeping records for the establishment of patterns and trends in outbreaks of pests. Additional information includes identification of pest species, population size, distribution pattern, complete plan of treatment action, and future prevention recommendations.

(3) *Action Levels*. Pests cannot be eliminated to a full extent. An action level is that population size for which actions are required.

(4) *Prevention*. Whether existing or designs of new structures, both must include preventive measure at a priority basis.

(5) *Tactics Criteria*. Use of chemicals is avoided to a maximum extent. Least toxic pesticides are applied if required.

(6) *Evaluation*. Evaluation programs are essential to be carried out in order to determine the success percentages of pest management strategies.

Least toxic pesticides include boric acid, microbe-based pesticides, desiccant dust (diatomaceous earth and silica gel), pesticides manufactured with essential oils (no pyrethrums) without toxic synergists, nonvolatile insect and rodent baits in tamper-resistant containers or for crack and crevice treatment only, and materials for which the inert ingredients are nontoxic and disclosed.

4.8.5   OPTIMIZE PESTICIDE USE

Over a few decades, the use of pesticides has been increased to target the wide-ranging variety of pests. However, it attacks the nontarget areas and nontarget arthropods (such as earthworms, predators, and pollinators) which cause an imbalance in the natural pest population level. This section will focus on control measures for optimization of pesticide use.

4.8.5.1   ALTERNATIVE METHODS OF PEST CONTROL

A large number of fish species died in the past due to the presence of pesticides in the water bodies, such as endosulfan in Germany, zinophos in England, and parathion in USA (Edwards, 2013). Therefore, alternative methods are needed to eliminate the effect of pesticides in the environment. Alternative methods of pest control include biological control, development of resistant crops and animals, use of attractants or repellents, cultural control, chemical control, sterile insect technique, and use of biological pesticides. Some of the alternative methods of pesticides are discussed below.

4.8.5.1.1   CULTURAL CONTROL

It is an old pest control technique used by the farmers throughout the world due to its cost-effectiveness. Some of the methods which are generally used for cultural control of pest include the use of resistant crops, crop rotation technique, soil solarization, certified seed, sanitation, and timely planting and harvesting. Other agricultural practices like plowing, weeding, and hoeing can destroy a large population of the pest.

4.8.5.1.2   BIOLOGICAL CONTROL

Biological control is a process in which natural enemies of the pest were used to reduce the population of the pest. One of the biological pest control techniques is a sterile insect technique (SIT) in which sterilize insects (usually males) are released in the wild to compete with another male to mate with the female. This technique reduces the next generation's pest population. Lanouette et al. (2017) used the SIT technique to control the population of *Drosophila suzukii* Matsumura, a pest of berries stone fruits.

4.8.5.1.3   CHEMICAL CONTROL

Chemical control is a preventive technique to reduce the pest problem when cultural and biological control method is insufficient. In this technique, the nontarget organism is free from the effect of chemicals. The quantity, density, and dosage time are optimized to provide an adequate result without any environmental concerns.

4.8.5.2   USE OF BIOPESTICIDES

Pesticides that are derived from the natural materials are known as biopesticides. Biopesticides can be classified broadly into three types mainly biochemical pesticides, microbial pesticides, and plant-incorporated-protectants. Biochemical pesticides use nontoxic mechanism by using naturally occurring substances to control the pest population. In microbial pesticides, the specific microbe is used as an active ingredient to control the target pests whereas, in plant-incorporated-protectants, the plant is genetically modified by introducing pesticidal substances. Lovett and Leger (2018) used genetically engineered fungi to control agricultural pests and vector diseases.

4.9   FUTURE PROSPECTS

The increasing food demand along with rising environmental concerns demands a novel approach for pest control. In this section, some new techniques have been discussed which is reported in the literature all around the world.

4.9.1   GARBAGE ENZYME

Garbage enzyme is a complex mixture made out of brown sugar/molasses, organic waste, and water in ratio 1:3:10 (Arun and Sivashanmugam, 2017). With 30 years of experience in the enzyme research, Dr Rosukon from Thailand invented garbage enzyme. Within a shorter span of time, garbage enzyme can achieve a greater extent of degradation. High removal of pollutants like chemical oxygen demand, biological oxygen demand, nitrate, sulfate, and alkalinity were observed in the past studies (Tang and Tong, 2011). Apart from the pollutant removal agent, it can be used as a pesticide and fertilizer as reported by Arun and Sivashanmugam (2015). Due to the low pH of garbage enzyme, the dilution is an important aspect. The specific amount of dilution and the application of garbage enzyme as a pesticide is still needed to be scientifically verified.

4.9.2   NEW APPROACHES

In the last few years, researchers had been isolating novel native microbial species for the bioremediation of soil contaminated with pesticides. Capellán et al. (2019) isolated novel bacterial strain from soil enriched with chlorpyrifos pollutant. A novel Zr-MOF-based adsorbent was used by Yang et al. (2018) for the removal of organophosphorus pesticides which are widely used in the agricultural field. Lu et al. (2018) used amorphous metal boride for development of acetylcholinesterase biosensor for detection of organophosphorus. However, further studies should be carried out to find effectiveness in more persistent pesticides.

4.10   CONCLUSION

Increased productivity and the ability to eliminate the pests and the infectious disease made pesticides popular worldwide. With its prolonged use, their adverse effects can be clearly seen in the environment; and they have overweighing the benefits associated with their application. The above discussion clearly highlights the major challenges of pesticide resistance, pest resurgence, and pest residues. The persistent nature of pesticides has entered them into our ecosystem followed by the food chain. There is a necessity of exigency from the application of pesticides. There is availability of many management options and plenty of solution strategies, but the preventive measures must be taken on a priority basis. Adoption of the innovative IPM strategies can lead to the reduction of deleterious impacts of harmful pesticides. Garbage enzyme can be a reliable alternative for pest's species management. It can be considered as a recommendation for future treatment options. Further research work can be done to explore its application in croplands and efficiency in the elimination of target and harmful pest species.

KEYWORDS

- **pesticide resistance**
- **pest resurgence**
- **integrated pest management**
- **biopesticides**
- **garbage enzyme**

REFERENCES

Agrawal, A.; Pandey, R.S.; Sharma, B.O. Water pollution with special reference to pesticide contamination in India. *Journal of Water Resource and Protection* **2010**, *2*(5), 432.

Arun, C.; Sivashanmugam, P. Solubilization of waste activated sludge using a garbage enzyme produced from different pre-consumer organic waste. *RSC Advances* **2015**, *5*(63), 51421–51427.

Arun, C.; Sivashanmugam, P. Study on optimization of process parameters for enhancing the multi-hydrolytic enzyme activity in garbage enzyme produced from preconsumer organic waste. *Bioresource Technology* **2017**, *226*, 200–210.

Ashton, F.M. Movement of herbicides in soil with simulated furrow irrigation. *Weeds* **1961**, *9*(4), 612–619.

Carriger, J.F.; Rand, G.M.; Gardinali, P.R.; Perry, W.B.; Tompkins, M.S.; Fernandez, A.M. Pesticides of potential ecological concern in sediment from South Florida canals: An ecological risk prioritization for aquatic arthropods. *Soil and Sediment Contamination* **2006**, *15*(1) 21–45.

Capellán, J.V.; Moreno, A.L.; Amador, E.H. Isolation of a novel bacterial degrader strain from a contaminated soil by the insecticide chlorpyrifos, using the enrichment culture technique. *Biosaia* **2019**, (8), 47. https://www.upo.es/revistas/index.php/biosaia/article/view/3937/3169

Dhaliwal, G.S.; Singh, R.; Chhillar, B.S. Essentials of Agricultural Entomology. **2016**, Kalyani Publishers.

Drinking Water Standards and Health Advisories. EPA 822-F-18-001. Office of Water U.S. Environmental Protection Agency Washington, DC March **2018**.

Edwards, C.A. Pesticide Residue in Soil and Water, in Edwards, C.A., 2013, Environmental pollution by pesticides. *Springer Science & Business Media* **2013**, *1*, 409–458.

Germany, P.A.N. Pesticides and Health Hazards Facts and Figures. Hamburg, Germany: PAN Germany—Pestizid Aktions-Netzwerk eV **2012**.

Gill, H.K.; Garg, H. Pesticides: Environmental Impacts and Management Strategies, Pesticides-Toxic Aspects, Dr. Sonia Soloneski (Ed.), **2014**. DOI: 10.5772/57399. In. Tech. DOI, 10, p. 57399.

Graham-Bryce, I.J. Diffusion of organophosphorus insecticides in soils. *Journal of the Science of Food and Agriculture* **1969**, *20*(8), 489–494.

Guenzi, W.D.; Beard, W.E. Movement and persistence of DDT and lindane in soil columns 1. *Soil Science Society of America Journal* **1967**, *31*(5), 644–647.

Hargrave, B.T.; Harding, G.C.; Vass, W.P.; Erickson, P.E.; Fowler, B.R.; Scott, V. Organochlorine pesticides and polychlorinated biphenyls in the Arctic Ocean food web. *Archives of Environmental Contamination and Toxicology* **1992**, *22*(1), 41–54.

Harriott, N. Bed bug policy. *Beyond Pesticides* **2010**, *30*, 20. Retrieved from: https://www.beyondpesticides.org/assets/media/documents/infoservices/pesticidesandyou/documents/Winter10-11vol.30no.4

Ibrahim, L.; Preuss, T.G.; Ratte, H.T.; Hommen, U. A list of fish species that are potentially exposed to pesticides in edge-of-field water bodies in the European Union—a first step towards identifying vulnerable representatives for risk assessment. *Environmental Science and Pollution Research* **2013**, *20*(4), 2679–2687.

IRAC. Resistance Management for Sustainable Agriculture and Improved Public Health; **2013**. Retrieved from: http://www.irac-online.org/

Karpouzas, D.G.; Walker, A. Factors influencing the ability of *Pseudomonas putida* epI to degrade ethoprophos in soil. *Soil Biology and Biochemistry* **2014**, *32*(11–12),1753–1762.

Lanouette, G.; Brodeur, J.; Fournier, F.; Martel, V.; Vreysen, M.; Cáceres, C.; Firlej, A. The sterile insect technique for the management of the spotted wing drosophila, Drosophila suzukii: Establishing the optimum irradiation dose. *PLoS One*, **2017**, *12*(9), e0180821.

Lovett, B.; St. Leger, R.J. Genetically engineering better fungal biopesticides. *Pest Management Science*, **2018**, *74*(4),781–789.

Lu, X.; Li, Y.; Tao, L.; Song, D.; Wang, Y.; Li, Y.; Gao, F. Amorphous metal boride as a novel platform for acetylcholinesterase biosensor development and detection of organophosphate pesticides. *Nanotechnology* **2018**, *30*(5), 055501.

Shearer, R.C.; Letey, J.; Farmer, W.J.; Klute, A. Lindane diffusion in soil 1. *Soil Science Society of America Journal* **1973**, *37*(2),189–193.

Tang, F.E.; Tong, C.W. A study of the garbage enzyme's effects in domestic wastewater. *International Journal of Environment, Chemical, Ecological, Geological and Geophysical Engineering* **2011**, *5*, 887–892.

Yang, Q.; Wang, J.; Chen, X.; Yang, W.; Pei, H.; Hu, N.; Li, Z.; Suo, Y.; Li, T.; Wang, J. The simultaneous detection and removal of organophosphorus pesticides by a novel Zr-MOF based smart adsorbent. *Journal of Materials Chemistry* **2018**, *6*(5), 2184–2192.

Water Pollution from Agriculture: A Global Review. Food and Agriculture Organization of the United Nations Rome, **2017**.

Zhang, W.; Jiang, F.; Ou, J. Global pesticide consumption and pollution: with China as a focus. *Proceedings of the International Academy of Ecology and Environmental Sciences* **2011**, *1*(2),125.

CHAPTER 5

Pesticides: Impact on Environment and Various Possible Remediation Measures

ANJALI PATHAK,[1] MAHENDRA K. GUPTA,[1*] MIR SAJAD RABANI,[1]
RACHNA SINGH,[2] SHIVANI TRIPATHI,[1] and SADHNA PANDEY[3]

[1]*Microbiology Research Lab, School of Studies in Botany, Jiwaji University, Gwalior 474011, Madhya Pradesh, India*

[2]*Microbiology Research Lab, School of Studies in Microbiology, Jiwaji University, Gwalior-474011, Madhya Pradesh, India*

[3]*Department of Botany, Government Kamla Raja Girls Post Graduation Autonomous College, Gwalior 474001, Madhya Pradesh, India*

Corresponding author. E-mail: mkgsac@yahoo.com

ABSTRACT

Unprecedented industrial and anthropogenic activities have led to release of enormous pollutants such as pesticides, heavy metals, hydrocarbons, etc., into the environment. The pesticides include a group of chemicals such as herbicides, insecticides, fungicides, etc., used worldwide to control pests, weeds, and crop diseases. The toxic effect of a pesticide varies with time, dosage, organism characteristics, pesticide type, and environmental conditions. The persistence of a pesticide in the environment leads to a life risk; the more persistent a pesticide is, the worse is its impact on ecosystem and mankind. Pesticides in agriculture were generally considered advantageous and no concern regarding the risks associated with these chemicals to environment and mankind existed until the publication—*Silent Spring* came in which the problems attributed to pesticides were addressed that could arise as a result of indiscriminate use of pesticides. The present chapter also highlights the impact of pesticides on environment and mankind with their possible remediation measures. For the effective management of pesticides other remediation measures such as bioremediation need to be paid more attention for broad-spectrum applications.

5.1   INTRODUCTION

A pesticide is defined as "any substance or a mixture of substances intended to prevent, destroy, repel, or mitigate any kind of pests" (Eldridge, 2008). The definition of pesticide may vary from time to time in different countries. However, the pesticides are basically poisonous mixed substances that are efficient to particular target species and safe to nontarget living organisms and environment (Zhang et al., 2011). Pesticides are the compounds that control pests, weeds, etc.. The pesticides generally include insecticides herbicides, termiticides, etc. (Randall et al., 2014). Majorly, pesticides include different classes like organochlorines, organophosphates, carbamates, pyrethroids, and neonicotinoids (Pesticide 101-A Primer, n.d.). These pesticides are produced synthetically or may be found naturally. In India, agricultural and industrial activities increased throughout 20th century which led to considerable contamination of water, soil, and air resources through different hazardous pollutants. For pesticide production, India has been ranked 12th in the world (Boricha and Fulekar, 2007). More than 70% of Indian economy is dependent on agriculture and related sectors. In recent times, feeding and clothing for the projected population of 1.3 billion by 2020 are becoming a great challenge (Kanekar et al., 2004).

In comparison to natural pesticides, synthetically produced pesticides are easy, fast and cheaper in use. It has large production and many applications in pest management (Daly et al., 1998). Over 60 years ago, the first generation of pesticides was introduced which had two compounds namely, organophosphate and organochlorine. It has been reported that about 2.3 million tons of pesticides are being used for pest control globally. The use of pesticides after world war second has increased enormously (Miller, 2002). Excluding the applications of synthetic pesticides in the socio-economic upliftment of farming, many disadvantages including its harmful effects on plants also came into existence. The inappropriate and unplanned usage of these chemical pesticides also shows a serious and devastating effect on environment (Kullman and Matsumura, 1996). Recently, over 2500 pesticides are being used worldwide in agricultural practices (Singh et al., 2006).

Nowadays, plant protectionists have taken a necessary step with the excessive use of synthetic pesticides, a most common tools used in agriculture all over the world. Due to the pest infestation, about 45% of the annual production of food is lost. Pesticides have made a significant impact on the farmer's economy by reducing the agricultural losses caused by pests, thereby improving crop yield and quality of food. The use of synthetic pesticides has significantly enhanced agricultural productivity, but simultaneously on the other hand these pesticides have created a general concern with respect to their toxicity as their residues are found in environment, food, and fodder. Thus, they are harmful to the mankind, livestock, and ecosystem (Sharma et al., 2016).

Organic and inorganic pollutants have the history of destroying world's large amount of soil and water resources. The pollutants reported in soil include polycyclic aromatic hydrocarbons (Rabani et al., 2018), pesticides, polychlorinated biphenyls (PCB), explosives, metals, etc. (Zhu and Shaw, 2000; Kumar et al., 2012; Testiati et al., 2013; Vane et al., 2014; Rehman et al., 2019; Singh et al., 2019). Nowadays, the use of chemical pesticides has become a common practice in agriculture. These synthetic chemicals from soil directly or indirectly come in contact with groundwater and surface water by leaching, runoff; thus, pollutes water bodies. The presence of these compounds in water and soil resources is said to be undesirable due to their toxic nature and anomalies they create. Therefore, pesticides and hydrocarbons when present in the soil adversely affect seed germination, plant growth, and soil microbial activities (Smith et al., 2006; Ahmad et al., 2012, 2013, 2014; Guo et al., 2012; Alrumman et al., 2015). Nie et al (2011) revealed that the efficiency of water and nutrients absorption from the soil by different plant species and microbes has got deteriorated due to these pollutants. In many reports, it was studied by the scientists that the pesticides are toxic and hazardous toward human health (Anderson and Meade, 2014). Different diseases in humans have been seen due to the exposure of these chemical pesticides present in plants (Isoda et al., 2005). Therefore, the importance of pesticide remediation is a major issue in the field of environmental and agricultural sciences.

The residues of chemical pesticides present in soil and water environment when enter the food chain, cause harmful effects to the living organisms. These residues after coming in contact with human body directly or indirectly affect the health of an individual and cause severe diseases. In human beings, these residues get accumulated mainly in the blood, adipose tissue, and lymphoid organs. The pesticides generally cause different acquired autoimmunity and hypersensitivity reactions like allergic respiratory diseases, eczema, dermatitis, etc., in animals and humans. Chauhan and Singhal (1997) had observed that different pesticides are mutagenic and teratogenic in nature and also known for causing mutations, carcinoma of liver and lungs in human beings.

Due to increased use of pesticides to meet the food demand of the growing population, a variety of chemical pests are introduced which ultimately affects the environment. Therefore, the production of different xenobiotics has led to the implementation of new technologies to remediate these harmful chemicals from the environment. Different technologies have been tried earlier which includes landfills, pyrolysis, recycling, etc., but these methods are not appropriate and had adversely affected ecosystem like the formation of toxic intermediates (Deberati et al., 2005). It is also reported that the methods used to eliminate these toxic substances from environment proved to be expensive and difficult to execute especially in the case of chemical pesticides (Jain et al., 2005). Recently, in many countries, efforts have been made to control the release of such pollutants and breakdown of existing pollutants by appropriate remediation techniques (Schnoor et al., 1995). Ex-situ and in-situ are the common methods used for remediation of different chemical pollutants including extraction and treatment by activated carbon adsorption, microbes or air, stimulation of aerobic and anaerobic microbial activities, etc.. Bioremediation is considered to be a very effective and promising technology to combat various types of pollution in the environment with the help of different microorganisms. Through the utilization of soil microorganisms the remediation of toxic substances is being practiced these days. It is an eco-friendly, versatile, and economical technology (Finley et al., 2010). Different researchers have reported various health issues in living beings; biodiversity and environment are also facing serious problems due to the overuse and improper use of pesticides (Gavrilescu, 2006; Hussain et al., 2009).

Pesticides degrade the soil quality and while reaching water bodies, they severely affect the aquatic ecosystem. Thus, decontamination of chemical pesticides from the areas which are polluted is a very complex or complicated process that needs a serious concern of researchers (Gavrilescu, 2006). The scientific community has special attention toward the indiscriminate use of pesticides for saving the plants from pests and vectors (Ahmed and Ahmed, 2014). The risk associated factors with the use of pesticides have surpassed its beneficial effects. Usually, there are many chemical pesticides that kill and harm nonselective targeted plants along with targeted ones. This chapter highlights the beneficial and harmful effects of pesticides along with the different remediation strategies and importance of pesticide remediation for the sustainable environment.

5.2   TYPES OF PESTICIDES WITH THEIR TARGET SPECIES

Pesticides include a variety of chemicals such as herbicides, insecticides, fungicides, rodenticides, and others (Table 5.1). On the basis of mode of action and their target sites the pesticides can be classified into various types (Frazer, 2010). Pesticides are beneficial in different ways such as for food preservation, crop protection, and also in prevention of vector-borne diseases (Basu et al., 2014). According to, different target groups, pesticides can be classified as fungicides (that kill fungi), insecticides (that kill insects), and herbicides (that kill weeds).

The chemical pesticides can be natural or synthetic, by which different organisms like insects, weeds, rodents, and fungi are killed and help in the reduction and repelling of these species. According to a report by United Nations Organization for Food and Agriculture (FAO), "pesticide is any compound or a mixture of substances which when applied inhibit the growth, kill, and help in controlling or managing any type of pests." These are also helpful in preventing or controlling vectors of human diseases, undesired plants, and animals that generally cause serious harm or interference in the production, storage, or marketing food stuffs (FAO, 2002).

As per the Code of Federal Regulations, the pesticides are known as chemical compounds used as plant regulators, defoliant, or many a times desiccant (CFR, 2004). These chemical pesticides are used all over because of the warm climatic conditions (Bhat and Padmaja, 2014). Some important measures for the regulation of pesticide usage are: pesticide safety, proper application of technologies, and integrated pest management (IPM). These are some key strategies that have been adopted for reducing human exposure to pesticides and also help to retain the fertility of the soils for better crop yield.

TABLE 5.1   Pesticides with Their Particular Target Organism

Types of Pesticides	Target Species/Pests
Algicides (Applied to control algae in water tanks, lakes, swimming pools, and canals other sites).	Algae
Avicides (Used for killing birds)	Birds
Bactericides (compounds used to kill or inhibit bacteria on plants as well as in soil) These are isolated from or produced by a microorganism, a related chemical, i.e., composed artificially.	Bacteria
Fungicides (Applied to kill molds or fungi), when applied to the wood, they are also called wood preservatives).	Fungi
Insecticides (Used to kill insects) include organochlorines, organophosphates, and carbamates.	Insects
Molluscicides (Applied to kill slugs and snails)	Snails
Nematicides (Kill nematode worm-like organisms that feed on plant roots)	Nematodes
Virucides (An agent having capacity to destroy an inactivated viruses)	Viruses
Rodenticides (Controls mice, rodents, etc.)	Rodents
Miticides or Acaricides (It is used to kill or destroy mites that generally feed on plant and animal).	Mites

5.3   CHARACTERISTICS OF DIFFERENT GROUPS OF PESTICIDES

The optimization of agricultural outputs is basically associated with the usage of agrochemicals in the agricultural practices. However, this statement is undermined with respect to the damages faced by the ecosystem because of the indiscriminate use of these chemicals (Paula et al., 2011). Therefore, it is important to know the characteristics of these agrochemical pesticides. The interactions of these pesticides within the plant–soil atmosphere system is quite complicated and usually depended on their physical and chemical characteristics (Capel, 1993).

TABLE 5.2   Characteristics of Some Pesticide Groups, Their Physical and Chemical Properties

Pesticide Group	Major Characteristics	Physical and Chemical Nature
Organophosphates	It shows solubility in water and organic solvents. After reaching to groundwater these pesticides get infiltrate, and are less persistent than chlorinated hydrocarbons. They affect the part of brain, i.e., central nervous system (CNS). This group of pesticides is absorbed by plants, transferred to leaves and stems which is the supply of leaf-eating insects. Examples: diazinon, methyl parathion, and malathion	Its organic compounds contain phosphorus. The properties of these pesticides vary with the size and structure. In general, these are more soluble in organic solvents.
Organochlorines	They can be soluble in lipids, they also get accumulated in animals (fatty tissue), and then transfer through the food chain. They are known to be toxic to a variety of animals, its persistence known for long term. Examples: dichlorodiphenyltrichloroethane (DDT), aldrin, lindane, and chlordane	This group of pesticides contains a cyclodiene ring. The fat-soluble compounds are persisting in human body as well as in the environment. Most of the organochlorines are semivolatile and sparingly soluble.
Pyrethroids	These affect the nervous system and are less persistent, safe in terms of their use than other pesticides. Some are used as household insecticides. Example: pyrethrins	Contain an acid moiety, an alcohol moiety, and a central ester bond in their structure. Generally, have low vapor pressure, low Henry law constant, and large octanol.
Carbamates	The carbamates are acid derivatives that kill insects in a limited spectrum and are highly toxic with relatively low persistence. Examples: sevin and carbaryl	This pesticide is an ester derivative. A wide range of melting points (50 °C to 150 °C) is found in this group of pesticides. They commonly have low vapor pressures and poor volatility.

The application of pesticides has positive aspect which includes enhanced production of food crops. Various plant growth regulators are also used for controlling crops from weeds, pests, and diseases (Agrawal et al., 2009; Damalas, 2009). Some common groups of pesticides are known to be very harmful having different human health issues such as organophosphorus, organochlorines, pyrethroids, carbamates, etc. (Table 5.2).

5.4 CLASSIFICATION OF PESTICIDES

On the basis of chemical and physical nature, the pesticides can be classified with respect to their effects on health of crops and humans as well. It is a well-known fact that pesticides are generally regarded as the effective and beneficial chemical agents to prevent crop plants from different types of diseases. Due to overuse of pesticides different insects and pests have become resistant to various commercial pesticides. Nowadays such kind of pesticides have been developed, that target multiple types of species which are responsible for different crop destruction and diseases (Speck-planche et al., 2012). IPM has taken a strong progress in eliminating different kinds of pests by the use of chemical pesticides and acts as a "comprehensive prophylactic remedial treatment" (Gentz et al., 2010). The effect of a particular pesticide on a crop plant depends on its exposure and toxicity.

TABLE 5.3   Showing Classification of Pesticides with Examples and Their Effects

Pesticide	Class	Examples	Effects on Human Health
Insecticides	Organophosphates	Parathion, malathion, methyl parathion, chlorpyrifos, diazinon, dichlorvos, phosmet, and azinphos methyl.	Its exposure causes neuropathy, myopathy, irritability, paralysis, convulsions; it inhibits the enzyme acetylcholinesterase.
	Organochlorines (dichloro diphenyl thanes and cyclodienes)	Aldrin, chlordane, DDT, dicofol, dieldrin, endosulfan, endrin, heptachlor, mirex, and pentachlorophenol.	It stimulates the nervous system through disrupting the sodium/potassium balance of the nerve fiber, tremors, irritability, convulsions, hyperexcitable state of the brain. And other effect includes cardiac arrhythmia tic and many reproductive problems.

TABLE 5.3   *(Continued)*

Pesticide	Class	Examples	Effects on Human Health
	Carbamates	Fenoxycarb, aldicarb, carbofuran (Furadan), carbaryl, ethienocarb, and fenobucarb.	Inhibition of acetylcholinesterase enzyme and paralysis.
	Pyrethroids and pyrethrenes	Cypermethrin, cyfluthrin, deltamethrin, etofenprox, fenvalerate, permethrin, phenothrin,	After exposure it generally causes allergic respiratory reactions (rhinitis and asthma) and contact dermatitis.
Herbicides	Triazines, phenoxy derivatives, benzoic acid, and urea.	Different examples of herbicides are chlorophenoxy acids, hexachlorobenzene (HCB), picloram, atrazine, simazine, propazine, diquat, paraquat, etc.	Dermal toxicity, carcinogenic effect, damages liver, thyroid, affects nervous system, bone-related problems, affects kidneys, blood, and immune system.
Fungicides	Substituted benzenes, thiocarbamates, organomercury compounds, thiophthalimides, etc.	Common fungicides include chloroneb, chlorothalanil, hexachlorobenzene, ferbam, thiram, metam sodium, ziram, ethyl mercury.	Damages liver, thyroid, and nervous system. Affects the bones, kidneys, blood, and immune system.
Rodenticides	Coumarins, 1,3-indandions	Warfarin, coumatetralyl, difenacoum, drodifacoum, flocoumafen, bromadiolone. diphacinone, chlorophacinone, pindone	Rodenticides are extremely poisonous. After its exposure its symptoms include nausea, headaches, eye irritation, etc.
Nematicides		Aldicarb, dibromochloropropane	It causes respiratory damages.

Insecticides like organophosphate and methylcarbamates cause different health issues like disruption of nerve impulse by the formation of covalent bonds with serine residues of acetylcholinesterase thereby alter its functioning (Table 5.3). Various insecticides and fungicides affect different organs in body, like damages liver, thyroid, nervous system, bones, etc. (Casida and Durkin, 2013).Van Djik (2010) had observed that the use of neonicotinoid pesticides is increasing day by day. These pesticides are responsible for different types of toxicities. Different classes of pesticides are discussed below.

5.4.1 ORGANOPHOSPHATES

This is the large class of pesticides, which binds to acetylcholinesterase and disrupts the nerve impulses in brain. For example, diazinon is one type of pesticide that works by damaging acetylcholinesterase in the brain. This pesticide is used commercially in agriculture for control and management of crops. Organophosphates also affect physiological functions of insects. They are also widely used inside homes to control other pests like termites and ants, but its usage has been discontinued now. It has an addictive toxic effect on ecosystem. Organophosphorus group includes acephate, azinphos methyl, chlorethoxyfos, chlorpyrifos, diazinon, dicrotophos, dimetshoate, disulfoton, fenitrothion, fenthion, fosthiazate, methamidophos, monocrotophos, and malathion (Ortiz Hernandez et al., 2013).

5.4.2 ORGANOCHLORINES

Organochlorines were the extensively used pesticides from 1940s to 1960s in the field of agriculture and for mosquito control. These pesticides are basically chlorinated hydrocarbons which include DDT, endosulfan, atrazine, vinclozolin, tributyltin, etc., Many derivatives of this kind of pesticides show their effect on environment. Although numerous organochlorine pesticides have been banned in the United States, but a few are still registered for use. Organochlorine pesticides can get easily accumulated in the environment and are very persistent, can move long distances through surface runoff or groundwater. These insecticidal pesticides are primarily formed by carbon, hydrogen, and chlorine. Some examples of organochlorines are endosulfan, DDT, benzene hexachloride (BHC), α-BHC, β-BHC, λ-BHC, etc. These types of pesticides generally break down slowly and endure in the environment for a longer period of time after their application and exposure (Ortiz Hernandez et al., 2013).

5.4.3 CARBAMATES

Carbamates are the chemical pesticides which are closely related to organophosphorus compounds in their mechanism of action and resistance development. However, they are said to be easily degradable than organo-phosphates. They cause similar effects as that of orgnophosphates, as these generally targets human melatonin receptors by inhibiting acetylcholines-terase (Popovaska et al., 2017; Colovic et al., 2013).

Some examples of carbamate insecticides are bendiocarb, carbofuran, carbaryl, dioxacarb, fenoxycarb, isoprocarb, and methomyletc (Ortiz-Hernandez et al., 2013). Carbamates usually kill insects by reversibly inhib-iting the enzyme acetylcholinesterase (Fukuto, 1990).

5.4.4 PYRETHROIDS AND PYRETHRENES

Pyrethroids and pyrethrenes are the natural compounds used as insecticides for management of various crop diseases which originated from the plant, that is, *Chrysanthemum cinerariaefolium, C. coccineum*, and *Tanacetum cinerariaefolium*.

These types of insecticides are photocatalytic in nature, decay rapidly in the presence of light. In 1950, the synthetic production of pyrethroids overcomes the disadvantages of natural pyrethrins (Ellenhorn and Barce-loux, 1988). The exposure of humans to this compound led to allergic or respiratory reactions (mainly rhinitis and asthma) and contact dermatitis (Reigart and Roberts, 1999). Pyrethroid, a commercial household insec-ticide is an organic compound similar to the natural pyrethrins, produced by the flower pyrethrins. This chemical compound was developed to mimic the insecticidal activity of pyrethrum. These are less toxic in comparison to organophosphates and carbamates and are nonpersistent. These compounds are used against household pests. Different examples of pyrethroids include cypermethrin, cyfluthrin, deltamethrin, etofenprox, fenvalerate, permethrin, phenothrin, prallethrin, resmethrin, tetramethrin, etc. (Ortiz Hernandez et al., 2013).

5.4.5 TRIAZINES

Triazine pesticides are used as herbicides. These are inexpensive and effec-tive compounds such as atrazine and simazine. This pesticide is used to control selective weeds of crops like sorghum, sugarcane, and corn. Also, some members of triazine family are useful for the weed control in orchards, horticulture, and perennial crops. It combats large variety of weeds through inhibiting the electron transport chain and photosynthesis in plants. The effects of these compounds when exposed to human beings are still unknown. But sometimes, exposure of triazines to humans can cause disruption of endocrine and is associated with carcinogenicity (Sathiakumar et al., 2011).

Gupta (2017) reported that triazine pesticides have been used for 40 years for preventing weeds in many crops worldwide. Triazines are differentiated into symmetric and asymmetric pesticides. Symmetrical triazines include chloro-s-triazines (simazine, atrazine propazine, cyanazine, etc.), the thiomethyl-s-triazines (ametryn, prometryn, etc.) and metribuzin are the examples of asymmetrical triazines. Triazines are less toxic and pose acute hazards with dermal introduction Studies about its chronic toxicity have primarily revealed that it reduces body weight gain (Costa, 2008). The most popular and studied triazine herbicide used is atrazine; it showed carcinogenic potential in mice and rats and particularly increases the incidence of mammary carcinoma in Sprague Dawley rats (Mayhew et al., 1986; Zeljezic et al., 2006). Triazines sometimes may be carcinogenic and teratogenic in nature; its contamination also causes sensory motor polyneuropathy in humans (Ellenhorn and Barceloux, 1988).

5.4.6   DERIVATIVES OF PHENOXY HERBICIDES

The phenoxy derivatives contain an aliphatic carboxylic acid moiety that is attached to chloride or methyl-substituted aromatic ring. Phenoxy herbicides need to be co-formulated with ioxynil or bromoxynil and they are more toxic than the other herbicides. Bradberry et al. (2004) revealed that at high temperatures, these herbicides during the fabrication give rise to some other harmful chemicals such as chlorinated dibenzo dioxin and chlorinated dibenzofuran. The toxicity of these phenoxy salts and ester derivatives depend on acid form of the pesticides as they undergo rapid hydrolysis or dissociation. The effect of these pesticides shows variation from 12 to 72 h but long half-lives occur after prolonged and huge doses of its exposure (Reigart and Roberts, 2014).

Bradberry et al. (2005) concluded that these pesticide derivatives generally exhibit a variety of toxic mechanisms which include uncoupling of oxidative phosphorylation, acetyl coenzyme-A disruption, neuronal membrane transport, and dose-dependent damage to cell membrane. The toxicity of this pesticide on the CNS is also dose dependent. In recent studies, many scientists revealed that these phenoxy herbicides are also known for causing burning sensation in the nasopharynx, dizziness, spontaneous abortion, infertility, and birth defects in females by their exposure and prolonged inhalation (Reigartand Roberts, 1999).

5.4.7   SOME OTHER CLASSES OF PESTICIDES (FUNGICIDES, RODENTICIDES, NEMATICIDES)

Different other classes of chemical pesticides exist such as, chloroacetanilide which is usually used in agriculture. Chloroacetanilide includes alachlor, acetochlor, metolachor, and propachlor, etc., and are carcinogenic in nature (Dearfield et al., 1999). Another important class of pesticide is benzimidazoles. The imidazole, a derivative of benzo is commonly referred to as benzimidazole (Bansal, 2002). The benzimidazole derivatives when binds with free β-tubulin monomers through colchicine binding site inhibits the formation of microtubule (Ermler et al., 2013). With respect to new technologies, nanoplant and nanopesticide protection products represent an emerging technology with respect to pesticide use. These pesticides have good effectiveness and durability with the use of small amount of active ingredients (Kookana et al., 2014). Another class of pesticides is fungicides; it constitutes different classes such as thiocarbamates, thiophthalimides, organomercury compounds. Examples of some fungicides are chloroneb, hexachlorobenzene, ferbam, metam sodium, thiram, ziram, chlorothalanil, etc. (Costa, 2008; Alonzo and Correa, 2014).

5.5   EFFECTS OF PESTICIDES ON ENVIRONMENT OR ECOSYSTEM BIOTA

Pesticides are being used by agricultural industries since first quarter of nineteenth century. They show major risk associated factors toward environment and ecosystem biota. It has been observed that pesticides show ill effects on untargeted species and therefore, affect the biodiversity may it be flora or fauna (Mahmood et al., 2016). Nowadays, pesticides are used all over the world to control a wide variety of crops from different insidious plant pathogens, for example, dicofol (DCF) is widely used for crops like cotton, citrus, vegetables, nuts, etc., as preharvest miticides (Osman et al., 2008). Approximately 80%–90% of pesticides when applied on the crops get volatilize in a few days (Majewski and Capel, 1995). Over 98% of pesticides are sprayed across the entire agricultural fields and sometimes affect other nontarget species in the environment (Miller, 2004). The volatilized pesticides evaporate into the air and consequently may cause harm to nontarget species. For instance, the vapors from volatilization of pesticides of treated plants are sufficient enough to cause severe damage to other plants (Straathoff, 1986). Indiscriminate use of pesticides may result in reduction of different aquatic and terrestrial biota. Pesticides have also been reported to affect several birds like bald eagle, peregrine falcon, and osprey (Helfrich et al., 2009). Along with this all the air, soil and water bodies are being contaminated with these harmful and toxic chemicals. Moreover, pesticides play a pivotal role in agricultural fields to reduce the losses caused by weeds, mites, agricultural pests microbial disease, etc., and thereby enhances yield and production (Ghorab and Khalil, 2015; Gupta, 2006).

It is quite evident from different studies that the biodiversity is facing serious problems because of the pesticide pollution in the environment. Pesticide pollution takes place because of their potential toxicity, slow degradation rate, and high persistence. But in recent years, a serious concern is given by researchers and various biotechnological approaches are sought for remediation of these harmful compounds and combat such pollution. Pesticides like organophosphate or organochlorine residues mainly affect target species but sometimes they also affect nontarget species in a large quantity. Many soil microorganisms mainly the bacteria have been widely reported playing a vital role in the biodegradation of pesticides such as organphosphate, profenofos, etc. Water is a very important resource on earth, which is also contaminated by pesticides via runoff or leaching. It is important to develop environmental friendly techniques to eradicate these pollutants from water. Water treatment helps in reducing the hazards associated to the various pollutants which affect the surrounding environment as well as human health (Routt and Roberts, 1999). To treat contaminated water; various new trends have been developed by the biotechnological approaches like microbial biofilms. These biofilms are very dynamic and applied for treating the organic and inorganic, polluted and effluent materials (Ghorab and Khalil, 2016).

Pesticides also affect the aquatic life forms as they may enter into the lakes and rivers via different routes such as excess of water flow, runoff, leaching through soil or sometimes these pesticides are applied directly onto surface water for the protection of mosquitoes. It has been reported that the lawn care pesticides are basically found on surface water and in different water bodies like lakes, ponds, rivers and streams. It decreases the dissolved oxygen in the water and affects aquatic life forms, therefore, causing physiological and natural behavioral changes in the aquatic population especially in fishes (Mahmood et al., 2016). The aquatic plants provide 80% of dissolved oxygen to the aquatic living organisms for their survival. Because of overuse of herbicides, the number of aquatic plants is diminishing day by day because of their harmful effects. This result in decrease of O_2 level in the water bodies drastically and ultimately causes suffocation to fishes and reduces fish productivity (Helfrich et al., 2009).

A pesticide also pollutes groundwater by reaching underground during the seepage of polluted water, inappropriate dumping, spillages, etc. (Pesticides in Groundwater, 2014). The terrestrial insect's like bees and beetles are also harmed by the pesticides; their number is also getting reduced day by day because of the indiscriminate uses of pesticides like organophosphates and pyrethroids. The chemical fungicides like triazole, imidazole, and pyrethroids have also negative effects on honey bees, butterflies, etc. (Pilling and Jepson, 2006).

5.6   FATE OF PESTICIDES: ENVIRONMENT AND ECOSYSTEM

The pesticides are found in all three phases, that is, liquid, solid, and gaseous (Marino et al., 2002). Their various applications depend on the constant of adsorption, desorption, and volatilization techniques. The fate of pesticides is strongly associated with the soil absorption processes which control their transfer and bioavailability in the environment (Besse-Hoggan et al., 2009). Adsorption is the most significant mode of interaction among soil and chemical pesticides. When pesticides are applied in the field they get bind to the soil particles or sediments and plants (Gebremariam et al., 2012). The adsorption process generally varies according to the soil properties and its composition. It may be purely physical as like vander Waals forces or chemical in nature like electrostatic interactions (Bailey and White, 1970).

The pesticides show their harmful effects to both the environment as well as public health. Soil contamination through pesticides is the accidental release of chemicals into the field surface and groundwater, etc. (Singh and Walker, 2006).

Several biological or environmental factors in soil that is moisture, temperature, along with physiochemical properties of soil, etc., influence the rate of biodegradation (Besse-Hoggan et al., 2009).

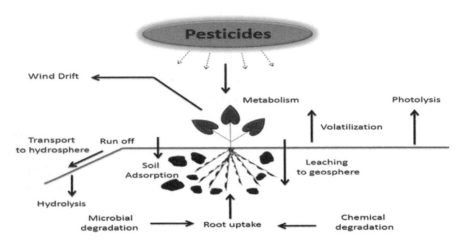

FIGURE 5.1   Illustration of fate of pesticides in environment.

The fate of pesticides is getting much affected because of their own physiochemical properties and also due to the environmental conditions, management practices, and the characteristics of the soil (Halimah et al., 2010). In a natural environment, the pesticide after their accumulation leads to release of metabolic dead-end products and get mix with the soil humus, also it enters the food chain leading to increase the concentration of toxic substances or biomagnification (Figure 5.1).

5.7   COMMON PESTICIDES USED IN AGRICULTURE IN INDIA AND THEIR EFFECTS ON HUMAN HEALTH

The first pesticide was developed by the Germans in the year 1930s. The first one introduced in 1965 in USA as an insecticide was the chloropyrifos [O,O-diethyl O-(3,5,6-trichloro-2-pyridinyl)-phosphorothioate]. It is most broadly used organophosphate pesticide in USA peoples especially in their homes and gardens (Worthing, 1979). In a study by Fang et al. (2006), it is reported that chloropyrifos is used against majority of insect pests for protecting economically important plants, and it is known as a nonsystemic insecticide. It is also harmful for health of humans when one gets exposed to it. In agriculture, the common pesticides used are fungicides, insecticides, and herbicides (Table 5.4).

TABLE 5.4   Toxic Effects of Pesticides on Human Health

Category of Toxic Effects	Target System	Symptoms Found
Respiratory	Nose, trachea, lungs	Irritation, tight chest, coughing, choking
Gastrointestinal	Stomach, intestines	Nausea, vomiting, diarrhea
Renal	Kidney	Back pain, urinating more or less than usual
Neurological	Brain spinal cord	Headache, dizziness, confusion, behavior, depression, coma, convulsions
Dermatological	Skin, eyes	Rashes, itching, redness, swelling
Hematological	Blood	Anemia (tiredness, weakness)
Reproductive	Ovaries, testes, fetus	Infertility, miscarriage

In India, insecticides form the highest share among total pesticides used around the country (Subhash et al., 2017). The chlorogenic compounds like aldrin, heptachlor, endosulphan, DDT, have deep concern, because of their ability to accumulate in tissues and cause serious diseases in humans (Das and Chandran, 2011). These pesticidal compounds have effect on soil flora and fauna (Bhat and Padmaja, 2014). Different workers had illustrated that the pesticide residues cause different neuropsychiatric diseases such as anxiety, mood disorders, depression, etc. Some pesticides bring about the changes in the function of central nervous system, for example, cholinergic crisis; this effect is ultimately followed by suicide attempts (Freire and Koifman, 2013). In a report of World Health Organization, it was stated that in many Asian and Latin American countries suicide by consuming pesticides is a common issue. These chemicals have been used indiscriminately and also are widely available in low and middle income countries (Sarchiapone et al., 2011).

The production as well as use of pesticides in India had started in 1952. Currently, India is the second largest manufacturer country of chemical pesticides in the world (Mathur, 1999). Pesticide usage pattern in India is different from that of the other countries.

5.8   VARIOUS PESTICIDE REMEDIATION TECHNIQUES

Pesticides are extensively applied to most of the crops all over the world for protecting them from pests. But many times these pesticides enter the environment directly by spillages during transportation or storage which harms the nontarget species (Khan et al., 2004). When these chemical pesticides are applied to the crop plants, through various processes it reaches to the soil such as by rain, irrigation water, and wind (Hamza et al., 2016). Odukkathil and Vasudevan (2013) had studied that many pesticides are ubiquitous in nature and due to their less bioavailability, they generally persist in soil and sediments. Hence, these chemical compounds are highly toxic for human health including biodiversity and environment (Jariyal et al., 2015). The presence of chemical pesticides in soil is undesirable because of their toxic nature and anomalies they normally create. From different studies the scientists had observed that the presence of pesticides and other pollutants like heavy metals and hydrocarbons residues in the soil adversely affects seed germination, plant growth (Smith et al., 2006; Ahmad et al., 2012, 2013, 2014), capability of plants and different microorganisms to absorb nutrients, water, minerals, etc., from the soil microbial activities (Nie et al., 2011; Guo et al., 2012; Alrumman et al., 2015). Pesticides are not toxic for living beings only (Anderson and Meade, 2014; Isoda et al., 2005), But these pesticides negatively affects other forms of life also; these include arthropods, plants, butterflies and many soil microbes (Desneux et al., 2007; Siddiqui and Ahmed, 2006; Longley et al., 1997), etc.. Environment and soil is of prime importance, for better health and sustainability. It is important to take serious remedial measures to eliminate these contaminants from the nature. In the past few decades, many methods have been successfully carried out for separating pollutants from the soil. Mainly, biological, chemical, and physical remediation techniques are used (Marican and Lara, 2018).

There are various environmental regulations for water and air pollution developed in most of the countries. But only a few countries have a regulation for removing soil pollutants. In the last few decades, the methods used to deal with the soil pollutants like excavating, put it in a landfill, or isolated by many types of barriers that are known by avoiding contamination from the neighboring sites (Scullion, 2006). Before applying any remediation method to a given polluted site, assessment of safety measures should be taken. Andreoni and Gianfreda (2007) had revealed that to obtain a successful bioremediation strategy, some measures must be taken like no production of any toxic by-product, the absence of inhibitory chemicals, contaminants must be bioavailable and also the optimized conditions are provided which supports the microbes for better growth and activity. Different remediation techniques are adopted including biological, physical, and chemical for the sustainable ecosystem (Table 5.5).

TABLE 5.5 Various Biological, Physical, and Chemical Techniques are Being Used to Treat Pollutants/Contaminants

Biological Remediation Technique	Elucidation
Bioremediation	In this process pollutants are treated, including water, soil, and subsurface material.
Bioaugmentation	Isolation and reinoculation of enriched bacteria capable to mobilize or immobilize pollutants.
Biostimulation	Nutrient elements like nitrogen and phosphorus incorporated to stimulate indigenous bacteria.
Phytoremediation	This technology uses living plants to clean up soil, air, and water contaminants.
Biopiles and land farming	The nonhalogenated volatile and semivolatiles are treated under this technique. Some other compounds that have been treated successfully are PCV and pesticide residues.
Composting	Composting is a process that involves the mixing of the polluted soil in a pile with a solid organic substrate, which serves as a carbon source for indigenous aerobic soil microorganisms.
Bioventing	Insertion of some oxidants (oxygen, sulfate, nitrate, etc.) into the wells to stimulate resident bacteria usually sulfate oxidizing/reducing bacteria
Physio-Chemical Remediation Techniques	
Soil vapor extraction	Used to treat volatile organic compounds
Solvent extraction	Used to treat sediments, sludge, and soils containing mainly organic compounds
Solidification/stabilization	Inorganic compounds, including radionuclides
Air sparging	In situ air stripping or in situ volatilization, it is less applicable to diesel fuel and kerosene
Soil flushing	Used to treat inorganic compounds

5.8.1 BIOLOGICAL TECHNIQUES

Remediation through the biological techniques shows profound results in the cleaning of environment and it is a very efficient technology. In this process, it allows the conversion of organic compounds into less harmful end products (i.e., CO_2 and H_2O). In comparison to other physical and chemical methods, this technology is said to be low cost and environment friendly (Nwankwegu and Onwosi 2017; Rajiv et al., 2009).

5.8.2 BIOREMEDIATION

Bioremediation is a biological process commonly used for remediating pollutants. It is a promising technology that includes existing microorganisms to eliminate contaminants from the sites (Samanta et al., 2002; Singh et al., 2004). For this process mainly bacteria, fungi, or yeast like biological agents are used (Strong and Burgess, 2008). With respect to physio-chemical techniques bioremediation process has high public acceptance to remediate pollutants from different sites (Vidali, 2001). The advantages of bioremediation process like being environmental friendly have been listed by several workers (Hussain et al., 2009).

"Bioremediation" mainly deals with any biological strategy used to eliminate or remediate undesirable contaminants from the environment. The important parameters such as pH, humidity content, dietary state, nature of pollutants, diversity of microbes, temperature and oxidation reduction potential, etc. (Dua et al., 2002). In the process of microbial bioremediation, microorganisms use contaminants as their nutrient or energy source. Bioremediation is based on two different methods that are rhizoremediation (i.e., plant and microbe interaction) and phytoremediation (includes plants for remediation) (Ma et al., 2011). Bioremediation process has many limitations; it can be only used for biodegradable substances. Sometimes, biodegraded products are said to be more persistent than the parent compounds itself, for example, 1,2,3-trichloropropane (in respect to chlorpyrifos), the biodegraded products of chlorpyrifos is examined more toxic and persistent. The biological processes are also said to be highly specific. Bioremediation is suitable for remediating many prevalent pollutants like polycyclic aromatic hydrocarbons (PAHs), petroleum hydrocarbons (PHC), solvents, and heavy metals, etc. (Damalas, 2009).

Approximately 90% of pesticides are used in agriculture, and these pesticides while reaching their target species, they get dispersed into soil, air, and water. This results in pesticide accumulation in the vegetables, and other crops, ultimately affects human beings and environment (Gamon et al., 2003). Pesticides may easily pass into the living tissues of organisms and it gives rise to bioaccumulation process. Therefore, the bioremediation techniques have received more attention to treat the pesticides from soil, by owing to its eco-friendliness behavior. These techniques are said to be an efficient tools used for eliminating the pollutants from contaminated sites in the existing environment (Mervat, 2009).

5.8.3   BIOAUGMENTATION

Bioaugmentation is also the most important biological technique; it is carried out for the improvement of contaminated soil and water by the help of unambiguous competent strains of microbes (Thierry et al., 2008). By the use of different microorganisms bioaugmentation process has been carried out for the several pesticide degradation processes (Silva et al., 2004). In the biological processes mainly microorganisms play a crucial role in enhancing degradation process, when these microbes are imported to the polluted sites it leads to the bioaugmentation. Herrero and Stuckey (2015) explained that bioaugmentation has good efficiency in waste water treatment, they reported different strategies for this process which includes on the ecological basis (for acquiring knowledge about the metabolic process of microbes), on monitoring techniques (i.e., PCR a thermal cycler to see the activity or existence of added microbes), on the management of plant species (antici-pating the changes in season changes or effects on bacterium or consortium communities), etc. The degradation of pesticides through the bioaugmenta-tion is the important process but at the same time it has some limitations too, these are (1) not suitable for all soils, (2) do not able to achieve complete degradation, (3) difficult to control natural conditions like temperature for optimum degradation (Chawla et al., 2013). Bioaugmentation is generally paired with biostimulation to strengthen the pesticide degradation potential of a contaminated medium or sites (Kadian et al., 2008).

5.8.4   BIOSTIMULATION

Biostimulation is the microbial ecological method which enhances the biore-mediation process, it includes the addition of trace minerals, soil nutrients, and oxygen, etc.(Scow and Hicks, 2005; Li et al., 2010). Biostimulation practice involves the modification of environment by the stimulation of existing bacteria which carry out bioremediation. Currently, number of products has been introduced by the bioremediation processes which are known to be used as biostimulation. However, it has been observed by the scientists that inorganic nutrients when added for the degradation of atrazine pesticide, they showed the good significant behavior toward it by the reduction of half-life of herbicides in the soil (Hance, 1973). Activated soil is directly used for degradation of pollutants when combined with inoculum (Barbeau et al., 1997; Runes et al., 2001).

5.8.5   PHYTOREMEDIATION

The phytoremediation method is the beneficial technology that uses plants for remediation purposes. It is an innovative, cost-effective, and socio-economic process. This is a plant assisted bioremediation which has been used for over 300 years generally used to clean and restore soil and wastewater by plants. This is a promising technology and useful in the elimination of a number of soil pollutants. It is an in situ, inexpensive, and environmentally friendly method (Brooks, 1998; Chaney et al., 1997; Utmazian and Wenzel, 2006; Zavoda et al., 2001). Overall it is a very easy process in which recovery and reuse of valuable products can be utilized. This includes several other techniques such as phytotransformation, phytoremediation, and rhizodegradation. (Sessitsch et al., 2002).

5.8.6   BIO PILES AND LAND FARMING

Bio piles and land farming is the biological pollutant remediation process. This includes a bed treatment, that is, an irrigable/nutrient system, mound of contaminated soil, an air exposure system, and a leachate collection system. A bio pile is an in-situ type method generally constructed into a ground where piles are driven into it. In this process different natural conditions are maintained to enhance biodegradation such as humidity, high temperature, oxygen, nutrients, and pH. The soil piles constructed are about 20 feet high and covered with a layer of plastic for controlling overflow, evaporation, etc. (Wu and Crapper, 2009). In land farming, the soil is dug up or excavated and mechanically separated by sieving process. After this the soil which is contaminated is positioned in layers over the uncontaminated soil by allowing natural remedies to detoxify and degrade contaminants (USEPA, 2006). Artificial or man-made concrete and membranes of clay are used to cover the polluted layer of soil. With the help of plowing, harrowing, or milling, oxygen is mixed well. Also the nutrients and moisture are, further-more, added to enhance the remediation of polluted soil. The pH (near 7.0) of the soil can be regulated by using crushed limestone or agricultural lime, (Van Deuren et al., 2002). Landfarming is predominantly intended for reme-diation of pesticide polluted soils (Felsot et al., 1992).

5.8.7 COMPOSTING

This is also a biological process of remediation, in which organic wastes are degraded through microbes by elevating the temperature (usually at the range of 55 °C–65 °C). In the composting practices, temperature is increased that results in the solubility of pollutants at a great metabolic rate. This process involves the microbial population which becomes more numerous and diverse in nature. It includes the excavated and screening of contami-nated soil; it removes the large debris and rocks (Blanca et al., 2007; Bouwer and Zehnder, 1993). After that soil is transported to composting pad, hence, for the carbon source and bulking agents different substrates are used, that is, alfalfa, manure, straw, agricultural wastes, etc. These amendments are layered into a pile which is known as windrows. With the help of windrow turning machine this windrow is mixed thoroughly. All the natural condi-tions are maintained same as in case of the other remediation techniques like pH, moisture, temperature, etc. After the completion of composting period, windrows are differentiated into exterior and static piles (Chawla et al., 2013). Mohamed and co-workers have reported that oxyfluorfen type of herbicide is degraded through the composting technique (Mohamed et al., 2011). Complete disappearance of many pesticides like malathion, chlorpyrifos-methyl, and lindane and a partial degradation of endosulfan pesticide residues have been observed after the composting within eight days of treatment (Frenich et al., 2005).

5.8.8 BIOVENTING

Bioventing method is an in-situ process that generally stimulates natural biodegradation of the pollutants in soil by providing oxygen to it for existing microbes. In this technique, oxygen and nutrients are injected into the soil in order to maintain the bioremediation (Shanahan, 2004). Phosphorus and nitrogen are the two most common nutrients applied in this method (Rockne and Reddy, 2003). Bioventing process requires less air flow rate and supply adequate amount of oxygen to microbes to sustain them in vadose zone (USEPA, 2006). This means of biological remediation basically suits drained, intermediate, and coarse-textured soil. It consists of a well and a blower, which usually pumps air into soil and well (Lee et al., 2006). A recent study reported that 93% of phenanthrene-based pesticides are reduced after seven months of treatment through bioventing technique (Frutos et al., 2010).

5.8.9   *PHYSIO-CHEMICAL TECHNIQUES*

The physio-chemical methods of remediating pesticides are based on physio-chemical phenomena. Some of these important techniques used for soil decontamination with their applications are discussed below.

5.8.10   *SOIL-VAPOR EXTRACTION*

Soil vapor extraction is one of the widely used physio-chemical methods. This technique is non-biological which is used to treat volatile organic compounds, semivolatile organic compounds, PCBs, etc. This process involves a source of vacuum applied to the soil matrix that creates a pressure gradient originated due to air present in wells of extraction (Sparks and Corn, 1993; Johnson et al., 1990; Beckett and Huntley, 1994). Through this procedure of remediation, the application is obtained in effervescent phase and then promoted for processing before launching it into the environment such as, adsorption of activated coal (Johnson et al., 1994; Wilson and Clarke, 1994).

5.8.11   *SOLVENT EXTRACTION*

This is also a physio-chemical technique, in which organic chemicals as solvents are used. This is generally used for water treatment. This technique is also used on commercial scale units. Solvent extraction technique does not carry out the actual remediation of pollutants, but helps to separate these from the soil, therefore, it is called as pretreatment technique and is similar to soil flushing (Raghavan et al., 1991; Sahle-Deessie et al., 1998). It is used in association with the other bioremediation techniques. The successes of this method directly depend upon the selection of solvent we use, it must be kept in consideration what kind of pollutants are to be extracted (Sparks and Corn, 1993).

5.8.12   SOLIDIFICATION/STABILIZATION

Pesticide remediation of the soil can also be done by the solidification or stabilization technique. It consists of mixing of different reactive materials (e.g., cement or concrete along with the solids, semi-solids, or sludge) which help in immobilization of pollutants. This technique consists of some variants which are cement-based solidification (in this the direct mixing or adding of cement in soil is done), second one is silicate-based solidification (in this process the material or leached ashes are mixed with the cement for stabilizing agents), microencapsulation (Sparks and Corn, 1993; Castelo-Grande and Barbosa, 2013).

5.8.13   AIR SPARGING

Air sparging is an in-situ, air stripping or in-situ volatilization type of method, which employs amplified type of oxygen to reach the subsurface of the saturated zone. This enables the hydrocarbons to get converted from its dissolved phase to its vapor phase. After that this air is vented from the unsaturated zone (Norris, 1994; Miller, 1996; Brown and Fraxedas, 1998).

5.9   MICROBIAL DEGRADATION OF PESTICIDES

A microbe has a determinant role in the remediation of pesticides from the environment. Especially, the bacteria are extensively used in the bioremediation techniques. This process can take place either in anaerobic or aerobic conditions. Various enzymes produced by bacteria contribute in the degradation process of pesticides (Langerhoff et al., 2001). In agricultural practices, a large number of chemical pesticides are used all over the world for protecting crops which have deteriorated the soil quality and environment. Thus, bioremediation is sought to achieve successful remediation and restoration of such polluted soils. Research is going on a large scale on microbial degradation techniques; it exhibits high impact on crop yield, productivity, and maintains losses by killing different pests and insects. The microbial degradation of pesticides generally occurs by biological activity of the microbes (bacteria or fungi) present in the soil under natural conditions. Some microorganisms are reported which use pesticides as their carbon source like *Pseudomonas, Flavobacterium, Penicillium, Rhodococcus, Trichoderma*, etc. (Aislabie and Lloyd-jones, 1995).

A diverse group of microbes have been found and reported to play the role in the biodegradation of various pesticides that are illustrated in Table 5.6. A microorganism in biological system has the capacity to biotransform pesticides. Studies have revealed that the soil biota has degradation ability to degrade certain pesticides quickly. These chemicals serve as carbon and energy source, electron donors for the microbes in an adequate quantity. This establishes a proper way for the treatment of pesticide polluted sites (Qiu et al., 2007; Araya and Lakhi, 2004). Many research laboratories around the world have a collection of microorganisms with their characteristics, identification, and reported degradation ability of pesticides. These different bacterial species with degradation ability of pesticides are used to treat polluted environment. These isolates are also helpful in treating the waste products prior to their absolute deposition (Ortiz-Hernández et al., 2011).

TABLE 5.6   List of Microbial Species and the Pesticide They Degrade

Pesticides	List Includes Microbial Spp. Involved in Degradation
DDT in soil	*Escherichia coli, Enterobacter aerogenes, Enterobacter cloacae, Klebsiella pneumonia, Pseudomonas putida, Bacillus* sp.
Endosulfan	*Pseudomonas* sp., *Bacillus* sp., and *Flavobacterium* sp.
DDT (activated sludge)	*Pseudomonas* sp., *Pseudomonas aeruginosa, Micrococcus, Bacillus pumilus, Bacillus circulans, Bacillus* sp., and *Flavobacterium* sp.
Lindane	*B. thiooxidans* and *Baseathiooxidans, Sphingomonaspaucimobilis, Streptomyces* sp., and *Pleurotusostreatus*
Nitrobenzene	*Pseudomonas pseudoalcaligenes* JS45 and *Pseudomonas putida* HS12, *Comamonas* sp. JS765, *Pseudomonas mendocina* KR-1, *Pseudomonas pickettii* PKO1, *cyanobacterium, Microcystis aeruginosa, Arthrobacter* sp. NB1, *Serratia* sp. NB2, and *Stenotrophomonas* sp. NB3, *Comamonadaceae and Clostridium* spp., and *Rhodotorulamucilaginosa* strain Z1
Cypermethrin	*E. coli, S. aureus, P. aeruginosa, B. subtilis, Enterobacter asuburiae*, and *Pseudomonas stutzeri*.
Carbendazin	*Pseudomonas*

TABLE 5.6   *(Continued)*

Pesticides	List Includes Microbial Spp. Involved in Degradation
Carbofuran	*Pseudomonas, Flavobacterium, Achromobacterium, Sphingomonas*, and *Arthrobacter*
PCP	*Arthrobacter* sp. and *Flavobacterium* sp.
1,4- Dichlorobenzene	*Pseudomonas* sp.
Heptachlor	*Phanerochaetechrysosporium* and *Phlebia* sp.
Dieldrin	*Pseudomonas* sp.
Toxaphene	*Bjerkandera* sp.
Aldrin	*Trichoderma viridae, Pseudomonas* sp., *Micrococcus* sp., and *Bacillus* sp.
Endrin	*Trichoderma viridae, Pseudomonas* sp., *Micrococcus* sp., *Arthrobacter* sp., and *Bacillus* sp.
Fenvalerate	*Bacillus cereus* and *Pseudomonas viridiflava.*
Pentachloronitrobenzene	*Rhizoctoniasolani, Botrytis* spp., *Aspergillus* spp., *Penicillium* spp., *Fusarium* spp., *Sclerotinia* spp., and *Tilletia caries.*
Atrazine, monocrotophos, alachlor, and 4-chlorophenol	*Arthrobacter, Clavibacter, Nocardia, Nocardioides, Rhodococcus*, and *Streptomyces.*

A serious concern had been shown by scientists toward the bacterial degradation of pesticides. A study showed that pesticide utilizing bacteria in water and irrigation channels include bacterial species of genus like *Bacillus, Micrococcus, Pseudomonas*, and *Vibrio* (Perclich and Lockwood, 1978). It is generally said that *Pseudomonas* and *Bacillus* are the two major groups that mainly degrades pesticides. This is because these bacterial species contain a broad variety of enzymes that helps in degradation and transformation of these toxic products (Walker et al., 1993).

Grant et al. (2002) had illustrated that technical grade of cypermethrin may be degraded from 60 to 6 mg/L in 20 days with the help of *Pseudomonas* sp. When the concentration of cypermethrin increases then it also shows negative effect on degradation rate. This is because of the low nutrients rate, which are generally required for the growth of bacteria that cause degradation of cypermethrin. Different studies had revealed the various *Pseudomonas* sp. (which is isolated from the soil) can lead to vast variety of pesticides like DDT, dichlorodiphenyldichloroethane, dodecylphenyl hydrogen phosphorodithioate, and hydrogen cyanide.

Most of the degradation study has been observed under the specific laboratory conditions in which the proper and optimum conditions for the growth and metabolism for bacteria are provided. However, to practice it on large scale in the fields needs more research in this area (Nawab et al., 2003). Bending et al. observed that including all these genera of bacteria, several fungi sp. like *Agrocybese miorbicularis, Auricularia, Coriolus versicolor,* etc., also have the degradation capacity of many pesticides, for example, phenylamide, triazine, phenylurea, dicarboximide, and organophosphorus compounds (Bending et al., 2002). White rot fungi are reported to degrade different classes of pesticides, for example, lindane, atrazine, DDT, gamma-hexachlorocyclohexane (Y-HCH), aldrin, heptachlor, etc. (Quintero et al., 2007). About 90% degradation of endosulfan pesticide in maize root (rhizosphere) occurs by the fungal consortium which comprises *Aspergillus* and *Trichoderma* sp. (Gangola, 2014).

Some pesticides are very sensitive to the hydrolytic activities of microbial enzymes such as malathion, carbamate, pyrethroid, diazinon, dichloropicolinic acid, etc. Many microbial enzymes (which are extracellular in nature) have capability to cleave broad range of pesticides. Instead of ordinary structure of the pesticides, the adsorption and volatility potential toward the soil is also an essential factor that affects the sensitivity of biological cleavage, these factors are dependent on temperature, light, soil, moisture, and pH, etc. When there is high level of moisture, it ultimately eases the degradation rate of the pesticides which are soluble in water by microbes. Therefore, it reduces the volatility of the soil compounds. There are many pesticides such as diazinon which show sensitivity toward less pH range, thus degradation at this range occurs considerably (Muller and Korte, 1975; Freed et al., 1979).

5.10 ADVANTAGES AND IMPORTANCE OF BIOREMEDIATION

The importance of pesticide remediation is aimed at eliminating toxic contaminants from the environment as well as atmosphere (Table 5.5). Different biological processes involved in pesticide remediation show lots of advantages in comparison to physical and chemical methods. The strategy for bioremediation process to be successful needs proper environmental conditions with specific type of microbes for proper degradation. The most specific and important microbes used in biological remediation of pesticides are bacteria, fungi, and algae which show great physiological metabolism for the degradation of contaminants. Biological techniques offer many advantages over conventional ones which include physical and chemical techniques. Pesticide remediation is of very pivotal importance in the present days for both environment as well as agricultural perspectives. Bioremediation is a cost-effective, eco-friendly, and easy process. There are many chemical compounds that are legally considered very toxic in nature that can be changed or converted to harmless compounds by means of bioremediation techniques (Chawla et al., 2013).

5.11   LIMITATIONS OF BIOREMEDIATION

The biological processes like bioremediation and phytoremediation are sometimes considered as problematic because these can lead only to partial degradation of contaminants which can be more toxic and harmful than the parent compound itself (Table 5.7). These processes take longer time duration to achieve acceptable levels, in comparison to other possible detoxifying methods (Rockne and Reddy, 2003). It also requires extensive monitoring throughout the process (FRTR, 2006). The microbial pesticide degradation is also having advantages as well as disadvantages. It sometimes needs a specific bacterial strain for the degradation of a specific type of chemical pesticides. But in the local market, the specific strains for specific pesticides are not available easily. However, a single type of microbial strains cannot degrade many types of pesticides together. Hence, the research in the field also addresses the drawbacks of microbial degradation of pollutants.

TABLE 5.7   Advantage and Disadvantage of Bioremediation Technique

Advantage	Disadvantage
• Can be done on spot.	• There are some chemicals which cannot be bioremediated.
• Irreversible elimination of waste.	
• Biological method-based systems are cheaper.	• It requires extensive monitoring.
• It is positively accepted by public.	• It requires specific sites.
• Long-term liability risk reduced.	• Toxicity of pollutants.
• It has minimum site disruption.	• Many times it is scientifically intensive.
• It reduces transportation cost and liability.	• Potential production of unknown by-products.
• It can be coupled with other treatment techniques.	• Perception of unproved technology.

5.12   CONCLUSION AND FUTURE PROSPECTS

Pesticides have proven boon for the agricultural farmers all around the world as they increase the crop production and provides numerous benefits to the society indirectly. Simultaneously pesticides also have proven to be hazardous substances which cause different types of chronic diseases in humans and affect the environment as well. So, the concern is to be made toward the safety measures regarding the use of the pesticides. There should be a proper judicious use and indiscriminate use of these chemicals should be avoided. As we know that, we cannot eliminate these hazardous chemicals but we can avoid or circumvent the overuse and improper usage of these pesticides. When these pesticides are used in appropriate amount or on dire requirement basis, the risk associated with the pesticides could be minimized. The harmful effects associated with these pesticides after its exposure can be reduced by acquiring different means of alternative cropping patterns. Also it may be reduced by using well-maintained spraying equipment. Hence, the strategy should be formulated for using safe and environment friendly pesticides. The havoc can be curbed by applying low doses of toxic formulations. There is a broad list of pesticides which have been banned in many countries like organochlorines because of their serious harmful effects. These pesticides cause serious water and soil pollution thereby causes serious health and environmental hazards. The reason behind the inappropriate usage of pesticides is because of lack of knowledge among the farmers. They hardly know the potential toxicities, safety measures, or the level of poisoning of these pesticides. Due to this reason, the people associated with the agriculture directly are at more risk compare to the ones which receive them indirectly. Thus, awareness should be provided to farmers to reduce the use of these toxic chemical pesticides. Soil contamination has become a complicated and global problem because of the generation of large number of toxic chemicals. Different techniques of remediation, that is, biological, physical, or chemical involves different strategies that are although available and considered as the excellent remedial techniques used for the elimination of the various pollutants. However, the biological process involving the microbial strains of bacteria, fungi, and algae serves as an eco-friendly, cost-effective process and a far better solution to the problem associated with these pollutants. Chemical pesticides can be used in the combination with natural remedies in future. By doing so, this technique of using natural remedies could help in environment sustainability. It is the need of hour to focus and address the different attributes including environmental chemistry, population biology, toxicology, community ecology including strategies related to conservation of environment to understand direct and indirect effects of these toxic pollutants on ecosystem. Moreover, bioremediation of pesticides and other pollutants needs a large concern of researchers to make this a promising technology through advanced molecular techniques. This will help in cleaning up of environments contaminated with combined pollutants. However, this technology needs to be explored and developed to show maximum potential.

KEYWORDS

- **pesticides**
- **warfare**
- **combat**
- **bioremediation**
- **contamination**

REFERENCES

Abhilash, P. C.; Singh, N. Pesticide use and application: an Indian scenario. *Journal of Hazardous Materials*, **2009**, *165*, 1–12.

Agrawal, A.; Pandey, R. S.; Sharma, B. Water pollution with special reference to pesticide contamination in India. *Journal of Water Resource and Protection*, **2009**, *2*(5), 432–448.

Ahemad, M. Pesticides as antagonists of rhizobia and the legumes-rhizobium symbiosis: a paradigmatic and mechanistic outlook. *Biochemistry & Molecular Biology*, **2013**, *1*(4): 63–75.

Ahmad, I.; Akhtar, M. J.; Asghar, H. N.; Zahir, Z. A., Comparative efficacy of growth media in causing cadmium toxicity to wheat at seed germination stage. *International Journal of Agriculture and Biology*, **2013**, *15*, 517–522.

Ahmad, I.; Akhtar, M. J.; Zahir, Z. A.; Jamil, A., Effect of cadmium on seed germination and seedling growth of four wheat (*Triticumaestivum* L.) cultivars. *Pakistan Journal of Botany*, **2012**, *44*, 1569–1574.

Ahmad, I.; Akhtar, M. J.; Zahir, Z. A.; Naveed, M.; Mitter, B.; Sessitsch, A. Cadmium-tolerant bacteria induce metal stress tolerance in cereals. *Environmental Science and Pollution Research*, **2014**, *21*, 11054–11065.

Ahmed, M.; Ahmed, I. *Bioremediation of pesticides*, In: Biodegradation and Bioremediation, Studium Press LLC, USA, pp 125–165, 2014.

Aislabie, J.; Lloyd-jones, G. A review of bacterial degradation of pesticides. *Australian Journal of Soil Research*, **1995**, *33*, 925–942.

Alonzo, H. G. A.; Corrêa, C. L.; Praguicidas. In: Oga, S.; Camargo, M. M. A.; Batistuzzo, J. A. (eds.). *Fundamentos de Toxicologia*. 4th. Ed. Editora Atheneu, São Paulo, Brazil, **2014**, pp. 323–341.

Alrumman, S. A.; Standing, D. B.; Paton, G. I. Effects of hydrocarbon contamination on soil microbial community and enzyme activity. *Journal of King Saud University Science*, **2015**, *27*, 31–41.

Anderson, S. E.; Meade, B. J. Potential health effects associated with dermal exposure to occupational chemicals. *Environmental Health Insights*, **2014**, *8*, 51–62.

Andreoni, V.; Gianfreda, L. Bioremediation and monitoring of aromatic-polluted habitats. *Journal of Applied Microbiology and Biotechnology*, **2017**, *76*, 287–308.

Araya, M.; Lakhi, A. Response to consecutive nematicide applications using the same product in mussaAAAcv. *Grande naine* originated from in vitro propagative material and cultivated in virgin soil. *Nematologia Brasileira*, **2014**, *28*(1), 55–61.

Bailey, G. W.; White, J. L. Factors influencing the adsorption desorption and movement of pesticide in soil. *Residues Review*, **1970**, *32*, 29–92.

Bansal, R. K. *Heterocyclic Chemistry*. (Third ed.), New Age International Publisher, New Delhi, India,, **2002**.

Beckett, G. D.; Huntley, D. *Groundwater*, **1994**, *32*, 247.

Bending, G. D.; Friloux, M.; Walker, A. Degradation of contrasting pesticides by white rot fungi and its relationship with ligninolytic potential. *FEMS Microbiology Letters*, **2002**, *212*, 59–63.

Bernardes, M.F.F.; Pazin, M.; Pereira, L.C.; Dorta, D.J. Impact of pesticides on environmental and human health. In: *Toxicology Studies—Cells, Drugs and Environment*, IntechOpen Limited, London, UK, 2010, pp 195–233.

Besse-Hoggan, P.; Alekseeva, T.; Sancelme, M.; Delort, A.; Forano, C. Atrazine biodegradation modulated by clays and clay/humic acid complexes. *Environmental Pollution*, **2009**, *157*(10), 2837–2844.

Bhat, D.; Padmaja, P. Assessment of organic pesticides in ground and surface water in Bhopal India. *Journal of Environmental Science, Toxicology and Food Technology*, **2014**, *8*, 51–52.

Blanca, A. L.; Angus, J. B.; Spanova, K.; Lopez-Real, J.; Russell, N. J. The influence of different temperature programmes on the bioremediation of polycyclic aromatic hydrocarbons (PAHs) in a coal-tar contaminated soil by in-vessel composting. *Journal of Hazardous Materials*, **2007**, *14*, 340–347.

Boricha, H.; Fulekar, M. H. *Pseudomonas plecoglossicida* as a novel organism for the bioremediation of cypermethrin. *Biology and Medicine*, **2007**, *1*(4), 1–10.

Bouwer, E. J.; Zehnder, A. J. B. Bioremediation of organic compounds putting microbial metabolism to work. *Trends in Biotech*, **1993**, *11*, 287–318.

Bradberry, S. M.; Proudfoot, A. T.; Vale, A. Glyphosate poisoning. *Toxicology Reviews*, 2004, *23*, 159–167.

Brooks, R. R. Geobotany and hyperaccumulators. In: Brook R.R. (ed.). *Plants Thathyperaccumulate Heavy Metals*, CAB International, Wallingford, United Kingdom, **1998**, pp. 855–894.

Brown, L. A.; Fraxedas, R. *Proceedings of the Symposium on Soil Venting*. (USEPA, Office of Research and Development, Publication # EPA/600/R-92/174, 1992*)*, **1998**.

Capel, P. D. Organic chemical concepts. In: Alley, W. M. (ed.). *Regional Ground-Water Quality*. Van Nostrand Reinhold, New York, USA. **1993**, pp. 155–180.

Casida, J. E.; Durkin, K. A. Neuroactive insecticides: Targets, selectivity, resistance, and secondary effects. *Annual Review of Entomology* **2013**, *58*, 99–117.

Castelo-Grande, T.; Barbosa, D. Chemical industry and environment. IV. In: Macias-Machin A.; Umbria J. (eds.), EMA, Universidad de Las Palmas de Gran Cannarias, Spain, **2013**.

Chaney, R. L.; Malik, M.; Li, Y. M.; Brown, S. L.; Brewer, E. P.; Angel, J. S.; Baker, A. J. Phytoremediation of soil metals. *Current Opinion in Biotechnology*, **1997**, *8*, 279–283.

Chauhan, R. S.; Singhal, L. Harmful effects of pesticides and their control through cowpathy. *International Journal of Cow Science*. **1997**, *2*(1): 61–70.

Chawla et al. Bioremediation: An emerging technology for remediation of pesticides. *Research Journal of Chemistry and Environment*, **2013**, *17* (4).

Colovic, M. B.; Krstic, D. Z.; Lazarevic-Pasti, T. D.; Bondzic, A. M.; Vasic, V. M. Acetylcholinesterase inhibitors: Pharmacology and toxicology. *Current Neuropharmacology*, *2013, 11, 315–335*.

Cookson, J.T. *Bioremediation Engineering: Design and Application.* McGraw-Hill, New York, USA, **1995**, p. 554.

Costa, L.G. Toxic effects of pesticides. In: Klaassen, C.D. (ed.). *Cassarett and Doull's Toxicology. The Basic Science of Poisons,* 7th Edition, McGraw-Hill, New York, USA, **2008**, pp. 883–930.

Daly, H.; Doyen, J. T.; Purcell, A. H. *Introduction to Insect Biology and Diversity,* 2nd Edition, Oxford University Press, New York, USA, **1998**.

Damalas, C. A. Understanding benefits and risks of pesticide use. *Scientific Research and Essays,* **2009**, *4*(10), 945–949.

Das, N.; Chandran, P. Microbial degradation of petroleum hydrocarbon contaminants: An overview. *Biotechnology Research International,* **2011**, *2011*, https://doi.org/10.4061/2011/941810.

Dearfield, K. L.; McCarroll, N. E.; Protzel, A.; Stack, H. F.; Jackson, M. A.; Waters, M. D. A survey of EPA/OPP and open literature on selected pesticide chemicals. II. Mutagenicity and carcinogenicity of selected chloroacetanilides and related compounds. *Mutatation Research*, **1999**, *443*, 183–221.

Debarati, P.; Gunjan, P.; Janmejay, P.; Rakesh, V. J. K. Accessing microbial diversity for bioremediation and environmental restoration. *Trends in Biotechnology*, **2005**, *23*, 135–142.

Desneux, N.; Decourtye, A.; Delpuech, J. M. The sublethal effects of pesticides on beneficial arthropods. *Annual Review of Entomology*, **2007**, *52*, 81–106.

Dua, M.; Singh, A.; Sethunathan, N.; Johri, A. Biotechnology and bioremediation: Successes and limitations. *Applied Microbiology and Biotechnology*, **2002**, *59*(2–3), 143–152.

Eldridge, B. F. *Pesticide Application and Safety Training for Applicators of Public Health Pesticides.* California Department of Public Health, Vector-Borne Disease Section, 1616 Capitol Avenue, MS7307, P.O. Box 997377, Sacramento, CA, Unites States, **2008**.

Ellenhorn, M. J.; Barceloux, D. G. Medical toxicology. *Diagnosis and Treatment of Human Poisoning.* Elsevier, New York, USA. **1998**, p. 1512.

Environmental Protection Agency. http://www.epa.gov/oppfead1/safety/healthcare, **1999**.

EPA. *What is a Pesticide?* http://www.epa.gov/opp00001/about/ (accessed 16 July **2012**).

Ermler, S.; Scholze, M.; Kortenkamp, A. Seven benzimidazole pesticides combined at sub-threshold levels induce micronuclei in vitro. *Mutagenesis*, **2013**, *28*, 417–426.

Fang, H.; Yu, Y. L.; Wang, X.; Shan, M.; Wu, X. M.; Yu, J. Q. Dissipation of chlorpyrifos in pakchoi-vegetated soil in a greenhouse. *Journal of Environmental Sciences (China)*, **2006**, *18*(4), 760–764.

FAO—Food and Agricultural Organization of United Nations. *Internal Code of Conduct on the Distribution and Use of Pesticides*, Rome, 36. [Cited on 2018 April 7] available from http://www.fao.org/docrep/005/y5544e/y4544e00htm **2002**.

Federal Remediation Technology Roundtable (FRTR). *Remediation Technologies Screening Matrix and Reference Guide*, **2006**, version 4.0.

Felsot, A. S.; Mitchell, J. K.; Bicki, T. J.; Frank, J. F. Experimental design for testing land farming of pesticide contaminated soil excavated from agrochemical facilities. *Pesticide Waste Management*, **1992**, *22*, 244–261.

Finley, S. D.; Broadbelt, L. J.; Hatzimanikatis, V. In silico feasibility of novel biodegradation pathways for 1,2,4-trichlorobenzene. *BMC Systems Biology*, **2010**, *4*, 7.

Frazer, C. The bioremediation and phytoremediation of pesticide contaminated sites. Prepared for U.S E.P.A Washington, DC. **2010**, 45–50.

Freire, C.; Koifman, S. Pesticides, depression and suicide: a systematic review of the epidemiological evidence. *International Journal of Hygiene and Environmental Health*, **2013**, *216*(4), 445–460.

Frenich, A. G.; Rodriguez, M. J. G.; Vidal, J. L. M.; Arrebola, F. J.; Torres, M. E. H. A study of the disappearance of pesticides during composting using a gas chromatography-tandem mass spectrometry technique. *Pest Management Science*, **2005**, *61*, 458–466.

Freed, V. H.; Chiou, C. T.; Schmedding, D. W. Degradation of selected organophosphorus pesticides in water and soil. *Journal of Agricultural and Food Chemistry*, **1979**, *27*, 706–708.

Frutos, F. J. V.; Escolano, O.; García, F.; Babín, M.; Fernández, M. D. Bioventing remediation and ecotoxicity evaluation of phenanthrene-contaminated soil. *Journal of Hazardous Materials*, **2010**, *183*(1–3), 806–813.

Fukumori, F.; Hausinger, R. P. Purification and characterization of 2,4-dichlorophenoxyacetate alpha-ketoglutarate dioxygenae. *Journal of Biological Chemistry*, **1993**, *268*, 24311–24317.

Fukuto, T. R. Mechanism of action of organophosphorus and carbamate insecticides. *Environmental Health Perspectives, 1990, 87, 245–254.*

Gamon, M.; Saez, E.; Gil, J.; Boluda, R. Direct and indirect exogenous contamination by pesticides of rice farming soils in a Mediterranean wetland. *Archives of Environmental Contamination and Toxicology*, **2003**, *44*, 141–151.

Gangola, S., Biodegradation of endosulfan and imidacloprid using indigenous fungal isolates of agriculture fields of Kumaun region of Uttarakhand. M.Sc. Thesis, G.B. Pant University of Agriculture and Technology, Pantnagar, **2014**.

Garbi, C.; Casasús, L.; Martinez-Álvarez, R.; Robla, J. I.; Martín, M. Biodegradation of oxadiazon by a soil isolated *Pseudomonas fluorescens* strain CG5: Implementation in an herbicide removal reactor and modelling. *Water Research*, **2006**, *40*, 1217–1223.

Gavrilescu, M. Fate of pesticides in the environment and its bioremediation. *Engineer in Life Science*, **2006**, *5*, 497–526.

Gebremariam et al. *Adsorption and Desorption of Chloropyrifos to Soils and Sediments*, https://link.springer.com/chapter/10.1007/978-1-4614-1463-6_3, **2012**.

Gentz, M. C.; Murdoch, G.; King, G. F. Tandem use of selective insecticides and natural enemies for effective, reduced-risk pest management. *Biological Control*, **2010**, *52*(3), 208–215.

George, T. M. *Sustaining the Earth: An Integrated Approach.* Thomson/Brooks/Cole, **2014**.

Ghorab, M. A.; Khalil, M. S. Toxicological effects of organophosphates pesticides. *International Journal of Environmental Monitoring and Analysis*, **2015**, *3*, 218–220.

Ghorab, M. A.; Khalil, M. S. The effect of pesticides pollution on our life and environment. *Journal of Pollution Effects & Control*, **2016**, *4*, 2.

Grant, R. J.; Daniell, T. J.; Betts, W.B. Isolation and identification of synthetic pyrethroid degrading bacteria. *Journal of Applied Microbiology*, **2002**, *92*, 534–540.

Guo, H.; Yao, J.; Cai, M.; Qian, Y.; Guo, Y.; Richnow, H. H.; Blake, R. E.; Doni, S.; Ceccanti, B. Effects of petroleum contamination on soil microbial numbers, metabolic activity and urease activity. *Chemosphere*, **2012**, *87*, 1273–1280.

Gupta, P. K. Herbicides and pesticides. *Reproductive and Developmental Toxicology*, 2nd Edition, Academic Press Elsevier, USA, pp. 657–679, **2017**.

Gupta, R. C. *Toxicology of Organophosphate & Carbamate Compounds*, Academic Press Elsevier, USA, **2006**.

Hamza, R. A.; Iorhemen, O. T.; Tay, J. H. Occurrence, impacts and removal of emerging substances of concern from wastewater. *Environmental Technology & Innovation*, **2016**, *5*, 161–175. https://doi.org/10.1016/j.eti.2016.02.003.

Hance, R.J. The effect of nutrients on the decomposition of the herbicides atrazine and linuron incubated with soil. *Pesticide Science* **1973**, *4*(6), 817–822.

Helfrich, L. A.; Weigmann, D. L.; Hipkins, P.; Stinson, E. R. *Pesticides and Aquatic Animals: A Guide to Reducing Impacts on Aquatic Systems*. Virginia Polytechnic Institute and State University, **2009**, Available from https://pubs.ext.vt.edu/420/420-013/420-013.htm. Accessed Jan 17, 2015.

Herrero, M.; Stuckey, D. C. Bioaugmentation and its application in wastewater treatment: A review. *Chemosphere*, **2015**, *140*, 119–128. https:// doi.org/10.1016/j.chemosphere.2014.10.033.

Husain, Q.; Husain, M.; Kulshrestha, Y. Remediation and treatment of organopollutants mediated by peroxidases: A review. *Critical Reviews in Biotechnology*, **2009**, *29*(2), 94–119.

Hussain, S.; Siddique, T.; Arshad, M.; Saleem, M. Bioremediation and phytoremediation of pesticides: Recent advances. *Critical Reviews in Environmental Science and Technology*, **2009**, *39*, 843–907.

Isoda, H.; Talorete, T. P.; Han, J.; Oka, S.; Abe, Y.; Inamori, Y. Effects of organophosphorous pesticides used in china on various mammalian cells. *Environmental Science*, **2005**, *12*, 9–19.

Jain, R. K.; Kapur, M.; Labana, S.; Lal, B.; Sarma, P. M. Microbial diversity: Application of microorganisms for the biodegradation of Xenobiotics. *Current Science*, **2005**, *89*, 101–112.

Jariyal, M.; Gupta, V.; Jindal, V.; Mandal, K., Isolation and evaluation of potent Pseudomonas species for bioremediation of phorate in amended soil. *Ecotoxicology and Environmental Safety*, **2015**, *122*, 24–30. https://doi.org/ 10.1016/j.ecoenv.2015.07.007.

Johnson, P. C.; Baehr, A. L.; Brown, R. A.; Hinchee, R.; Hoag G. E. Innovative site remediation technology. In: Anderson, W.C. (eds.), American Academy of Environmental Engineers Monograph Series, Annapolis, MD, USA, **1994**, *8*, 224.

Johnson, P. C.; Stanley, C. C.; Kemblowski, M. W.; Byers, D. L.; Colthart, J. D. Ground water. *Monitoring Review*, **1990**, *10*, 159.

Kadian, N.; Gupta, A.; Satya, S.; Mehta, R. K., Malik, A. Biodegradation of herbicide (atrazine) in contaminated soil using various bioprocessed materials. *Bioresource Technology*, **2008**, *99*(11), 4642–4647.

Kanekar, P. P.; Bhadbhade, B.; Deshpande, M. N.; Sarnaik, S. S. Biodegradation of organophosphorous pesticides. *Indian National Science Academy*, **2004**, *70*, 57–70.

Karpouzas, D.; Fotopoulou, A.; MenkissogluSpiroudi, U.; Singh, B. Non-specific biodegradation of the organophosphorus pesticides, cadusafos and ethoprophos, by two bacterial isolates. *FEMS Microbiology Ecology*, **2005**, *53*, 369–378.

Kato, T.; Haruki, M.; Imanaka, T. Isolation and characterization of long-chain-alkane degrading *Bacillus thermoleovorans* from deep subterranean petroleum reservoirs. *Journal of Bioscience and Bioengineering*, **2011**, *91*, 64–70.

Khan, F. I.; Husain, T.; Hejazi, R. An overview and analysis of site remediation technologies. *Journal of Environmental Management*, **2004**, *71*(2), 95–122. doi.org/10.1016/j.jenvman.2004.02.003.

Kookana, R. S.; Boxal, A. B.; Reeves, P. T.; Ashauer, R.; Chaudhry, Q.; Cornelis, G. T. F.; Gan, J.; Kah, M.; Lynch, I.; Ranville, J.; Sinclair, C.; Spurgeon, D.; Tiede, K.; Van Den Brink, P. J. Nanopesticides: Guiding principles for regulatory evaluation of environmental risks. *Journal of Agriculture and Food Chemistry*, **2014**, *62*, 4227–4240

Kopittke, P. M.; Blamey, F. P. C.; Asher, C. J.; Menzies, N. W. Trace metal phytotoxicity in solution culture: A review. *Journal of Experimental Botany*, **2010**, *61*, 945–954.

Kullman, S. W.; Matsumura, F. Metabolic pathways utilized by the *Phanerochaete chrysosporium* for degradation of cyclodiene pesticide endosulfan. *Applied and Environmental Microbiology*, **1996**, *62*:593–600.

Kumar, B.; Gaur, R.; Goel, G.; Mishra, M.; Singh, S. K.; Prakash, D.; Sharma, C. S. Residues of pesticides and herbicides in soils from agriculture areas of Delhi region, India. *Journal of Environmental Agricultural and Food Chemistry*, **2012**, *11*, 328–338.

Kumar, R. et al. Bioremediation of contaminated sites: allow cost nature's biotechnology for environmental cleanup by versatile microbes, plants & earthworms. *Solid Waste Management and Environmental Remediation*. Nova Science Publishers, Incoparate, Hauppauge, **2009**, pp. 1–69.

Langerhoff, A.; Charles, P.; Alphenaar, A.; Zwiep, G.; Rijnaarts, H. Intrinsic and stimulated in situ biodegradation of hexachlorocyclohexane (HCH). *6th International HCl and Pesticides Forum*. Poland, **2001**, pp. 132–185.

Lee, T. H.; Byun, I. G.; Kim, Y. O.; Hwang, I. S.; Park, T. J., Monitoring biodegradation of diesel fuel in bioventing processes using in situ respiration rate. *Water Science and Technology*, **2006**, *53*(4–5), 263–272.

Longley, M.; Sotherton, N. W. Factors determining the effects of pesticides upon butterflies inhabiting arable farmland. *Agriculture Ecosystems and Environment*, **1997**, *61*, 112.

Ma, Y.; Prasad, M. N. V.; Rajkumar, M.; Freitas, H. Plant growth promoting rhizobacteria and endophytes accelerate phytoremediation of metalliferous soils. *Biotechnology Advances*, **2011**, *29*, 248–258.

MacLennan, N. P. A.; Mandel, J.; Delzell, E. A review of epidemiologic studies of triazine herbicides and cancer. *Critical Review in Toxicology*. **2011**, *41*, 1–34.

Macneale, K. H.; Kiffney, P. M.; Scholz, N. L. Pesticides, aquatic food webs, and the conservation of Pacific salmon. *Frontiers in Ecology and the Environment*, **2010**, *8*, 475–482.

Mahmood et al. Effect of pesticides on environment. *International Journal on Plant-Soil and Microbes,* **2016**, 254–256.

Majewski, M.; Capel, P. Pesticides in the atmosphere: Distribution, trends, and governing factors. *Pesticides in the Hydrologic System*. Vol. 1. Ann Arbor Press Inc., Boca Raton, FL, USA. **1995**, p. *118*.

Marican, A.; Lara, E. F. D. A review on pesticide removal through different processes. *Journal of Environmental Science and Pollution Research,* **2018**.

Marino et al. Pesticide transport modelling in soils, ground water and surface water. In: Schmitz, G.H. (ed.), *Water Research and Environment Research. Proceedings of ICWRER*, Vol. II. Dresden, Germany, **2002**.

Mathur, S.C. Future of Indian pesticides industry in next millennium. *Pesticide Information*, **1999** *XXIV*(4), 9–23.

Mayhew, D. A.; Taylor, G. D.; Smith, S. H.; Banas, D. A. Twenty-four month combined chronic oral toxicity and oncogencity Sathiakumar study in rats utilizing atrazine technical grade. Lab Study No.: 410-1102, Accession No. 262714-262727, American Biogenics Corp., Decatur, **1986**, 2–6.

Mervat, S. M. Degradation of methomyl by the novel bacterial strain *Stenotrophomonas maltophilia* M1. *Electronic Journal of Biotechnology*, **2009**, *12*(4), 1–6.

Miller, G. T. *Living in the Environment*. 12th Edition. Praeger Publishers, London, **2002**.

Miller, R. Technology Overview Report. *Ground Water Remediation Technologies*, Analysis Center, Pittsburgh, PA, USA, **1996**.

Mohamed, A. T.; El Hussein, A. A.; El Siddig, M. A.; Osman, A. G. Degradation of oxyfluorfen herbicide by soil microorganisms. *Biodegradation of Herbicides, Biotechnology.* **2011**, *10*, 274–279

Muller, W. P.; Korte, F., Microbial degradation of benzo-(a)-pyrene, monolinuron and dieldrin in waste composting. *Chemosphere*, **1975**, *4*, 195–198.

Nawab, A.; Alimand, A.; Malik, A. Determination of organochlorine pesticides in Agricultural soil with special reference to alpha-HCH degradation by *Pseudomonas* Strains. *Bioresource Technology*, **2003**, *88*, 41–46.

Nie, M.; Zhang, X.; Wang, J.; Jiang, L.; Yang, J.; Quan, Z.; Cui, X.; Fang, C.; Li, B. Rhizosphere effects on soil bacterial abundance and diversity in the Yellow River Deltaic ecosystem as influenced by petroleum contamination and soil salinization. *Soil Biology and Biochemistry,* **2011**, *41*, 2535–2542.

Norris, R. *Handbook of Bioremediation,* Lewis Publishers, Boca Raton, FL, USA, **1994**.

Nwankwegu, A. S.; Onwosi, C. O. Bioremediation of gasoline contaminated agricultural soil by bioaugmentation. *Environmental Technology & Innovation,* **2017**, *7*, 1–11. https://doi.org/10.1016/j.eti.2016.11.003.

Odukkathil, G.; Vasudevan, N. Toxicity and bioremediation of pesticides in agricultural soil. *Reviews in Environmental Science and Bio/Technology*, **2013**, *12*(4), 421–444. https://doi.org/10.1007/s11157-013-9320-4.

Ortiz-Hernandez, M. L.; Sánchez, S. E.; González, E. D.; Godínez, M. L. C. Pesticide biodegradation: Mechanisms, genetics and strategies to enhance the process. *IntechOpen— Open Science Open Minds*, **2013**, 251–287.

Ortiz-Hernández, M. L.; Sánchez-Salinas, E.; Olvera-Velona, A.; Folch-Mallol, J. L. Pesticides in the environment: Impacts and its biodegradation as a strategy for residues treatment, pesticides—formulations, effects, fate. In: Margarita Stoytcheva (ed.), *InTech*, **2011**, DOI: 10.5772/13534.

Osman, K. A.; Ibrahim, G. H.; Askar, A. I.; Rahman, A.; Alkhail, A. Biodegradation kinetics of dicofol by selected microorganisms. *Pesticide Biochemistry and Physiology*, **2008**, *91*, 180–185.

Overview of Ecological Risk Assessment process in the office of pesticides programme, *U.S. Environmental Protection Agency, Endangered & Threatened species Effect Determination*. Office of Pesticide Programmes, Washington, DC, USA, January, **2004**.

Paula, J. M.; Vargas, L.; Agostinetto, D.; Nohatoo, M. A. Manejo de Conyzabonariensis resistenteao herbicida glyphosate. *Planta Daninha*, **2011**, *29*(1), 217–227. http://dx.doi. org/10.1590/S0100-83582011000100024.

Perclich, J. A.; Lockwood, J. L. Interaction of Atrazine with soil microorganisms: Population changes and accumulation. *Canadian Journal of Microbiology*, **1978**, *24*, 1145–1152.

Pesticides 101-A Primer. *Pesticide Action Network North America*. Available from http:// www.panna.org/issues/pesticides-101-primer. Accessed January 10, **2015**.

Pesticides in Groundwater. *The USGS Water Science School*. Available from http://water.usgs. gov/edu/pesticidesgw.html. Accessed December 17, **2014**.

Pilling, E. D.; Jepson, P. C. Synergism between EBI fungicides and a pyrethroid insecticide in the honeybee (*Apis mellifera*). *Pesticide Science*, **2006**, *39*, 293–297.

Popovska-Gorevski, M.; Dubocovich, M. L.; Rajnarayanan, R. V. Carbamate insecticides target human melatonin receptors. *Chemical Research in Toxicology*, **2017**, *30*, 574–582.

Porto, A. L. M.; Melgar, G. Z.; Kasemodel, M. C.; Nitschke, M. *Biodegradation of Pesticides, Pesticides in the Modern World—Pesticides Use and Management*. Dr. Margarita Stoytcheva (*ed.*), **2011**. ISBN: 978-953-307-459-7.

Qiu, X.; Zhong, Q.; Li, M.; Bai, W.; Li, B. Biodegradation of p-nitrohenol by methyl parathion-degrading *Ochrobactrum* sp. B2. *International Biodeterioration and Biodegradation*, **2007**, *59*, 297–301.

Quintero, J. C.; Lu-Chau, T.; Moreira, M. T.; Feijoo, G.; Lema, J. M. Bioremediation of HCH present in soil by the white-rot fungus *Bjerkandera adusta* in a slurry batch bioreactor. *International Biodeterioration & Biodegradation*, **2007**, *60*, 319–326.

Rabani, M. S.; Sharma, R.; Gupta, M. K. Bioremediation of naphthalene and other PAH contaminants: An approach for cleaning of environment. *Journal of Environmental Research and Development*, **2018**, *12*(3), 292–297.

Raghavan, R.; Cles, E.; Dietz, D. *Journal of Hazardous Materials*, **1991**, *26*, 81.

Randall, C. et al. Pest Management. *National Pesticide Applicator Certification Core Manual, (2nd Edition)*, National Association of State Departments of Agriculture *Research Foundation, Washington, USA, 2014.*

Rehman, A.; Singh, R; Rabani, M. S.; Sharma, R.; Gupta, M. K. Characterization of chromium (Cr(VI) reducing bacteria from soil and waste water. *International Journal of Advance and Innovative Research*, **2019**, *6*(1), 92–98.

Reigart, J. R.; Roberts, J. R. Recognition and Management of Pesticide Poisonings, 5th Edition, United States Environmental Protection Agency, **1999**, http://www.epa.gov/ oppfead1/safety/healthcare/handbook/handbook.pdf (accessed 4 September 2014).

Rockne, K.; Reddy, K. *Bioremediation of Contaminated Sites*. University of Illinois at Chicago. **2003**.

Routt, R. J.; Roberts, J. R. *Recognition and Management of Pesticide Poisonings*. EPA, Washington DC, USA, **1999**, 223.

Runes, H. B.; Jenkins, J. J.; Bottomley, P. J. Atrazine degradation by bioaugmented sediment from constructed wetlands. *Applied Microbiology and Biotechnology*, **2001**, *57*, 427.

Sahle-Deessie, E.; Meckes, M. C.; Richardson, T .L. *Environmental Progress*, **1998**, *15*, 293.

Samanta, S. K.; Singh, O. V.; Jain, R. K. Polycyclic aromatic hydrocarbons environmental pollution and bioremediation. *Trends in Biotechnology*, **2002**, *20*, 243–248.

Sarchiapone, M.; Mandelli, L.; Iosue, M.; Andrisano, C.; Roy, A. Controlling access to suicide means. *International Journal of Environmental Research and Public Health*, **2011**, *8*, 4550–4562.

Sathiakumar, N.; MacLennan, P. A.; Mandel, J.; Delzell, E. A review of epidemiologic studies of triazine herbicides and cancer. *Critical Review in Toxicology*, **2011**, *41*, 1–34.

Schnoor, J. L.; Licht, L. A.; McCutcheon, S. C.; Wolfe, N. L.; Carreira, L. H. Phytoremediation of organic and nutrient contaminants. *Environmental Science & Technology*, **1995**, *29*, 318A–323A.

Scow, K. M.; Hicks, K. A. Natural attenuation and enhanced bioremediation of organic contaminants in groundwater. *Current Opinion in Biotechnology*, **2005**, *16*(3), 246–253.

Sessitsch, A.; Reiter, B.; Pfeifer, U.; Wilhelm, E. Cultivation independent population analysis of bacterial endophytes in three potato varieties based on eubacterial and Actinomycetes-specific PCR of 16S rRNA genes. *FEMS Microbiology Ecology*, **2002**, *39*, 3–32.

Shanahan, P. Bioremediation, *Waste Containment and Remediation Technology*. Spring, Massachusetts Institute of Technology, MIT Open Course Ware, **2004**.

Sharma, D. R.; Thapa, R. B.; Manandhar, H. K.; Shrestha, S. M. Use of pesticides in Nepal and impacts on human health and environment. *Journal of Agriculture Environment*, **2012**, *13*, 67–72.

Sharma et al. Microbial degradation of pesticides for environmental cleanup. *Bioremediation of Industrial Pollutants*. **2016**, 178–205.

Siddiqui, Z. S.; Ahmed, S. Combined effects of pesticide on growth and nutritive composition of soybean plants. *Pakistan Journal of Botany*, **2006**, *38*, 721–733.

Silva, E.; Fialho, A. M.; Sa-Correia, I.; Burns, R. G.; Shaw, L. J. Combined bioaugmentation and biostimulation to cleanup soil contaminated with high concentrations of atrazine. *Environmental Science & Technology*, **2004**, *38*: 632.

Singh, B. K.; Walker, A. Microbial degradation of organophosphorus compounds. *FEMS Microbiology Reviews*, **2006**, *30*(3), 428–471.

Singh, B. K.; Walker, A.; Morgan, J. A.; Wright, D. J. Biodegradation of chlorpyrifos by Enterobacter strain B-14 and its use in bioremediation of contaminated soils. *Applied and Environmental Microbiology*, **2004**, *70*, 4855–4863.

Singh, R.; Gupta, M. K. Exploring the role of selected bacterial strains in chromium (Cr (VI) reduction from soil. *International Journal of Scientific Research in Biological sciences*, **2019**, *6*(1), 226–232.

Smith, D.; Crowley, D. E. Contribution of ethylamine degrading bacteria to atrazine degradation in soils. *FEMS Microbiology Ecology*, **2006**, *58*, 271–277.

Smith, M. J.; Flowers, T. H.; Duncan, H. J.; Alder, J. Effects of polycyclic aromatic hydrocarbons on germination and subsequent growth of grasses and legumes in freshly contaminated soil and soil with aged PAHs residues. *Environment Pollution*, **2006**, *141*, 519–525.

Sparks, D. L.; Corn, M. *Soil Decontamination—Handbook of Hazardous Materials*. Academic Press, San Diego, CA, USA, **1993**.

Speck-Planche, A.; Kleandrova, V. V.; Scotti, M. T. Fragment-based approach for the in silico discovery of multi-target insecticides. *Chemometrics and Intelligent Laboratory Systems*, **2012**, *111*, 39–45.

Straathoff, H. Investigations on the phytotoxic relevance of volatilization of herbicides. *Mededelingen* **1986**, *51*(2A), 433–438.

Strong, P. J.; Burgess, J. E. Treatment methods for wine related ad distillery wastewaters: A review. *Bioremediation Journal*, **2008**, *12*, 70–87.

Subhash, S. P. et al. *Pesticide Use in Indian Agriculture: Trends, Market Structure and Policy Issues*. National Centre for Agricultural Economics and Policy Research, **2017**.

Teresa et al. Remediation of soils contaminated with pesticides: A review. *International Journal of Environmental Analytical Chemistry,* **2010**, *90*, 438–467.

Testiati, E.; Parinet, J.; Massiani, C.; Laffont-Schwob, I.; Rabier, J.; Pfeifer, H. R.; Prudent, P. Trace metal and metalloid contamination levels in soils and in two native plant species of a former industrial site: evaluation of the phytostabilization potential. *Journal of Hazardous Material,* **2013**, *248*, 131–141.

Thatheyus, A. J.; Selvam, A. D. G. Synthetic pyrethroids: Toxicity and biodegradation. *Applied Ecology and Environmental Sciences,* **2013**, *1*(3), 33–36.

Thierry, L.; Armelle, B.; Karine, J. Performance of bioaugmentation-assisted phytoextraction applied to metal contaminated soils: A review. *Environmental Pollution,* **2008**, *153*, 497–522.

United States Environmental Protection Agency. *A Citizen's Guide to Bioremediation,* **2006**.

Utmazian, M. N.; Wenzel, W. W. Phytoextraction of metal polluted soils in Latin America. *Environmental Applications of Poplar and Willow Working Party* **2006**.

Van Deuren, J.; Lloyd, T.; Chhetry, S.; Raycharn, L.; Peck, J. Remediation technologies screening matrix and reference guide. *Federal Remediation Technologies Roundtable,* **2002**, *4*.

Van Djik, T. C. *Effects of Neonicotinoid Pesticide Pollution of Dutch Surface Water on Non-target Species Abundance,* **2010**.

Vane, C.H.; Kim, A. W.; Beriro, D. J.; Cave, M. R.; Knights, K.; Moss-Hayes, V.; Nathanail, P. C. Polycyclic aromatic hydrocarbons (PAH) and polychlorinated biphenyls (PCB) in urban soils of Greater London, UK. *Applied Geochemistry,* **2014**, *51*, 303–314.

Vidali, M. Bioremediation. An overview, *Pure and Applied Chemistry,* **2001**, *73*(7), 1163–1172.

Walker, A.; Perekh, N. R.; Roberts, S. J.; Welch, S. J. Evidence of the enhanced biodegradation of napropamide in soil. *Pesticide Science,* **1993**, *39*(1), 55–60.

WHO—World Health Organization. *Suicide Prevention (SUPRE).* http://www.who.int/ mental_health/prevention/suicide/suicideprevent/en/ *(accessed 16 August* **2014***).*

Wilson, D. J.; Clarke, A. N. *Soil Vapour Stripping in Hazardous Waste Site Soil Remediation,* Marcel Dekker, New York, USA, **1994**.

Worthing, C.R. *The Pesticide Manual: A World Compendium,* 6th Edition, The British Crop Protection Council, Croydon, England, **1979**, p. 655.

Wu, T.; Crapper, M. Simulation of biopile processes using a hydraulics approach. *Journal of Hazardous Materials,* **2009**, *171*(1–3), 1103–1111.

Zavoda, J.; Cutright, T.; Szpak, J.; Fallon, E., Uptake, selectivity and inhibition of hydroponic treatment of contaminants. *Journal of Environmental Engineering,* **2001**, *127*, 502–508.

Zeljezic, D.; Garaj-Vrhovac, V.; Perkovic, P. Evaluation of DNA damage induced by atrazine and atrazine-based herbicide in human lymphocytes in vitro using a comet and DNA diffusion assay. *Toxicology In Vitro,* **2006**, *20*, 923–935.

Zhang, W.; Jiang, F.; Feng, Ouj. Global pesticide consumption and pollution: with China as a focus. *Proceedings of the International Academy of Ecology and Environmental Sciences,* **2011**, *1*(2), 125–144.

Zhu, Y. G.; Shaw, G. Soil contamination with radionuclides and potential remediation. *Chemosphere,* **2000**, *41*, 121–128.

CHAPTER 6

Biopesticides: Importance and Challenges

MUHAMMAD FAHAD SARDAR,[1*] TANVEER ABBAS,[2]
MUHAMMAD NAVEED,[3] SULMAN SIDDIQUE,[3] ADNAN MUSTAFA,[2]
BILAWAL ABBASI,[2] and KHURAM SHEHZAD KHAN[4]

[1]*Agricultural Clean Watershed Research Group,*
Institute of Environment and Sustainable Development in Agriculture,
Chinese Academy of Agricultural Sciences, Beijing 100081, PR China

[2]*Institute of Agricultural Resources and Regional Planning,*
Chinese Academy of Agricultural Sciences, Beijing 100081, China

[3]*Institute of Soil and Environmental Sciences, University of Agriculture,*
Faisalabad, Pakistan

[4]*Department of Ecological Science and Engineering, College of Resource*
and Environmental Sciences, China Agriculture University, China

*Corresponding author. E-mail: fahadsardar16@yahoo.com

ABSTRACT

Over the past 10 decades, the world population has grown rapidly and to fulfill the human needs ultimately pressures come on agriculture with intensive production systems. Due to intensive development in world, agriculture sector faces several challenges in maintaining the trend of increasing production and countering the abiotic and biotic stresses. These stresses involve attack of different pests and diseases. Pests and plant diseases are among the major human competitors of agricultural products, especially those grown under high productivity regions. The damage by these pests and associated risks are major factors in reducing the crop production, either before harvest or after harvest losses. Increasing public concern about adverse effect of chemical pesticides on agriculture products has prompted industry to develop new products that must be less harmful or having higher biodegradability and effective at lower rate than the existing chemical pesticides and they will also nontoxic for workers and consumers. Biopesticides contain an array of several living and nonliving entities that vary considerably from each other based on the characteristic composition, mechanisms of action, and fate and behavior in environment after application.

6.1 INTRODUCTION

Over the past 10 decades, the world population has grown rapidly and to fulfill the human needs ultimately pressures come on agriculture with intensive production systems (Kohler and Triebskorn, 2013). Due to intensive development in world, agriculture sector faces several challenges in maintaining the trend of increasing production and countering the abiotic and biotic stresses. These stresses involve attack of different pests and diseases. Pests and plant diseases are among the major human competitors of agricultural products, especially those grown under high productivity regions (Oerke and Dehne, 2004; Oliveira et al., 2014). The damage by these pests and associated risks are major factors in reducing the crop production, either before harvest or after harvest losses. On average, 35% of potential crops in the world are destroyed by pests before harvest, and postharvest losses can reach 35% or more (Oerke, 2006; Popp et al., 2012).

As the world hurtles to achieve maximum global agricultural production, practices have been categorized to get the desired goal. This production method is often referred to as "alternative agriculture" that relies on the use of fertilizer to enhance or sustain soil fertility and agrochemicals to control pests, diseases, and weeds. The overuse of chemicals affects natural biodiversity or ecosystem functioning, which gradually become unsuccessful due to pest resistance (George et al., 2014; Rosner and Markowitz, 2013). In addition, the continued use of chemical pesticides in agriculture has resulted in the accumulation of pesticide residues in soil and groundwater. In fact, ~99.7% of pesticides remain and accumulate in the environment (Tejada et al., 2017). These pesticides also affect natural enemies' population, either by dying or drifting into the alternative ecosystem (Thacker, 2002). It is predicted that only 0.1% of the pesticides are expected to have reached the target pests, while 99.9% have been released into the environment (Goswami et al., 2017; Goulson et al., 2015). Overall, synthetic pesticides significantly decrease the potential losses for a short period, but they also pose additional threats to human health, and their efficiency drops as resistance developed in pests and weeds. According to market research, humans are also at risk of unintentional pesticide intoxication, which is strongly linked to their misuse and excessive environmental impact (Jones et al., 2014). Thus, too much dependence on agrochemicals as the only form of crop protection is therefore not a sustainable long-term strategy (Shaner, 2009). In addition, the development of poorly regulated products, particularly in developing countries, requires low-cost alternatives that do not meet international quality standards and pose risks to ecosystems and health human (Popp et al., 2012).

However, pesticides can significantly affect the environment because of their high usage. The overuse of persistent traditional insecticides such as organophosphates and organochlorines are harmful to the environment and microbial diversity. This has directed to a growing interest in the use of an alternative to chemicals in agriculture (Devine and Furlong, 2007; Kumar et al., 2017). When a pesticide is applied through the spray to soil, some of its parts are directly adsorbed in the soil as and present for the longer time period as residues. Some of the applied pesticides go into the atmosphere via evapotranspiration that causes atmospheric pollution, and some parts are leached down with irrigation water. Most dangerous component of pesticide is directly absorbed by plants that enter into human and animal food chain (Figure 6.1)

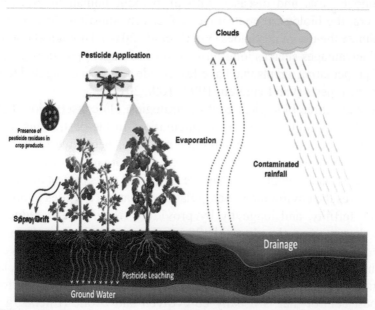

FIGURE 6.1   The fate of applied pesticide and its interaction with the atmosphere and soil environment.

6.2   BIOPESTICIDES

Increasing public concern about adverse effect of chemical pesticides on agriculture products has prompted industry to develop new products that must be less harmful or having higher biodegradability and effective at lower rate than the existing chemical pesticides (Saxena and Pandey, 2001) and they will also nontoxic for workers and consumers (Weidemann et al., 1995). Unlike most conventional chemical pesticides, Steinke and Giles (1995) have defined biopesticides as products originated from living entity such as bacteria, viruses, fungi, etc. On the basis of active ingredients, biopesticides are divided into two categories, that is, biochemical and microbial pesticides. Biochemical pesticides are natural substances that counter the pests attack by adopting nontoxic mechanisms. Microbial pesticides consist of beneficial microorganisms (such as bacteria, fungi, viruses, or protozoa) as an active ingredient or metabolite they produce (Bonaterra et al., 2011).

Biological control has been defined many times, but Eilenberg et al. (2001) proposed a generally accepted definition: "The use of living organisms to suppress population density or the effects of a particular pest that makes it less abundant or less harmful than it would otherwise be." Longe (2016) recently defined biological control agents as natural organisms that can remove pests and diseases with or without human involvement. In recent era, the biological control can effectively substitute or help farmers to minimize the use of pesticides (Puech et al., 2014). Biological control has several advantages such as long-term control of target pests through the use of self-perpetuating agents that have lesser side effects and targeted to one or other related pests (van Lenteren, 1992; Toth, 2012).

There are three biological control strategies: (1) conservation, (2) classical biological control, and (3) augmentation. In the past, focus has been on classical biological control, although much has been invested recently to strengthen augmentative control (Toth, 2012). In organic farming, conservation and classical biological control are important parts of pest management, that involves the environment manipulation of natural enemies to boost their survival, fertility, and longevity by providing resources and reducing the factors that interfere (Landis et al., 2000). These techniques include the use of cover crops to establish an ecological infrastructure that improves their food, habitat, and reproduction sites (Franco et al., 2006). While, classical biological control includes the intentional use of biological agents for long term pest control that is usually co-evolved and exotic (Eilenberg et al., 2001), is often used against introduced pests which arrive in a new area and becomes permanently established without an associated natural enemy complex (Amaro, 2003).

Whereas in augmentation, natural enemy's population is increased by the mass production in a laboratory and liberating it into the field at a time when the pest is most vulnerable. The augmentation does not offer a permanent solution and also needs continuous human involvement, unlike conservation and classical approach. It includes two basic approaches: inoculation and inundation techniques (Luck and Forster, 2003; Longe, 2016). This technique is in practice worldwide, and more than 230 species are used as biological control agents and are commercially available (van Lenteren and Bueno, 2003; van Lenteren, 2012).

New parasites are continuously emerging accidentally or intentionally, leaving behind their natural competitors. When they become pests, the introduction of some of their natural competitors can be an important way to minimize the risk of damage they can cause. For best results, a biological control agent must have a colonizing ability to keep pace with the spatio-temporal disturbances of natural habitat caused by the pest (Driesche et al., 2009; Longe, 2016).

6.2.1   TYPES OF BIOPESTICIDES

Against pests and diseases, different control methods can be used, such as regulation (regulatory measures to limit the spread of insects and pathogens), cultural (practices that modify the behavior and environmental conditions of the pests, genetic (selection of plants cultivars that are resistant to certain diseases and pests), physical (use of equipment, machinery or other physical means of pest control), biological (use of natural competitors of pests and organisms pathogens and promotion of their growth), and biotechnology (use of physiological mechanisms or environmental behavior of insects that minimize their survival rate and reproduction efficiency). On the basis of origin, biopesticides are classified as (1) microbial, (2) plant-based, and (3) insect-based

6.2.1.1   MICROBIAL BIOPESTICIDES

Microbial pesticides are composed of living organisms such as a bacterium, fungi, virus, or protozoan and active ingredient produced by microorganisms. Chemical pesticides are relatively specific for the target insects, but microbial pesticides can control different species of pests (Bravo et al., 2007). This feature alleviates concerns about disease or pest resistance and its harmful effects on human health (Ragsdale and Sisler, 1994; Montesinos and Bonaterra, 2009). On the basis of microorganisms, microbial biopesticides are classified into the following types.

6.2.1.2   FUNGI BASED BIOPESTICIDES

Fungi are natural entomopathogens that can spread faster and have significant epizootic potential. Entomopathogenic fungi are the most abundant organisms used as a biological control against pests due to their ability to cause epizootic outbreaks, ease of application with a small amount inoculum (Faria et al., 2009; Hajek and Delalibera, 2010). Bio-fungicides have the best biological control characteristics such as environmentally hardy, highly reproductive, and antagonistic. Fungus thrives in the host's and attacks the host's organs possibly by producing toxins. Fungus infected host usually takes several days to die (Dent, 2000).

The fungal-based biopesticides can be used to control insects, weeds, and plant diseases caused by other pathogenic fungi or bacteria. Similar bacteria may act by out-competing the targeted pathogen or producing toxins. About 1500 species of bacteria, fungi, viruses, and protozoa have been discovered that infect arthropods species. In 1884 Elie Metchnikoff, a Russian Scientist first used fungi (*Metarhiziumanisopliaeto*) as bioinsecticide on two invertebrate pest's cereal cockchafer and sugar-beet weevil. In the 1930s, *Bacillus thuringiensis* (Bt) was the first commercially available bioinsecticide and meanwhile, many bioinsecticides have been developed, which are mostly hyphomycete fungi that are entomopathogenic. Many of the present biopesticides comprise of species of the similar genus but they work in different ways to achieve greater levels of control. *Trichoderma harzianumis* a fungus that is used as a fungicide against *Fusarium*, *Pythium*, and *Rhizoctonia*. Currently, there are only a few microbial pesticides available commercially that are comprised of fungi, but the genera *Metarhizium*, *Beauveria*, *Verticillium*, *Hirsutellaare*, and *Nomuraea*, are of greatest interest for biological control. Some species are very selective to insects and some act on a wide range of insects, like *Beauveria bassiana* that is used against thrips, weevils, aphids, and whiteflies (Longe, 2016). *Metarhizium anisopliae* focused most due to root dynamics interaction and plant defense, know reclassified as *Metarhizium brunneum* (Bischoff et al., 2009).

Entomopathogenic fungi are also helpful for sucking insects like whitefly and aphids because fungi penetrate the insect body through the cuticle and do not need to be consumed by insects. Naturally occurring fungi also act as entomopathogenic to some insect species. However, several fungi species that are the pathogen for arthropods are also considered as plant growth promotor, rhizosphere colonizers, and plant endophytes (Lacey et al., 2007).

Bio-fungicides suppress pathogens attack through resource competition, production of antibiotics, and hyperparasitism. There are multiple products that are available in the market for crop/pathogen systems. Many agents are highly competitive for space and nutrients but usually have a lower capability of displacing established enemies. These characteristics make bio-fungicides preventative against pests (Table 6.1).

TABLE 6.1 Fungal Pesticides, Target Pest (s) and Mode of Action

Fungi Species	Category	Target Pest (s)	Mode of Action	References
Beauveria bassiana	Insecticide	Borers, thrips, foliar feeding insects, wireworm	Digestion of insect cuticle, white muscadine disease	Ladurner et al. (2008)
Metarhizium anisopliae	Insecticide	Cabbage root fly, wireworm	Parasitism, sporulation	Bruck et al. (2005)
Trichoderma harzianum	Fungicide Nematicide	Fungal diseases root rot, botrytis rot, powdery mildew, nematodes	Mycoparasitic, pathogens cell wall digestion	Costa et al. (2016)
Muscodoralbus	Biofumigant	Bacterial soil and seed-borne diseases	Release volatile toxins	Costa et al. (2016)
Verticillium lecani	Insecticide	Aphids, jassids, whiteflies, and mites	Parasitism, sporulation	Costa et al. (2016)
Paecilomyces lilacinus	Nematicide	All types of nematodes	Parasitism, sporulation	Costa et al. (2016)

6.2.1.3 BACTERIAL-BASED BIOPESTICIDES

Biopesticides are used to counter insects, nematodes, weeds, and plant diseases based on bacteria or bacterial products termed as bacterial-based biopesticides. *B. thuringiensis* (Bt) is the most important Gram-positive bacteria used for insect control, naturally occurring in soils throughout the world. From an agricultural point of view, bacterial biopesticides increase crop production by suppressing the phytopathogen sand plant pest's growth (Longe, 2016). According to Aravind et al. (2009) bacteria significantly increase the growth of the potato, black pepper, and wheat, also suppress fungal pathogen growth. The bacteria secrete antibiotics, siderophore, toxin, antimicrobial volatiles compounds, and wall hydrolytic enzymes that act antagonistically with phytopathogens (Sheoran et al., 2015). Endophytic bacteria also target both fungal and bacterial associated pathogens (Lodewyckx et al., 2002). The most commonly reported bacteria against phytopathogens and antimicrobial activity are belonging to genera *Bacillus, Actinobacteria, Enterobacteor, Pseudomonas, Paenibacillus,* and *Serratia* (Liu et al., 2010; Lodewyckx et al., 2002) (Table 6.2).

TABLE 6.2   Bacterial Biopesticides, Target Pest (s) and Mode of Action

Bacteria Species	Category	Target Pest (s)	Mode of Action	References
Bacillus thuringiensis	Insecticide	Black fly, fungus gnats, diamondback moth, cabbage looper, and beetles	Digestive system damage	Khater (2012)
Bacillus subtilis	Bactericide/ fungicide/ insecticide	Bacterial and fungal pathogens	Host root colonization, toxin production	Deng et al. (2011), Dong et al. (1994), Costa et al. (2016)
Bacillus sphaeicus	Insecticide	Mosquito larvae	Production of proteinaceous toxins	Costa et al. (2016)
Pantoea vagans	Fungicide	Used against fire blight disease	Colonization	Smits et al. (2011)
Pseudomonas fluorescens	Fungicide/ bactericide/ viricide	Many fungal, bacterial, and viral diseases	Control the plant growth	Costa et al. (2016)
Serratia marcescens	Insecticide	Sugarcane borer (*Eldana saccharina*)	Toxin production	Downing et al. (2000)
Pasteuria spp.	Nematicide	Nematodes and microscopic worms	Sporulation within nematode body	Costa et al. (2016)
Bacillus megaterium	Nematicide	*Radopholus similis*	Sporulation	Aravind et al. (2009)

TABLE 6.2   *(Continued)*

Bacteria Species	Category	Target Pest (s)	Mode of Action	References
Streptomyces lydicus	Fungicide	*Rhizoctonia, Fusarium,* Pythium, Aphanomyces, Verticillum, Botrytus, Ervinia and many other fungal pathogens	Root and above-ground plant parts colonization, competition for nutrition, and parasitism	Costa et al. (2016)
Streptomyces avermitilis	Insecticide	Several kinds of leaf miners	Toxin production	Grewal et al. (2008)
Curtobacterium luteum	Nematicide	*Radopholus similis*	Sporulation in nematode pathogen body	Aravind et al. (2009)

The mechanisms used by plant beneficial bacteria against bacterial, fungal, and viral pathogens to protect their host are called induced systemic resistance (ISR) (Alvin et al., 2014). Bacteria induce ISR using networks of signaling pathways that are interconnected and mediated by jasmonic acid (JA), salicylic acid (SA), and ethylene (ET) pathways (Pieterse et al., 2012). The prime function of ISR is to protect unexposed plants parts from further pathogenic attack. This defense system also is seen in endophytic bacteria of genus *Pseudomonas, Bacillus,* and *Serratia* (Kloepper and Ryu, 2006; Pieterse et al., 2012). ISR may involve both the SA and JA/ET pathways. Niu et al. (2011) found that *Bacillus cereus* AR156 led an additive effect in *Arabidopsis* by activating both the SA and JA/ET pathways to induce ISR. Likewise, Conn et al. (2008) showed that inoculation of endophytic *Actinobacteria* up-regulates defense pathways against bacterial *Erwinia carotovora* and fungal *Fusarium oxysporum* pathogen in *A. thaliana* plant. However, the resistance to *E. carotovora* was triggered by the JA/ET Pathway and *F. oxysporum* by SA Pathways. Hence, the same bacterium is capable to prime two different pathways to confer resistance mechanism to the different pathogen (Conn et al., 2008).

Bacterial biological control is difficult to control for sucking insects like aphids and scale insects because bacteria infect insect pests through their digestive tract. Only a few numbers of bacterial products have been commercially available that are entomopathogenic. About 2% of the market successfully used Bt-based products against many insects (Bravo et al., 2011).

This bacterium produces toxins that bind the specific protein in the gut of target insect species causing the insect larvae to starve and ultimately death. Bacterial-based biopesticide suppresses fungal pathogens by releasing fungal wall degrading enzymes including glucanase, protease, and chitinases (Zhang et al., 2012). Bacterial biological control agents including species of the genera *Photorhabdus* and *Xenorhabdus* can also control nematodes-based disease (Chaston et al., 2011). Nematodes are killed by bacteria present in the gut symbiotically when they attack plants, bacteria released into the hemocoel of the host when they attack the plants and ultimately kill the host (Ruiu et al., 2013). *P. fluorescent* spp. is a valuable biocontrol agent present naturally in the rhizosphere that protects plants from biotic stress (Kloepper et al., 1980).

Plant growth promoting rhizobacteria (PGPR) with biocontrol activities can be considered as an efficient alternative to the higher doses of pesticides applied on plants for remediation of the pathogens. Multiple approaches are used by PGPR which are directly or indirectly involved in control of diseases such as antagonistic pathway via production of antibiotics, siderophores, HCN, hydrolytic enzymes (chitinases, proteases, lipases, etc.), or indirect mechanisms in which the biocontrol agents act as a competitor against pathogen for a specific habitat (Lugtenberg and Kamilova, 2009). Among all the PGPRs strains, *Bacillus* and *Pseudomonas* are the two most important genera, studied extensively for antibiosis mechanisms in the disease management practices (Jayaprakashvel and Mathivanan, 2011; Dominguez-Nunez et al., 2016). Many bacterial species are available are for biological control of diseases (Table 6.3).

TABLE 6.3   PGPR as Biopesticides/Biocontrol Agents Against Various Plant Diseases

Bacterial species	Crop	Disease	References
Azospirillum spp.	Rice	Rice blast	Naureen et al. (2009)
Bacillus amyloliquefaciens	Tomato, cucumber	Tomato mottle virus, cucumber mosaic virus	Murphy et al. (2000), Zehnder et al. (2000)
Bacillus pumilus	Cucumber, tobacco	Cucumber mosaic virus, bacterial wilt, blue mold	Zehnder et al. (2000)
Bacillus cereus	Tomato	Foliar diseases	Silva et al. (2004)
Bacillus subtilis	Peral millet	Downy mildew, soil-borne diseases	Chung et al. (2008)
Burkholderia spp.	Maize	Maize rot	Hernandez-Rodringuez et al. (2008)
Enterobacter sp.	Chickpea	*Fusarium avenaceum*-based disease	Hynes et al. (2008)

TABLE 6.3 *(Continued)*

Bacterial species	Crop	Disease	References
Paenibacillus polymyxa	Sesame	Fungal disease	Ryu et al. (2006)
Pseudomonas fluorescens	Banana, white clover, and medicago	Banana bunchy top virus	Kavino et al. (2010), Kempster et al. (2002)
Streptomyces marcescens	Tobacco	Blue mold	Zhang et al. (2002)

Most of the PGPRs can synthesize antifungal metabolites such as antibiotics, fungal cell wall-lysing enzymes, or hydrogen cyanide, which effectively suppress the growth of fungal pathogens. PGPR and endophytic bacteria could play a key role in the management of various fungal diseases but one of the major constraints faced with biocontrol agents is lack of suitable delivery system. Thus, PGPRs as biocontrol agents can be potentially used by doing genetic modifications which could further sustain and also enhance agricultural output. Therefore, bacterial strains can be used practically for the plant growth promotion or as a biocontrol inoculant, together with other several growth-promoting microbes (Laslo et al., 2012).

6.2.1.4 NEMATODE-BASED BIOPESTICIDES

Nematodes are colorless, nonsegmented, and elongated roundworms. Nematodes that parasitize insects, known as entomopathogenic nematodes, have been described from 23 nematode families (Koppenhofer and Fuzy, 2008). Among all the nematodes studied for biological control of insects, the families *Steinernematidae* and *Heterorhabditidae* have received remarkable attention because they have many of the characteristics of active biological control agents (Grewal et al., 2008; Lacey et al., 2015). They have the capability to kill their hosts in a relatively shorter time and have been used as classical, conservational, and augmentative biological control agents. Many nematodes have a parasitic relationship with plants and can cause serious damage to crops and other types of plants. However, some are beneficial and attack wide range of insect pests (corn rootworm, caterpillars, crown borers, grubs, cutworms, thrips, fungus gnats, and various beetles (Kaya and Lacey, 2007; Grewal et al., 2008). Mostly nematodes are effective against insect pests and affect insect pests and their symbiotic microorganisms at the same time. Firstly, nematodes enter into insect body then break the cell wall and release symbiotic bacteria from the insect body. Preferably entomopathogenic nematodes attack on the intestine of insect pests (Costa et al., 2016).

6.2.1.5   INSECTS-BASED BIOPESTICIDES

Herbivorous insects have a remarkable ability to counter the plant defense system, and their diversity makes them the most challenging pests in agriculture (Kupferschmied et al., 2013). Insects living belowground are the most important in agricultural and natural ecosystems (van Dam, 2009) and their deteriorating effects are often underestimated, probably because they are not much explored and are difficult to study.

Insects-based biopesticides mostly kill the insects by predation or feeding upon other insects (Cruz-Rodriguez et al., 2016). As compared to chemical insecticides, these pathogens take longer time span (several days) to kill or debilitate the target pest (Table 6.4). Although they kill, minimize the reproduction and growth rate, or shorten the life of pests, their effectiveness largely depends on environmental conditions or host abundance. The efficiency and pest control ratio by naturally occurring pathogens is still unpredictable.

TABLE 6.4   Insects Based Biopesticides, Target Pest (s) and Mode of Action

Insect Species	Category	Target Pest (s)	Mode of Action	References
Typhlodromus pyri	Insecticide	On phytophagous mites and thrips	Predation	van Maanen et al. (2010)
Orius spp	Insecticide	Many other insects and mites	Predation	Cruz-Rodriguez et al. (2016)
Coccinellidae	Insecticide	Aphids, whiteflies, mites, mealybugs, and scale insects	Predation	Obrycki and Kring, (1998), Obrycki et al. (2009), Michaud (2012)
Lacewings	Insecticide	Aphids, other small insects, and mites	Predation	Pappas et al. (2011), Loru et al. (2014)

6.3   FUTURE CHALLENGES OF BIOPESTICIDES

At present, there are about 1400 biopesticides products being sold globally (Marrone, 2007), representing only 1% of the agrochemicals applied in crop production worldwide (Hajek, 2004). About 90% of which are based on just one commercial formulation containing *Bacillus thuringiensis* (Advisory Committee on Pesticides, 2004). According to an estimate by (Marrone, 2007) biopesticide market currently, have a 5 years annual growth rate of 16% as compared to chemical pesticide market (3%) and is expected to make a role in the global market of up to $10 billion by 2017. There are many reasons and explanations on why take up of biopesticides in the global market for crop production is at such a low level? Some of these reasons and challenges faced by biopesticides formulations and developments are categorized as (1) challenges faced in production and formulation, (2) challenges faced in regulation, and (3) challenges faced in the technical performance of biopesticides and are discussed under;

6.3.1 CHALLENGES FACED IN PRODUCTION AND FORMULATION

Effective mass production and formulation is a crucial factor in biopesticide research. This part is also dependent on the application method for instance, seed treatment requires less microbial load as compared to drench product (Jackson, 2003). Therefore, cost-effectiveness for commercial productions should be taken into account. This has been achieved for many biopesticides by producing large volumes of inoculum based on waste/cheaper materials and some by-products of industrial origin (Costa et al., 2001; Visnovsky et al., 2008; Brara et al., 2005). Commercial manufacturing of such biological control agents (BCAs) requires suitable formulation and production in such a way as to ensure the biocontrol activity of fresh biocontrol agents (Powell and Jutsum, 1993; Burgues and Jones, 1998). Different types of carriers both solid and liquid are being used in order to produce and ensure the shelf life of the biocontrol agents. Liquid formulations are mass cultured suspensions in water, emulsions, and oils retaining viability and efficacy of the microbial loads for considerable time duration and are generally not stored in the refrigerator (Abadias et al., 2013). Whereas, dry or solid formulations are applied as wettable powders, granules, and dust (Schisler et al., 2004). Long shelf life and potent efficacy limit the use and commercialization of bacterial biopesticides (Morales, 1996). In case of fungal biocontrol agents, a special emphasis has to be given during spore formulation in order to ensure higher shelf life that in turn ensures consistent biocontrol activity under field conditions (Jackson et al., 2010). However, in order to maximize the virulence of fungal agent surface chemistry and ecological and environmental factors should be taken into account (Jackson et al., 2010). In a recent study by Grewal (2002), on nematode-based biocontrol agents advocate the use of several formulations including wettable granular suspensions but have shown inconsistency on a larger scale as compared to clay and gel-based pesticide formulation (Georgis, 2002). In the case of viruses, easy cultivation of virus-based agents on commonly available dried formulations make their use attractive than other liquid formulated control agents (Burges and Jones, 1998).

All of the dry formulations need an additional step of dehydration to achieve a stable formulation before transportation and storage that makes it somehow tricky because cell damage can occur during dehydration that loses viability of the live cells (Rhodes, 1993). Thus, additional protections have to be applied in dry formulations in order to avoid dehydration. Therefore, it is an expensive alternative particularly when the aim is to produce formulations on a large scale which deter manufacturers to commercialize biopesticides.

6.3.2   REGULATORY CHALLENGES IN BIOPESTICIDE DEVELOPMENT

Biopesticides contain an array of several living and nonliving entities that vary considerably from each other based on the characteristic composition, mechanisms of action, and fate and behavior in the environment after application. Therefore, the prime priority should be given to those species which are more selective, less/nontoxic and casts no negative impression on the environment (Gullino and Kuijpers, 1994). These biocontrol agents are grouped together based on their properties by the officials in several countries to regulate and authorize their use given that (1) human and environmental safety and (2) to ensure a reliable and consistent supply of these biopesticides to the market (Harman, 2000; Montesinos and Bonaterra, 2009). In European Union the use of biopesticides is restricted due to inadequate legislation measures and only the authorized and registered products can be used legally (Fravel et al., 1999; Mathre et al., 1999). The biopesticides registration process generally requires detailed scientific data on microbial identity, modes of action, ecological adaptations, host range, specific methods to culture, residues, toxicological and ecotoxicological testing, and potential impacts on the environment and human health (Bonaterra et al., 2012).To put the record straight provision of such information is too expensive for the manufacturers which hampers the commercialization of biopesticides (Chandler et al., 2011).

Biosafety and environmental risks associated with microbial biopesticides are due to their wide host range, toxic nature, pathogenicity, and negative impacts on nontarget entities (Cuddeford and Kabaluk, 2010). However, scientific publications hitherto indicated minimal or no toxic effects of biopesticides on humans and the environment (Vestergaard et al., 2003). Though these test bioherbicides have little detectable risks, lack of evidence indicating negative effects of BCAs does not necessarily mean that the novel biocontrol product should not proceed for biosafety testing because there may exist a variety of strains of several species differing in host range, pathogenic, and toxic characteristics; hence, potentially affecting environmental safety (Chandler et al., 2008). Efficient methodologies need to be incorporated with the development of BCAs which determine their effects on nontarget species (Bidochka et al., 1991). As far as the impacts of these bio-based pesticides on nontarget species are concerned, these can be direct or indirect. BCAs with high host-specific nature are considered safe having no negative effects on nontarget species. In a few studies, however, host-specific BCAs can have effects on unwanted species (Pearson and Callaway, 2005). The magnitude and strength of such effects depend on the ability of BCAs to persistence and eco-ecological adaptations (e.g. interactions with the biotic and abiotic environmental components) (Chandler et al., 2008). It is worth mentioning that the knowledge on the natural enemies of the introduced BCAs should be in place because if the enemies are well adapted to the native environment this could have profound implications for biocontrol of pests and effects on the nontarget species (Dybdahl and Storfer, 2003). Therefore, manufacturers are under market pressure to formulate BCAs with wider host ranges which can cast negative impressions on nontarget entities and environment (Jackson, 2003; Chandler et al., 2008). On the other hand, environmentalists are not much interested to see broad-spectrum formulations based on strains isolated from the native area (Waage, 2001). All these conditions put commercial pressure on manufacturers to produce such formulations; hence, hindering their development.

Assessing the fate and interactions of the introduced microbes in the environment is of vital importance to gather information on residue analysis, traceability, and impacts on the exposed environment for smooth registration of formulated biocontrol agents (De Clercq et al., 2003; Montesinos, 2003). A variety of techniques are available to track and to assess the fate of applied BCAs such as immunological or morphological analysis (ELISA), molecular analysis (probe-hybridization of the 16S/18S rDNA sequences), and determination of microsatellite markers (Ryder, 1995; Plimmer, 1999). Longer time durations and higher costs associated with these tests are among the key constraints to develop and commercialize biopesticides.

6.3.3 CHALLENGES IN TECHNICAL PERFORMANCE

Technical issues in commercialization of biopesticides may be categorized as (1) improvements in the efficacy of formulated products, (2) novel field delivery methods, and (3) improved compatibility (Glare et al., 2012; Martin et al., 2012). There exist a number of technical gaps in effective commercialization of biopesticides that need to be narrowed in order to improve the utilization of BCAs in crop protection measures (Kumar, 2015). Development of biopesticides has long been using the synthetic chemical model which does not exploit the biological potency of the BCAs (Kumar, 2015). The efficacy of the BCAs is governed by the way they are applied to control pests or diseases. Effective delivery of these agents to the target host is, therefore, important (Nuyttens et al., 2009). Biopesticide delivery is not much complicated; however, the application of BCAs requires proper training and knowledge on host-pathogen relationships (Kumar, 2015).

There is also a need to understand about the proper effective concentration of the target biopesticide in case of microbial agents, for instance (per unit leaf surface area/ per unit volume of the substrate) to ensure higher efficacy (Montesinos and Bonaterra, 1996; Frances et al., 2006). In the case of bacterial agents, a concentration of 108 CFU mL^{-1} and 107 for fungi and yeast are required to ensure high efficacy (Bonaterra et al., 2012). More or less similar concepts are applied to botanicals and semiochemicals. Honestly speaking, there are only a few studies where the application of BCAs has been on the utilization of effective concentration (Jaronski, 2010). Despite the fact that delivery methods had significant effects on the efficacy of BCAs, a little attention has been given to optimizing efficient delivery of biopesticides (Gan-Mor and Matthews, 2003; Gwynn, 2011). Inefficiency in understanding such methods creates market entry barriers of biopesticides.

Similarly, determining how long the applied biopesticide will remain active in the target area is also important to optimize application requirements, that is, when and how frequently they should be applied. The shelf life and activity of any BCA is directly related to its rate of production and biological characters of the disease or pests (Scheepmaker and Butt, 2010). It is a matter of misfortune that most of the research has been done on the relationship between the persistence of biopesticides and its environmental fate after application to figure out environmental risk assessment rather than finding ways to improve efficacy (Mudgal et al., 2014). Inadequate knowledge on pesticide persistence to growers is also a drawback of neglected commercialization of biopesticidal microorganisms. Proper information on compatibility of produced biopesticide with other agents is highly important as, there are some synthetic chemicals which are antagonistic to microbial pesticides (Jaronski, 2010). Studies to find such compatibility relations are only studied using in vitro assays under a test synthetic chemical therefore, these tests often mimic the kind of exposure that occurs in fields and hence give different results (Chandler et al., 2005). It seems to be difficult on manufacturer side to conduct and test such relationships to find compatibility with other crop protection agents which leads to failure in regulation and hence production and commercialization of biopesticides.

6.4   CONCLUSION

Inadequate knowledge on pesticide persistence to growers is also a drawback of neglected commercialization of biopesticidal microorganisms. Proper information on compatibility of produced biopesticide with other agents is highly important as there are some synthetic chemicals which are antagonistic to microbial pesticides. Studies to find such compatibility relations are only studied using in vitro assays under a test synthetic chemical, therefore, these tests often mimic the kind of exposure that occurs in fields and hence give different results. It seems to be difficult on manufacturer side to conduct and test such relationships to find compatibility with other crop protection agents which leads to failure in regulation and hence production and commercialization of biopesticides.

KEYWORDS

- **biopesticides**
- **fungicides**
- **pesticides**
- *Bacillus thuringiensis*
- *bio-safety*

REFERENCES

Advisory Committee on Pesticides. Final report of the subgroup of the Advisory Committee on Pesticides on alternatives to conventional pest control techniques in the UK: a scoping study of the potential for their wider use. **2004**.

Amaro, P. A. ProteccaoIntegrada. ISA Press, Lisboa, Portugal. **2003**.

Aravind, R.; Kumar, A.; Eapen, S.; Ramana, K. Endophytic bacterial flora in root and stem tissues of black pepper (*Piper nigrum* L.) genotype: isolation, identification and evaluation against Phytophthora capsici. *Lett. Appl. Microbiol.* **2009**, *48*, 58–64.

Bidochka, M. J.; Walsh, S. R. A.; Ramos, M. E.; St. Leger, R. J.; Silver, J. C.; Roberts, D. W. Fate of biological control introductions: monitoring an Australian fungal pathogen of grasshoppers in North America. *Proc.Natl. Acad. Sci. U. S. A.* **1991**, *93*, 918–921.

Bischoff, J. F.; Rehner, S. A.; Humber, R. A. A multilocus phylogeny of the *Metarhizium anisopliae* lineage. *Mycologia.* **2009**, *101*, 512–530.

Bonaterra, A.; Badosa, E.; Cabrefiga, J.; Francés, J.; Montesinos, E. Prospects and limitations of microbial pesticides for control of bacterial and fungal pomefruit tree diseases. *Trees.* **2011**, *26*, 215–226.

Bonaterra, A.; Badosa, E.; Cabrefiga, J.; Francés, J.; Montesinos, E. Prospects and limitations of microbial pesticides for control of bacterial and fungal pomefruit tree diseases. *Trees.* **2012**, *26*, 215–226.

Brar, S. K.; Verma, M.; Tyagi, R. D.; Valero.; J. R.; Surampalli, R. Y. Starch industry wastewater-based stable *Bacillus thuringiensis* liquid formulations. *J. Econ. Entomol.* **2005**, *98*, 1890–1898.

Bravo, A.; Gill, S. S.; Soberon, M. Mode of action of *Bacillus thuringiensis* Cry and Cyt toxins and their potential for insect control. *Toxicon.* **2007**, *49*, 423–435.

Bravo, A.; Likitvivatanavong, S.; Gill, S. S.; Soberon, M. *Bacillus thuringiensis*: a story of a successful bioinsecticide. *Insect Biochem. Mol. Biol.* **2011**, *41*, 423–431.

Bruck, D. J.; Snelling, J. E.; Dreves, A. J.; Jaronski, S. T. Laboratory bioassays of entomopathogenic fungi for control of *Delia radicum* (L.) larvae. *J. Invertebr. Pathol.* **2005**, *89*, 179–183.

Burges H. D.; Jones, K. A. Formulation of bacteria, viruses and protozoa to control insects. In: Burges HD, (Ed). Formulation of Microbial Pesticides: Beneficial Microorganisms, Nematodes and Seed Treatments. Kluwer, Dordrecht. **1998**, 33–127.

Chandler, D.; Bailey, A. S.; Tatchell, G. M.; Davidson, G.; Greaves, J.; Grant, W. P. The development, regulation and use of biopesticides for integrated pest management. *Philos. Trans. R. Soc. B: Biol. Sci.* **2011**, *366*, 1987–1998.

Chandler, D.; Davidson, G.; Grant, W. P.; Greaves, J.; Tatchell, G. M. Microbial biopesticides for integrated crop management: an assessment of environmental and regulatory sustainability. *Trends Food Sci. Technol.* **2008**, *19*, 275–283.

Chandler, D.; Davidson, G.; Jacobson, R. J. Laboratory and glasshouse evaluation of entomopathogenic fungi against the two-spotted spider mite, *Tetranychusurticae* (Acari: Tetranychidae) on tomato, *Lycopersicon esculentum. Biocontrol Sci. Technol.* **2005**, *15*, 37–54.

Chaston, J. M.; et al. The entomopathogenic bacterial endosymbionts *Xenorhabdus* and *Photorhabdus*: convergent lifestyles from divergent genomes. *PLoS One,* **2011**, *6*, e27909.

Conn, V. M.; Walker, A.; Franco, C. Endophytic actinobacteria induce defense pathways in *Arabidopsis thaliana. Mol. Plant-Microbe Interact.* **2008**, *21*, 208–218.

Costa, E.; Teixido, N.; Usall, Atares, E.; Vinas, I. Production of the biocontrol agent *Pantoea agglomerans* strain CPA-2 using commercial products and by-products. *Appl. Microbiol. Biotechnol.* **2001**, *56*, 367–371.

Costa, C. I.; de V.P.A.; da. Integrated Pest Management and the (un) Sustainable Use of Pesticides **2016**. www.repository.utl.pt/handle/10400.5/12017 (verified 31 October 2016).

Cuddeford, V.; Kabaluk, J. T. Alternative Regulatory Models for Microbial Pesticides. In: Kabaluk, J. T.; Svircev, A. M.; Goettel, M. S.; Woo, S. G. (eds.). The Use and Regulation of Microbial Pesticides in Representative Jurisdictions Worldwide. IOBC Global, **2010**, 94–98.

De Clercq, D.; Cognet, S.; Pujol, M.; Lepoivre, P.; Jijakli, M. H. Development of a SCAR marker and a semi-selective medium for specific quantification of *Pichia anomala* strain K on apple fruit surfaces. *Postharvest Biol Technol.* **2003**, *29*, 237–24.

Deng, Y.; Zhu, Y.; Wang, P.; Zhu, L.; Zheng, J.; Li, R.; Ruan, L.; Peng, D.; Sun, M. Complete genome sequence of *Bacillus subtilis* BSn5, an endophytic bacterium of *Amorphophallus konjac* with antimicrobial activity for the plant pathogen *Erwinia carotovora* subsp. *carotovora. J. Bacteriol.* **2011**, *193*, 2070–2071.

Devine, G. J.; Furlong, M. J. Insecticide use: contexts and ecological consequences. *Agric. Hum. Values.* **2007**, *24*, 281–306. doi:10.1007/s10460-007-9067-z

Dong, Z.; Canny, M. J.; McCully, M. E.; Roboredo, M. R.; Cabadilla, C. F.; Ortega, E.; Rodes, R. A nitrogen-fixing endophyte of sugarcane stems (a new role for the apoplast). *Plant Physiol.* **1994**, *105*, 1139–1147.

Downing, K. J.; Leslie, G.; Thomson, J. A. Biocontrol of the sugarcane borer *Eldana saccharina* by expression of the *Bacillus thuringiensis* cry1Ac7 and *Serratia marcescens chiA* genes in sugarcane-associated bacteria. *Appl. Environ. Microbiol.* **2000**, *66*, 2804–2810.

Driesche, R.; van Hoddle, M.; Center, T. Control of Pests and Weeds by Natural Enemies: An Introduction to Biological Control. John Wiley & Sons. **2009**.

Dybdahl, M. F.; Storfer, A. Parasite local adaptation: Red Queen vs. Suicide King. *Trends Ecol. Evol.* **2003**, *18*, 523–530.

Eilenberg, J.; Hajek, A.; Lomer, C. Suggestions for unifying the terminology in biological control. *Biol. Control.* 2001, *46*, 387–400.

Faria, M.; Hajek, A. E.; Wraight, S. P. Imbibitional damage in conidia of the entomopathogenic fungi *Beauveria bassiana*, *Metarhizium acridum* and *Metarhizium anisopliae. Biol. Control.* **2009**, *51*, 346–354.

Frances, J.; Bonaterra, A. Moreno, M. C.; Cabrefiga, J.; Badosa, E.; Montesinos, E. Pathogen aggressiveness and postharvest biocontrol efficiency in *Pantoea agglomerans. Postharvest Biol. Technol.* **2006**, *39*, 299–307

Franco, J. C.; Ramos, A. P.; Moreira, I. Infra-estruturasecolo ´gicas e protecc¸a ˜o biolo ´gicacaso dos citrinos. ISA Press, Lisboa. **2006**.

Fravel, D. R.; Rhodes, D. J.; Larkin, R. P. Production and commercialization of biocontrol products. In: Albajes R., Gullino L. M., van Lenteren J. C., Elad Y. (eds.). Integrated Pest and Disease Management in Greenhouse Crops. Kluwer, Dordrecht, **1999**, 365–376.

Gan-Mor, S.; Matthews, G. A. Recent developments in sprayers for application of biopesticides—an overview. *Biosyst. Eng.* **2003**, *84*, 11, 9–12.

George, D. R.; Finn, R. D.; Graham, K. M.; Sparagano, O. A. E. Present and future potential of plant-derived products to control arthropods of veterinary and medical significance. *Parasit. Vectors.* **2014**, 710, https://doi.org/10.1186/1756-33057-28.

Georgis, R. The Biosys experiment: an insider's perspective. In: Gaugler, R. (Ed). Entomopathogenic Nematology. CAB International, Wallingford, UK; **2002**, 357–372.

Glare, T.; Caradus, J.; Gelernter, W.; Jackson, T.; Keyhani, N.; Kohl, J.; Marrone, P.; Morin, L.; Stewart, A. Have biopesticides come of age? *Trends Biotechnol*. **2012**, *30*, 250–258.

Goswami, L.; Kim, K. H.; Deep, A.; Das, P.; Bhattacharya, S. S.; Kumar, S.; Adelodun, A. A. Engineered nano particles: nature, behaviour, and effect on the environment. *J. Environ. Manag*. **2017**, *196*, 297–315.

Goulson, D.; Nicholls, E.; Botías, C.; Rotheray, E. L. Bee declines driven by combined stress from parasites, pesticides, and lack of flowers. *Science*. **2015**, *347*, 1255957.

Grewal, P. S. Formulation and application technology. In: Gaugler, R. (ed.). Entomopathogenic Nematology. CAB International, Wallingford, UK; **2002**, 265–287.

Grewal, P. S.; Ehlers, R. U.; Shapiro-Ilan, D. I. Nematodes as Biocontrol Agents. CABI Publishing, Wooster, USA. **2008**.

Gullino, M. L.; Kuijpers, L. A. M. Social and political implications of managing plant diseases with restricted fungicides in Europe. *Annu. Rev. Phytopathol*. **1994**, *32*, 559–579.

Gwynn, R. Review of literature and existing state of the art with respect to the application of microbial biopesticides in agriculture. Final report of Defra project PS2027. **2011**.

Hajek, A. Natural Enemies: An Introduction to Biological Control. Cambridge University Press, Cambridge, UK. **2004**.

Hajek, A. E.; Delalibera, J. I. Fungal pathogens as classical biological control agents against arthropods. *Biol. Control*. **2010**, *55*, 147–158.

Harman, G. E. *Myths* and *dogmas* of biocontrol. *Plant Dis*. **2000**, *84*, 377–393.

Jackson, T. A. Environmental safety of inundative application of a naturally occurring biocontrol agent, *Serratia entomophila*. In: Hokkanen, H. M. T.; Hajek, A.E. (eds.), Environmental Impacts of Microbial Insecticides. Kluwer Academic Publishers, Dordrecht. **2003**, 169–176.

Jackson, T. A. A novel bacterium for control of grass grub. In: Vincent, C.; Goettel, M. S.; Lazarovits, G. (eds.), Biological Control: A Global Perspective. CABI Publishing, Wallingford, UK. **2007**, 160–168.

Jaronski, S. T. Ecological factors in the inundative use of fungal entomopathogens, *Biol. Control*. **2010**, *55*, 159–185.

Jones, K.; Everard, M.; Harding, A. H. Investigation of gastrointestinal effects of organophosphate and carbamate pesticide residues on young children. *Int. J. Hyg. Environ. Health*. **2014**, *217*, 392–398.

Kaya, H. K.; Lacey, L. A. Introduction to microbial control. In: Lacey, L. A.; Kaya, H. K. (eds.), Field Manual of Techniques in Invertebrate Pathology. Springer, Netherlands. **2007**, 3–7.

Khater, H. F. Ecosmart biorational insecticides: alternative insect control strategies. Insecticides—Advances in Integrated Pest Management. In. Tech. **2012**, 17–60.

Kloepper, J. W.; Ryu. C. M. Bacterial endophytes as elicitors of induced systemic resistance. In Microbial Root Endophytes. Springer. **2006**, 33–52.

Kloepper, J. W.; Schroth, M. N.; Miller, T. D. Effects of rhizosphere colonization by plant growth-promoting rhizobacteria on potato plant development and yield. *Phytopathology*. **1980**, *70*, 1078–1082.

Kohler, H. R.; Triebskorn, R. Wildlife ecotoxicology of pesticides: can we track effects to the population level and beyond? *Science*. **2013**, *341*, 759–765.

Koppenhofer, A. M.; Fuzy, E. M. Early timing and new combinations to increase the efficacy of neonicotinoid entomopathogenic nematode (Rhabditida: *Heterorhabditidae*) combinations against white grubs (Coleoptera: *Scarabaeidae*). *Pest. Manag. Sci.* **2008**, *64*, 725–735.

Kumar, S. Biopesticides: a need for food and environmental safety. *J. Biofertil. Biopest.* **2012**, *3*, e107.

Kumar, S. Biopesticide: an environment friendly pest management strategy. *J. Biofertil. Biopest.* **2015**, *6*, e127.

Kumar, R.; Verma, H.; Haider, S.; Bajaj, A.; Sood, U.; Ponnusamy, K.; Nagar, S.; Shakarad, M. N. Negi, R. K.; Singh, Y.; Khurana, J. P.; Gilbert, J. A.; Lal, R. Comparative genomic analysis reveals habitat-specific genes and regulatory hubs within the genus *Novo sphingobium*. *Systems*. **2017**, *210*. https://doi.org/ 10.1128/mSystems.00020-17.

Lacey, L. A.; Grzywacz, D.; Shapiro-Ilan, D. I.; Frutos, R.; Brownbridge, M.; Goettel, M. S. Insect pathogens as biological control agents: back to the future. *J. Invertebr. Pathol.* **2015**, *132*, 1–41.

Lacey, L. A.; Unruh, T. R.; Simkins, H.; Thomsen-Archer, K. Gut bacteria associated with the Pacific Coast wireworm, Limoniuscanus, inferred from 16s rDNA sequences and their implications for control. *Phytoparasitica*. **2007**, *35*, 479–489.

Ladurner, E.; Benuzzi, M.; Fiorentini, F.; Franceschini, S. Beauveria bassiana strain ATCC 74040 (Naturalis®), a valuable tool for the control of the cherry fruit fly Rhagoletis cerasi. In: Ecofruit 13th International Conference on Cultivation Technique and Phytopathological Problems in Organic Fruit-Growing: Proceedings of the Conference from 18th February to 20th February 2008 at Weinsberg, Germany. **2008**, pp. 93–97.

Landis, D. A.; Wratten, S. D.; Gurr, G. M. Habitat management to conserve natural enemies of arthropod pests in agriculture. *Annu. Rev. Entomol.* **2000**, *45*, 175–201.

Liu, X.; Jia, J.; Atkinson, S.; Cámara, M.; Gao, K.; Li, H.; Cao, J. Biocontrol potential of an endophytic *Serratia* sp. G3 and its mode of action. *World J. Microbiol. Biotechnol.* **2010**, *26*, 1465–1471.

Lodewyckx, C.; Vangronsveld, J.; Porteous, F.; Moore, E. R.; Taghavi, S.; Mezgeay, M.; der Lelie, D.V. Endophytic bacteria and their potential applications. *Crit. Rev. Plant Sci.* **2002**, *21*, 583–606.

Longe, O. O. Biological control: A veritable natural pest management strategy. *Int. J. Agric. Environ. Res.* **2016**, *2*, 57–74.

Luck, R. F.; Forster, L. D. Quality of augmentative biological control agents: a historical perspective and lessons learned from evaluating *Trichogramma*. Quality Control and Production of Biological Control Agents: Theory and Testing Procedures. CABI Publishing, The Netherlands. **2003**, pp. 231–240.

Marrone, P. G. Barriers to Adoption of Biological Control Agents and Biological Pesticides. CAB Reviews: Perspectives in Agriculture, Veterinary Science, Nutrition and Natural Resources. CABI Publishing Wallingford, UK. **2007**, *2*.

Martin, L.; Marques, J. L.; Gonzalez-Coloma, A.; Mainar, A. M.; Palavra, A. M. F.; Urieta, J. S. Supercritical methodologies applied to the production of biopesticides: a review. *Phytochem. Rev.* **2012**, *11*, 413–431.

Mathre, D. E.; Cook, R. J.; Callan, N. W. From discovery to use. Traversing the world of commercializing biocontrol agents for plant disease control. *Plant Dis.* **1999**, *83*, 972–983.

Montesinos, E. Development, registration and commercialization of microbial pesticides for plant protection. *Int. Microbiol.* **2003**, *6*, 245–252.

Montesinos, E.; Bonaterra, A. Dose–response models in biological control of plant pathogens. An empirical verification. *Phytopathology*. **1996**, *86*, 464–472.

Montesinos, E.; Bonaterra, A. Microbial pesticides. In: Schaechter, M. (ed.), Encyclopedia of Microbiology, 3rd edn. Elsevier, New York. **2009**. pp. 110–120.

Mudgal, S.; De Toni, A.; Tostivint, C.; Hokkanen, H.; Chandler, D. Scientific support, literature review and data collection and analysis for risk assessment on microbial organisms used as active substance in plant protection products. *EFSA Tech. Rep.* **2014**, p. 156.

Nuyttens, D.; De Schampheleire, M.; Verboven, P.; Brusselman, E.; Dekeyser, D. Droplet size and velocity characteristics of agricultural sprays. *Trans. Am. Soc. Agric. Eng.* **2009**, *52*, 1471–1480.

Oerke, E. C. Crop losses to pests. *J. Agric. Sci.* **2006**, *144*, 31–43.

Oerke, E. C.; Dehne, H. W. Safeguarding production-losses in major crops and the role of crop protection. *Crop Prot.* **2004**, *23*, 275–285.

Oliveira, C. M.; Auad, A. M.; Mendes, S. M.; Frizzas, M. R. Crop losses and the economic impact of insect pests on Brazilian agriculture. *Crop Prot.* **2014**, *56*, 50–54.

Pearson, D. E.; Callaway, R. M. Indirect nontarget effects of host-specific biological control agents: implications for biological control. *Biol. Control.* **2005**, *35*, 288–298.

Pieterse, C. M.; Van der Does, D.; Zamioudis, C.; Leon-Reyes, A.; Van Wees, S. C. Hormonal modulation of plant immunity. *Annu. Rev. Cell Dev. Biol.* **2012**, *28*.

Plimmer, J. R. Analysis, monitoring, and some regulatory implications. In: Hall, F. R.; Menn, J. J.; (eds.) Methods in Biotechnology, vol 5. Biopesticides: Use and Delivery. Humana Press, Totowa, NJ, USA. **1999**, pp. 529–555

Popp, J.; Peto, K.; Nagy, J. Pesticide productivity and food security. A review. *Agron. Sustain. Dev.* **2012**, *33*, 243–255.

Puech, C.; Baudry, J.; Joannon, A.; Poggi, S.; Aviron, S. Organic vs. conventional farming dichotomy: does it make sense for natural enemies? *Agric. Ecosyst. Environ.* **2014**, *194*, 48–57.

Ragsdale, N. N.; Sisler, H. D. Social and political implications of managing plant diseases with decreased availability of fungicides in the United States. *Annu. Rev. Phytopathol.* **1994**, *32*, 545–557.

Rhodes, D. J. Formulation of biological control agents. In: Jones, D. G. (ed.), Exploitation of Microorganisms. Chapman and Hall, London. **1993**, pp. 411–439.

Rosner, D.; Markowitz, G. Persistent pollutants: a brief history of the discovery of the widespread toxicity of chlorinated hydrocarbons. *Environ. Res.* **2013**, *120*, 126–133.

Ruiu, L.; Satta, A.; Floris, I. Emerging entomopathogenic bacteria for insect pest management. *Bull. Insectol.* **2013**, *66*, 181–186.

Ryder, M. H. Monitoring of biocontrol agents and genetically engineered microorganisms in the environment: biotechnological approaches. In: Rudra, P. S.; Uma, S. S. (eds.), Molecular Methods in Plant Pathology. CRC Press, Boca Raton, FL, USA. **1995**, pp. 475–492.

Saxena, S.; Pandey, A. K. Microbial metabolites as ecofriendly agrochemicals for the next millennium. *Appl. Microbiol. Biotechnol.* **2001**, *55*, 395–403.

Scheepmaker, J. W. A.; Butt, T. M. T. Natural and released inoculum levels of entomopathogenic fungal biocontrol agents in soil in relation to risk assessment and in accordance with EU regulations. *Biocontrol Sci. Technol.* **2010**, *20*, 503–552.

Shaner, D. L.; Wiles, L.; Hansen, N. Behavior of atrazine in limited irrigation cropping systems in Colorado: prior use is important. *J. Environ. Qual.* **2009**, *38*, 1861–1869.

Sheoran, N.; Nadakkakath, A. V.; Munjal, V.; Kundu, A.; Subaharan, K.; Venugopal, V.; Rajamma, S.; Eapen, S. J.; Kumar, A. Genetic analysis of plant endophytic *Pseudomonas putida* BP25 and chemo-profiling of its antimicrobial volatile organic compounds. *Microbiol. Res*. **2015**, *173*, 66–78.

Smits, T. H.; Rezzonico, F.; Kamber, T.; Blom, J.; Goesmann, A.; Ishimaru, C. A.; Frey, J. E.; Stockwell, V. O.; Duffy, B. Metabolic versatility and antibacterial metabolite biosynthesis are distinguishing genomic features of the fire blight antagonist Pantoeavagans C9-1. *PLoS One*. **2011**, *6*, e22247.

Steinke, W. E.; Giles, D. K. Delivery systems for biorational agents. ACS Symposium Series. **1995**, 0097-6156/95/ 595-0080

Tejada, M.; Morillo, E.; Gomez, I.; Madrid, F.; Undabeytia, T. Effect of controlled release formulations of diuron and alachlor herbicides on the biochemical activity of agricultural soils. *J. Hazard. Mater.* **2017**, *322*, 334–347.

Thacker, J. R. M. An Introduction to Arthropod Pest Control. Cambridge University Press. **2002**.

Toth, P. Biological control. Sustainable Agriculture. Baltic University Press. Ecosystem Health and Sustainable Agriculture, **2012**, pp. 206–213.

Universidad Autenoma de Nuevo Leen. **1996**, pp. 157–77.

Van Lenteren, J. C. Biological control in protected crops: where do we go? *Pesticide Sci.* **1992**, *36*, 321–327.

Van Lenteren, J. C. IOBC Internet Book of Biological Control. IOBC Publication. International Organization for Biological Control, St. Paul, USA. **2012**.

Van Lenteren, J. C.; Bueno, V. H. Augmentative biological control of arthropods in Latin America. *Biol. Control*. **2003**, *48*, 123–139.

Vestergaard, S.; Cherry, A.; Keller, S.; Goettel, M. Safety of hyphomycete fungi as microbial control agents. In Hokkanen, H. M. T.; Hajek, A. E. (eds.), Environmental Impacts of Microbial Insecticides. Kluwer Academic Publishers, Dordrecht, **2003**, pp. 35–62.

Visnovsky, G. A.; Smalley, D. J.; O'Callaghan, M.; Jackson, T. A. Influence of culture medium composition, dissolved oxygen concentration and harvesting time on the production of *Serratia entomophila*, a microbial control agent of the New Zealand grass grub. *Biocontrol Sci. Tech*. **2008**, *18*, 87-100

Waage, J. K. Indirect ecological effects in biological control: the challenge and the opportunity. In Wajnberg, E.; Scott, J. K.; Quimby, P. C. (eds.), Evaluating Indirect Ecological Effects of Biological Control. CABI Publishing, Wallingford, UK. **2001**, pp. 1–12

Weidemann, G. J.; Boyette, C. D.; Templeton, G. E. Utilization criteria for mycoherbicides. ACS Symposium Series. **1995**, 0097-6156/95/595-0238

Zhang, D.; Spadaro, D.; Valente, S.; Garibaldi, A.; Gullino, M. L. Cloning, characterization, expression and antifungal activity of an alkaline serine protease of *Aureobasidium pullulans* PL5 involved in the biological control of postharvest pathogens. *Int. J. Food Microbiol*. **2012**, *153*, 453–464.

CHAPTER 7

Advances in Pesticide Bioremediation Technology

GULZAR A. RATHER

CSIR-Indian Institute of Integrative Medicine, Jammu 180001, India
E-mail: gulzargani.87@gmail.com

ABSTRACT

Over the past decade, increase in demand for food and vegetables has simultaneously forced injudicious use of pesticides which resulted in contamination of agricultural products as well as soil, water, air, etc., and have a long term effect on the humans and other life forms. The extensive use of pesticides results in their accumulation in the agricultural products and causes wide range of health problems. With the inception of industrial revolution and increased application of newly discovered pesticides has enhanced the yield of our agricultural products by protecting majority of our crops from the pest attack. Pesticides at present play an important role in the agricultural lands but their degradation is a major concern today. Accumulation of pesticides in the agricultural products poses a serious threat to humans and other life forms as well as to the ecosystem. Therefore, isolation, identification, and characterization of biota, having ability to degrade pesticides provide an opportunity to develop new strategies to restore contaminated sites or to treat wastes before their final discharge. Enzymatic degradation, mycoremediation, and phytoremediation, having exceptional abilities for decontamination of contaminated sites are nowadays promising environmentally sound technologies.

7.1 INTRODUCTION

The word pesticide covers a wide range of substances or mixture of chemical compounds used for repelling, destroying, preventing, extenuating, or incapacitates biological organisms including rodents, weeds, insects, and diseases in the crop fields. The introduction of synthetic organophosphate (OP) insecticides in the 1960s, pyrethroids in the 1980s, carbamates in the 1970s, and the introduction of fungicides and herbicides in the 1970s to1980s contributed greatly to pest management and agricultural output (Aktar et al., 2009). Although having benefits, there have been many problems associated with the use of pesticides. Application of pesticides reduces the pest attack, control dreadful diseases, and boost the agricultural production. Over the past decade, increase in demand for food and vegetables has simultaneously forced injudicious use of pesticides which resulted in contamination of agricultural products as well as soil, water, air, etc., and have a long term effect on the humans and other life forms. The extensive use of pesticides results in their accumulation in the agricultural products and causes wide range of health problems (Bhanti et al., 2007). From experimental evidence, pesticide exposures were more and more related to hormone disruption, immune suppression, neurologic impairment, diminished intelligence, reproductive disorders, cardiovascular diseases, and cancers (Corsini et al., 2008). The first chemical substance used to combat plant pathogens, fungi, and other harmful insects were inorganic materials containing sulfur, copper, arsenic, etc. Some naturally occurring organic insecticides like protenone were also used. However, due to advancement in technology, synthetic organic chemicals containing mercury, phosphorus, arsenic, and chlorine molecules have largely replaced these early pesticides. All organic compounds that occur naturally or produced via industrially are discharged into soil and decomposed by a combination of different chemical and biological methods but the rate of decomposition varies with the nature of compound (Pal et al., 2006). Organic pesticides are consumed by soil bacteria of several genera and eventually decomposed. However, it is not true for all synthetic chemicals. In recent years people have added large number of slowly degradable or nonbiodegradable compounds in the soil by using different pesticides in many parts of world. The chemical composition of insecticides and herbicides covers an extremely broad range of organic acids, organic compounds, chlorinated nitrophenols, and other substances (Adhikari et al., 2010).

Pesticide consumption around the globe has reached two million tons in which European countries utilize 45%, USA 24%, and 25% utilization occurs in rest of the countries. Consumption of pesticides in Asia is highly alarming. Among Asian countries, China exercises the highest percentage followed by Japan, Korea, Pakistan, and India (Su et al., 2014). India is the leading producer of pesticide chemicals and ranks 12th in the world market for the consumption of pesticides. Majority of Indian population is affianced with agriculture and average consumption of pesticides in India is far lower in comparison to other developed countries (Abhilash et al., 2009). Farmer community is always under pressure and tries to adopt advanced technology for enhanced food production. For management of plant pathogens use of pesticides or related synthetic chemicals has become the most essential component of modern agriculture as more than 40% of annual loss in food production is protected from the pest infestation. Pesticides have undoubtedly significant impact on farmer's economy via reducing the agricultural production loss, improving yield, and quality of food commodities in terms of cosmetic appeal. While on the other side, they have created a general concern as their residues are found in food, environment, humans, and animals. The presence of pesticide residue in several food products has also affected their export for the last few years. Against this background, regulation and management of pesticide use, its safety measures, integrated pest execution, and proper application of technologies are some of the key strategies to minimize their harmful impact on humans and other organism. However, there is a lack of studies related to harmful impacts of pesticides which should be explored using new tools and approaches. Pesticides differ in their physical properties, chemical nature and are mostly classified on the presence of chemical group, environmental perseverance, target organism, toxicity, or other characteristics. On the basis of multilateral environmental treaty Stockholm Convention on Persistent Organic Pollutants, 9 of the 12 persistent organic substances are pesticides. Classes of organic pesticides include organochlorine, organometallic, OP, pyrethroids, and carbamates (Table 7.1).

Therefore, it is important to study and predict the environmental affluence of pesticides, review the pesticide production technology, and recommend future degradation strategies which are efficient and cost-effective and would make rational use of pesticides. Bioremediation is an innovative technology having potential to assuage pesticide contamination. Bioremediation involves microorganisms like bacteria, fungi, algae, plants, or their enzymes to restore natural environment altered by toxic contaminants. Bioremediation process could be employed in a direction to attack specific environmental contaminants, such as degradation and detoxification of chlorinated hydrocarbons via bacteria or mineralization of chemical compounds using different plant species. Bioremediation mediated breakdown of pesticides is relatively a new technology which usually occurs in soil and has gone through intense study as of recent decades. Overall, from the past decade bioremediation technology has progressed much more as it provides a novel, eco-friendly, economical, and highly efficient method for detoxification of hazardous pesticides.

TABLE 7.1 Classification of Pesticides on the Basis of Chemical Nature

Pesticides (Chemical Group)	Characteristics	Examples
Organochlorines	These are highly toxic, exhibit slow degradation and bioaccumulation in lipids, and fatty tissues of animals, long-term persistent.	Dichlorodiphenyltrichloroethane, lindane, chlordane, aldrin, mirex. methoxychlor, toxaphene, mirex, kepone
OPs	These are esters of phosphoric acid which are soluble in water as well as in organic solvents. They infiltrate through soil and reach groundwater table, less persistent in the environment than chlorinated hydrocarbons, affect central nervous system. They are absorbed by leaves and stems tissues of plants and transferred to leaf-eating insects.	Methyl parathion, malathion, diazinon, dichlorovas, diptrex, demetox, oxydemeton-methyl, phosphomidon
Carbamates	These organic compounds are derivatives of carbamic acid, destroy a wide spectrum of insects as well as highly toxic to vertebrates, less persistence	Carbaryl, carbanolate, prupoxur, sevin, vernolate, pebulate, diallate, monilate
Pyrethroids	Affect the nervous system, show relatively low persistence than other pesticides, safe to use, some are effective against household insecticides	Allethrin, pyrethrins, bonthrin, tetramethrin, dimethrin, ptrethrin, cyclethrin, fenevelerate, furethrin, alphamethrin
Neonicotinoids	Neuro-active insecticides chemically having similar structure as that of nicotine. They bind to nicotinic acetylcholine receptors and stimulate a response signal.	Acetamiprid, imidacloprid, clothianidin, nitenpyram, thiacloprid, nithiazine, and thiamethoxam.
Spinosyns	The spinosyns are fermentation-derived potent insecticidal activity and lower environmental effect	Spinosyn A, spinosyn D
Diacylhydrazines	Diacylhydrazines are important nonsteroidal inducing agent used against lepidopteron. They show excellent insecticidal activity by inducing precocious molting	Tebufenozide, halofenozide, methoxyfenozide, chromafenozide

7.2 BACTERIA-MEDIATED REMEDIATION OF CONTAMINATED SOILS

With the inception of industrial revolution and increased application of newly discovered pesticides has enhanced the yield of our agricultural products by protecting majority of our crops from the pest attack. Pesticides at present play an important role in the agricultural lands but their degradation is a major concern today. Accumulation of pesticides in the agricultural products poses a serious threat to humans and other life forms as well as to the ecosystem. As a consequence of apparent persistence and toxicity of pesticides, remediation of contaminated lands is a priority. However, to lessen the impact of these dreadful chemicals, degradation process can be employed by using different bacterial strains which carry forward the process under normal atmospheric conditions. Different strains of bacteria were used to biotransform the pesticide residues. Some microorganisms have been isolated and characterized by a wide variety of contaminated sites having a potential of degrading pesticides. Currently, in different research laboratories around the globe there are ongoing collections of microorganisms which are further characterized by identifying their nature toward degradation of pesticides. Cypermethrin is a synthetic pesticide of the pyrethroid family that has high insecticidal activity. It is used to control many dangerous pests including lepidopterous pests of fruit, vegetable crops, and cotton. Due to excessive use, cypermethrin is now a common environmental pollutant with high toxicity and persistence (Zhang et al., 2009). From many studies, it has been revealed that *Pseudomonas aeruginosa, Micrococcus* sp. *Escherichia coli, Bacillus subtilis*, and *Staphylococcus aureus* were found to be effective in the degradation of cypermethrin (Tallur et al., 2015; Zhang et al., 2011). There are also various other microorganisms which use the pesticides as carbon source like *Flavobacterium, Rhodococcus, Alcaligenes, Gliocladium, Trichoderma,* and *Penicillium* (Aislabie and Lloyd-Jones, 1995). Neonicotinoid is a neurotoxic systemic pesticide widely used in crop protection. These are effective in killing honey bees and other natural pollinators including butterflies, wild bees, dragonflies, lacewings, bats, and ladybugs. Neonicotinoid is water soluble and having highly persistent nature. These pesticides are proficient in disrupting food chains and altering biogeochemical cycles (Hussain et al., 2016). *Pseudoxanthomonas indica* isolated from soil has revealed the fastest biodegradation of neonicotinoid imidacloprid (Ma et al., 2014). Similarly, *Stenotrophomonas maltophilia* transformed imidacloprid to olefin metabolite via hydroxylation and dehydrogenation process (Dai et al., 2010). It has been reported that biotransformation of thiamethoxam and imidacloprid has been carried out by *Pseudomonas* species in presence of glucose as a carbon source (Pandey *et al.*, 2009). Moreover, recent investigations on microbes have explored their role in biodegradation of neonicotinoids. Ethion [(*O,O,O',O'*-tetraethyl *S,S'*-methylene bis(phosphorodithioate))] is highly persistent in the soil as it undergoes a little biodegradation. Its persistence is a grave concern as current investigation has estimated nearly 1600 ethion contaminated sites in New South Wales. The mesophilic bacteria isolated from contaminated lands identified as *Azospirillum* and *Pseudomonas* species, were capable of biodegrading OP ethion when cultured in minimal salts medium (Foster et al., 2004). *Bemisia tabaci* and *Comomonadaceae* bacterial species biodegrade thiomethoxam in soil (Xie et al., 2012; Zhou et al., 2014). Moreover, *Botrytis cinerea* eliminates the metroburon and linuron herbicides almost completely (Ortiz et al., 2013). Therefore, isolation, identification, and characterization of bacterial species having ability to degrade pesticides provide an opportunity to develop new strategies to restore contaminated sites or to treat wastes before their final discharge.

7.3   PHYTOREMEDIATION

Phytoremediation is a promising technology that aims to make use of green plants for remediation of contaminated soil and waters. It has also been known as "botano-remediation," "agro remediation," "green remediation," and "vegetative remediation". Phytoremediation has been less investigated than those of other strategies like bacteria fungi, etc. This cost-effective plant-based approach can degrade pesticides and reduce their impact on environment. Nevertheless, it has been proved to be more efficient cleanup system for water, soil, and even for air pollution than bacteria (Andrew, 2007; Pascal et al., 2011). Phytoremediation involves several strategies like rhizodegradation, phytoextraction, phytodegradation, rhizofiltration, phytostabilization to concentrate compounds and elements and to metabolize them in their tissues (Sarma et al., 2011). *Chrysopogon zizanioides*a perennial bunchgrass is resistant to atrazine and having ability to absorb it from contaminated soil further degrades and metabolizes it. Similarly, *Zea mays* and *Setaria faberi* can biotransform some pesticides (Marcacci, 2004; Hatton et al., 1999). Crop plants like *Spinacea oleracea, Raphanus sativus, Solanum melongena,* and *Oryza sativa*can bioaccumulate pesticides like DDT and benzene hexachloride (Mishra et al., 2009). *Ocimum basilicum* and *O.*

minimum can bioremediate extremely toxic endosulfan from soil (Ramírez et al., 2011). Drainage ditch macrophytes like *Leersia oryzoides*, *Sparganium americanum* and *Typha latifolia* were most effective in mitigating atrazine [2-chloro-4-(ethylamino)-6-(isopropylamino)-striazine], diazinon [*O,O*-diethyl-*O*-(2-isopropyl-6-methyl-4-pyrimidinyl) phosphorothioate], permethrin [3-phenoxybenzyl-(1RS)-*cis*,*trans*-3-(2,2-dichlorovinyl)-2,2-dimethylcyclopropanecarboxylate], malathion, and dimethoate pollutants from agricultural runoff waters (Moore et al., 2013). Plants create a favorable microenvironment around their root zone possessing diverse microbial community that facilitates the degradation of harmful products to a greater extent. Degradation of toxic organic chemicals in soil by plant-associated bacteria involves endophytic and rhizospheric bacteria. The rhizosphere bacterial association could support plants in the rhizospheric zone to enhance their ability to degrade pesticides. In phytoremediation process, mostly used plant species include *Amaranthus caudate*, *Nasturtium officinale*, *Phaseolus vulgaris*, and *Lactuca sativa* species. These species could detoxify and degrade malathion and dimethoate during cultivation periods (Fahd and Ahmed, 2009). *Morus alba* and *Populus deltoids* trees have been used successfully to clean up the chlorophenols and chlorinated solvents such as trichloroethylene from the contaminated lands (Stomp et al., 1994).

The concept of phytoremediation can play a vital role in removal of noxious waste products and is widely recognized innovative technology among scientific community. However, direct uptake of pesticides by plant cells would be a common method for modifying physicochemical properties as well as biochemical characteristics of the pesticides.

7.4   MYCOREMEDIATION

Use of fungal strains for degradation of pesticides has shown a promise since 1985. The white rot fungus *Phanerochaete chrysosporium* is tolerant to high concentrations of polluting chemicals. *P. chrysosporium* is able to degrade a number of harmful environmental pollutants (Nyakundi et al., 2012). This ability is generally attributed to potent enzymes which are beneficial for detoxifying pesticides and other pollutants. Fungi secrete dioxygenases, peroxidases, and oxidases to remediate pesticides more efficiently than cytochromes. Lignin peroxidase, dichlorohydroquinone dioxygenase, and laccase are some of the important examples of biotransformation enzymes produced by fungal species like *Pleurotus ostreatus, P. chrysosporium, Ganoderma lucidum* (Kaur et al., 2016). *Fusarium ventricosu, Phanerochaete chrysosporium, and Trametes versicolor* are *growing on wood chips and can detoxify* simazine, trifluralin, and dieldrin pesticides (Fragoeiro et al., 2008). *F. ventricosum* and *P. chrysosporium* are members of soil microbial community. From previous investigations, it has been shown that *P. chrysosporium* mineralizes endosulfan (Kim, 2001). *F. ventricosum* and *Pandoraea* species are likely to be effective in the degradation of endosulfan and endosulfan sulphate (Siddique et al., 2003). Moreover, brown rot fungus *Gloeophyllum striatum* and *G. trabeum* produce dioxygenases and peroxidases which are involved in biodegradation of 2,4-dichlorophenol and pentachlorophenol (Fahr et al., 1999). Chlorpyrifos [O,O-diethyl-O-(3,5,6-trichloro-2-pyridinyl)phosphorothionate] an organophosphorus pesticide and acaricide widely applied for pest control on cotton, grains, fruits, and vegetable crops. These pesticides have slow rate of degradation and persist for longer periods in the soil. Toxicity of these chemicals affects the cardiovascular system, central nervous system, respiratory system as well as causes several skin and eye irritation problems (Yu et al., 2006). *Acremonium* genus also known as *Cephalosporium* belonging to Hypocreaceae family was found to be effective for the degradation of chlorpyrifos from the contaminated lands (Kulshrestha et al., 2011). Among different bioremediation processes of pesticide, fungal ones being innovative, eco-friendly, and cheap and have been investigated expansively because most of fungus strains are more tolerant of high chemical concentrations. Giving their exceptional abilities use of fungal strains for bioremediation process is a promising technology using their metabolic potential for decontamination.

7.5   ENZYMATIC BIOREMEDIATION

Enzymes are biological catalysts which can be used for detoxifying or degrading the contaminants into nontoxic or less-hazardous substances. Enzymatic bioremediation is efficient, robust, and fast method for removing pesticide residues from the environment and is an innovative technology for extraction of chemicals from contaminated areas. Its applications include the treatment of pesticide residues resulting from processing industries and agricultural products, such as the treatment of contaminated fruits, vegetables, waters, and spent dip liquors (Karigar et al., 2011). A variety of enzymes has been developed for detoxifying xenobiotics and other synthetical products including fungicides, insecticides, and mycotoxins (Sutherland et al., 2004). The OP degrading enzymes attack the phosphoester bonds of aromatic oxon and thion OPs, and degrading these with high turnover. Such enzymes form the basis of numerous remediation strategies to reduce the environmental impacts of pesticide residues. Different types of enzymes involved in detoxifying process are provided in Table 7.2. Phosphotriesterase enzyme shows a highly catalytic potential toward OP pesticides. It was first isolated from *Pseudomonas diminuta* species. The phosphotriesterases are encoded by OP-degrading gene. Phosphotriesterases isolated from *Rhizobium* and *Bradyrhizobium* strains hydrolyze some organophosphorus compounds. Phosphotriesterase like lactonase extracted from *Geobacillus kaustophilus* is catalytically efficient against OP pesticides including parathion, diazinon, and chlorpyrifos (Zhang et al., 2012). There has also been significant research into the use of vertebrate paraoxonases. Mammalian paraoxonases a calcium-dependent enzyme hydrolyze only chlorpyrifos-oxon (Carr et al., 2015). Carboxylic ester (CE) plays a significant role in bioremediation of pesticides and in recent years several CEs of plant, animal, and bacterial origin have been identified and studied. CEs hydrolyze organophosphorus pesticides, pyrethroids, carbamates, and even some other organochlorinated pesticides. They hydrolyze the ester bonds of carboxyl ester molecules, such as those present in malathion (Khan et al., 2016). Malathion is a well-known OP pesticide which is degraded by carboxyesterases. This enzyme is found in several species of fungus like *Penicillum glaucum, Aspergillus*, and *Rhizoctonia* species (Hasan et al., 1999). Not only the fungal cells but also the bacterial species are capable of bioremediation of malathion. From previous studies, it has been found that a strain of *Bacillus thuringiensis* isolated from agricultural waste transforms malathion into mal-monocarboxylic acid and mal-dicarboxylic acid (Xie et al., 2009; Kamal et al., 2008). Different microbial enzymes having catalytic potential to hydrolyze methyl parathion have been identified. For example, organophosphorus hydrolase encoded by the *mpd* gene, and novel phosphotriesterase hydrolysis of coroxon enzyme isolated from *Pseudomonas moteilli* encoded by the *hocA* gene enable to use coroxon as its sole source phosphorous (Horne et al., 2002). *Flavobacterium* bacterial strain is capable of using OP diazinon (*O,O*-diethyl *O*-[6-methyl-2-(1-methylethyl)-4-pyrimidinyl] phosphorothioate) as a carbon source (Sethunathan et al., 1973; Latifi et al., 2012). The other enzymes involved in biotransformation are mainly carboxylesterases, cytochrome p450s, GST, and translocases (Velázquez et al., 2012).

TABLE 7.2 Enzymes Involved in Detoxification of Pesticides

Enzyme	Organism	Pesticide
Oxidoreductases	*Pseudomonas* species *Agrobacterium* strains	Glyphosate
CYP1A1, CYP1A2, CYP2B6	Rats, *Homo sapiens*	Norflurazon, atrazine, isoproturon, metolachlor
Cyp76B1	*Helianthus tuberosus*	Linuron, isoproturon and chlortoluron
P450	*Pseudomonas putida*	Pentachlorobenzene and hexachlorobenzene
Laccase	*Coriolus versicolor*	Pentachlorophenol
Dioxygenases	*Pseudomonas putida*	Trifluralin
Glutathione *S*-transferase	*Arabidopsis*	Acetochlor, metolachlor, triazine
Phosphotriesterases	*Pseudomonas diminuta, Agrobacterium radiobacter Flavobacterium* sp.	Insecticides phosphotriester
Dehalogenases	*Sphingobium* sp. *Shingomonas paucimobilis*	Hexachlorocyclohexane
Hydrolase	*Arthrobacter sp. Achromobacter sp.*	Carbaryl, carbofuran
Haloperoxidase	*Rhodococcus erythropolis*	Thiocarbamate
Urease	*Agrobacterium radiobacter, A. tumefaciens, Variovorax paradoxus*	Atrazine

Hence, microbial enzyme-mediated bioremediation would contribute toward developing new advanced remediation technologies to lessen the toxic effects of pollutants and also to create novel compounds.

7.6 IMMOBILIZATION OF CELLS ENHANCE PESTICIDE DEGRADATION

Degradation of pesticides by using physical or chemical methods such as oxidation with ozone, photolysis, ultrasonic degradation, incineration, fenton degradation, and adsorption was found to be less effective and more expensive than bioremediation of toxic pollutants (Chen et al., 2012). The use of freely suspended cells of microorganisms for detoxification of various toxic substances has a number of drawbacks. It is mostly due to low density of cell cultures, low mechanical strength, and the difficulty in the biomass effluent separation (Wang et al., 2007). To overcome these hurdles, cell immobilization techniques have now been well established. Cell immobilization process made pesticides more susceptible to biodegradation by maintaining catalytic activity of enzymes over long periods of time (Kadakol et al., 2011). This technique of degradation provides several advantages over conventional biological approaches that use free cells. The benefits of using cell immobilization technique provide a possibility of employing high density cell cultures, at high dilution rates avoidance of cell washout, easy severance of growing cells from the reaction systems, continual use of cells, high productivity yields, and better protection of immobilized cells from unfavorable environments (Cassidy et al., 1996). From previous reports, it has been revealed that immobilized cell cultures show higher degradation of pesticides and their tolerance to higher chemical concentrations have been further enhanced via genetic modifications. There are many studies indicating that immobilized cells are much more efficient and tolerant of perturbation in the reaction environment together with better cell survivability and are less vulnerable to toxic molecules. These characteristics make immobilized cell cultures more attractive for the degradation of toxic substances like pesticides (Fernández et al., 2017). The enhanced degradation capability of immobilized cell cultures is primarily due to the protection of the cells from inhibitory compounds present in their surrounding environment. Immobilized microbial systems are highly versatile and have more advantages over freely suspended cells as well as immobilized enzymes in various matrices. In order to remove pesticides, it is imperative to search for efficient and relevant materials with flattering characteristics for the immobilization of cells, some of the aspects such as physical texture, sterilization properties, and the possibility of reusing it repeatedly. The immobilizing materials like agar, sodium alginate, and polyurethane foam (PUF) are inert, nontoxic to cells, and inexpensive (Mulla et al., 2012). Immobilized cell systems allow the operation of bioreactors at flow rates which are independent of cell cultures, tolerable to high concentrations of toxic substances in comparison to their nonimmobilized counterparts. The calcium alginate gelatinous beads operate like a slow delivery system wherein the substances are slowly moving across microbial cell cultures for mineralization without any significant impact on the surrounding environment. There are a number of studies on the potential use of immobilized bacterial cell systems in different matrices for the dispossession of numerous pesticide chemicals (Mulla et al., 2013; Zheng et al., 2009). The degradation of highly toxic cypermethrin by immobilized cell systems of *Micrococcus* species strain CPN 1 using various matrices such as, PUF, sodium alginate, agar, and polyacrylamide revealed that PUF-immobilized cells showed high degradation of cypermethrin s compared to freely suspended cells and cells immobilized in other matrices like polyacrylamide, sodium alginate and agar. PUF-immobilized cell systems of *Micrococcus* are also potentially involved in the remediation process of cypermethrin from contaminated water (Tallur et al., 2015). *Pseudoxanthomonas suwonensis* immobilized on sodium alginate, sodium alginate polyvinyl alcohol, and sodium alginate bentonite clay showed enhanced degradation of profenofos OP pesticide as compared to freely suspended cells (Talwar et al., 2015). Methyl parathion most commonly used OP pesticide was degraded by *Burkholderia cenocepacia. Further,* degradation potential of *B. cenocepacia* cells was evaluated on three different matrices *Opuntia, Agave,* and powdered zeolite fibers. A significant increase in methyl parathion and *p*-nitrophenol degradation was observed in immobilized cell systems as compared to free cell cultures (Fernández et al., 2017). Moreover, a material that has presented better results in the degradation process of pesticide mixtures is the tezontle rock that is a native of volcanoes of Morelos state of central Mexico. The texture of tezontle is highly porous which provides a large surface area for contact and also have a property of being sterilized via autoclaving and can be reused again and again. The presence of micropores allows the rich growth of bacterial colonies. The biofilm of bacterial colonies is formed in the micropores of tezontle rock which provides immobilization of bacterial cells. Subsequently, when a pesticide waste is allowed to pass through these micro-pores having immobilized microorganisms, the biodegradation process can be executed. This innovative strategy of halting cells has been really efficient and is an effective technique that can be exploited for the degradation of highly toxic pesticide wastes (Ortiz et al., 2013). Moreover, there are so many reports that discuss a variety of materials involved to immobilize microbial cells. For example, various plant fibers were used to immobilize bacterial cells or degradation of xenobiotics. The use of other structural matrices such as sheath of palm tree for the cell trapping has added another element to a number of immobilization matrices. Petiolar felt-sheath of palm is inexpensive, stable, and easily handle and available. In this material, the procedure of entrapment of microalga is simple and reliable. This biomatrix material is reusable, free from toxic problems, having mechanical strength for providing necessary support, open and wide spaces within the matrix providing better conditions for growing cells, and thus avoiding rupturing and diffusion problems. These new findings have paved way to search for other sources of biomaterials from the diverse plant species that may be used for cell immobilization (Iqbal et al., 1997). Hence, immobilized microbial cell system technology has a benefit of enhanced rate of biodegradation, tolerance of higher chemical concentrations, and their reusability. Thus, the immobilized microbial systems provide a highly versatile and cost-effective approach that can be utilized for degradation of highly toxic pesticides and contaminated wastewater.

7.7   GENETIC MANIPULATION

Microorganisms respond differently to various external stimuli such as stresses and are adapted accordingly in different environments. Plants, fungi, and bacteria use pesticides as a carbon source and metabolize them into nonhazardous products. However, the rate of detoxification of these harmful compounds is low. This remediation process can be enhanced by using different genetic engineering tools and techniques. The recombinant DNA approaches and other molecular cloning strategies have enabled isolation, disruption, regulation, and/or modification of the desired genes that encode the enzymes involved in pesticide degradation, enhancement of redox reactions, minimization of pathway bottleneck energy generation, and overexpression of heterologous genes to generate new characteristics (Johri et al., 1996). Various bimolecular engineering tools have been developed and employed to optimize the catalytic activity of enzymes, metabolic pathway modulation, and raise transgenic organisms for bioremediation of persistent chemicals (Jiang et al., 2007). Deciphering new information related to metabolic routes and bottlenecks of detoxification require the available molecular toolbox. Nevertheless, the characterized and functionally validated genes need to be integrated within metabolic pathways for proper expression (Kang et al., 2002). Detoxification of hazardous OP pesticides and other toxic chemicals was demonstrated by application of genetically engineered microorganisms in which genes encoding hydrolases, laccases, P450s, etc., have been transferred and expressed. Some of genetically modified organisms having ability to degrade highly toxic chemicals are *P. pseudoalcaligenes, Streptomyces lividans, Escherichia coli, Yarrowia lipolytica,* and *Pichia pastoris* (Adrio et al., 2010). Plants generally lack catabolic pathways for complete degradation of pesticides. To overcome this limitation, transgenic lines of plants were generated which expresses pesticides degrading genes. These plants have enhanced potential to degrade organic chemicals. From previous reports, an extracellular fungal enzyme such as lacasse from *Coriolus versicolor* was introduced into *Nicotiana tabaccum* plants. The transgenic tobacco plants were able to remove efficiently pentachlorophenol by secreting laccase into rhizosphere zone. Similarly, a bacterial organophosphorus hydrolase was expressed in tobacco plants for the detoxification of organophosphorus pesticide. The transgenic plants have the potential to degrade more than 99% of methylparathion within a period of 14 days of growth (Sonaki et al., 2005; Wang et al., 2008). Moreover, a number of transgenic plant species have been produced and genetically modified to overexpress glutathione S-transferase. These plants have gained immense attention in remediation process due to their critical role in detoxifying hazardous chemicals from contaminated lands. Transgenic poplars overexpressing GSH gene have shown the higher potential for uptake and detoxification of heavy metals and pesticides (Hussain et al., 2009). Hence, decontamination of agricultural lands using transgenic lines of biological organisms is eco-friendly and highly promising approach to enhance biodegradation.

7.8   CONCLUSION AND FUTURE PERSPECTIVE

For the degradation of pesticide residues, different techniques have been employed. Among the existing techniques there are chemical, physical treatments, for instance, adsorption, oxidations of mainly the hydroxyl radical, hydroxylations, group transfer, and so forth. These conventional physico-chemical approaches are generally expensive, less efficient, labor-intensive, and remediation process is often ineffective due to the partial conversion of the desired/targeted compounds to different products which are less persistent and equally nontoxic for organisms. Use of biological organisms is an efficient method for the removal of hazardous pollutants as it becomes the safest and easiest method. Detoxification of pesticides via biological agents is an important process and to understand the chemical nature pesticides, molecular mechanisms of enzymatic action will create new avenue to design tools and techniques for the degradation of pesticide residues which are accumulated in contaminated lands. Bioremediation process overcomes the limitations of traditional techniques for mitigation of toxic compounds, so it has allowed the destruction of many highly toxic organic chemicals at a reduced cost. Moreover, implementing several strategies to enhance degradation process, such as cell immobilization of bacteria or fungi, genetic engineering, enzymatic processing, and so forth, will reduce the danger of insecticides on the environment and human health. Further studies should be carried out to understand metabolism of pesticides in plants, fungi, and microorganisms. Consequently, from the past decade, biotechnological interventions of metabolic pathways of bacteria fungi and plants have improved remediation. Moreover, understanding enzymatic activities, especially conception related to pesticide mechanism of action, resistance, tolerance, selectivity, inhibition, and environmental fate, has advanced our perceptive of pesticide science, and of plant and microbial interaction biochemistry and physiology.

KEYWORDS

- genetic manipulation
- enzymatic bioremediation
- degradation
- organophosphates
- mycoremediation
- phytoremediation

REFERENCES

Abhilash, P.C.; Singh, N. Pesticide use and application: an Indian scenario. *Journal of Hazardous Materials.* **2009**, *165*(1–3), 1–2.

Adhikari, S. Bioremediatioii of Malathion from environment for pollution control. *Research Journal of Environmental Toxicology.* **2010**, *4*(3), 147–150.

Adrio, J.L.; Demain, A.L. Recombinant organisms for production of industrial products. *Bioengineered Bugs.* **2010**, *1*(2), 116–131.

Aislabie, J.; Lloyd-Jones, G. A review of bacterial-degradation of pesticides. *Soil Research.* **1995**, *33*(6), 925–942.

Aktar, W.; Sengupta, D.; Chowdhury, A. Impact of pesticides use in agriculture: their benefits and hazards. *Interdisciplinary Toxicology.* **2009**, *2*(1):1–2.

Al-Qurainy, F.; Abdel-Megeed, A. Phytoremediation and detoxification of two organophosphorous pesticides residues in Riyadh area. *World Applied Sciences Journal.* **2009**, *6*(7), 987–998.

Andrew, A. Phytoremediation: an environmentally sound technology for pollution prevention, control and remediation in developing countries. *Educational Research and Reviews.* **2007**, *2*(7), 151–156.

Bhanti, M.; Taneja, A. Contamination of vegetables of different seasons with organophosphorous pesticides and related health risk assessment in northern India. *Chemosphere*. **2007**, *69*(1), 63–68.

Carr, R.L.; Dail, M.B.; Chambers, H.W.; Chambers, J.E. Species differences in paraoxonase mediated hydrolysis of several organophosphorus insecticide metabolites. *Journal of Toxicology*. **2015**.

Cassidy, M.B.; Lee, H.; Trevors, J.T. Environmental applications of immobilized microbial cells: a review. *Journal of Industrial Microbiology*. **1996**, *16*(2), 79–101.

Chen, S.; Luo, J.; Hu, M.; Lai, K.; Geng, P.; Huang, H. Enhancement of cypermethrin degradation by a coculture of *Bacillus cereus* ZH-3 and *Streptomyces aureus* HP-S-01. *Bioresource Technology*. **2012**, *110*, 97–104.

Corsini, E.; Liesivuori, J.; Vergieva, T.; Van Loveren, H.; Colosio, C. Effects of pesticide exposure on the human immune system. *Human and Experimental Toxicology*. **2008**, *27*(9), 671–680.

Cui, Z.; Li, S.; Fu, G. Isolation of methyl-parathion-degrading strain M6 and cloning of the methyl-parathion hydrolase gene. *Applied and Environmental Microbiology*. **2008**, *67*, 4922–4925. DOI:10.1128/AEM.67.10.4922-4925.2001.

Dai, Y.J.; Ji, W.W.; Chen, T.; Zhang, W.J.; Liu, Z.H.; Ge, F.; Yuan, S. Metabolism of the neonicotinoid insecticides acetamiprid and thiacloprid by the yeast *Rhodotorula mucilaginosa* strain IM-2. *Journal of Agricultural and Food Chemistry*. **2010**, *58*(4), 2419–2425.

Fahr, K.; Wetzstein, H.G.; Grey, R.; Schlosser, D. Degradation of 2, 4-dichlorophenol and pentachlorophenol by two brown rot fungi. FEMS *Microbiology Letters*. **1999**, *175*(1), 127–132.

Fernández-López, M.G.; Popoca-Ursino, C.; Sánchez-Salinas, E.; Tinoco-Valencia, R.; Folch-Mallol, J.L.; Dantán-González, E.; Laura Ortiz-Hernández, M. Enhancing methyl parathion degradation by the immobilization of *Burkholderia* sp. isolated from agricultural soils. *Microbiology Open*. **2017**, *6*(5), e00507.

Foster, R.L.J.; Kwan, B.H.; Vancov, T. Microbial degradation of the organophosphate pesticide, Ethion. FEMS *Microbiology Letters*. **2004**, *240*(1), 49–53.

Fragoeiro, S.; Magan, N. Impact of *Trametes versicolor* and P*hanerochaete chrysosporium* on differential breakdown of pesticide mixtures in soil microcosms at two water potentials and associated respiration and enzyme activity. *International Biodeterioration and Biodegradation*. **2008**, *62*(4), 376–383.

Hasan, H.A.H. Fungal utilization of organophosphate pesticides and their degradation by *Aspergillus flavus* and *A. sydowii* in soil. *Folia Microbiologica*. **1999**, *44*, 77–84.

Hatton, P.J.; Cummins, I.; Cole, D.J.; Edwards, R. Glutathione transferases involved in herbicide detoxification in the leaves of *Setaria faberi* (giant foxtail). *Physiologia Plantarum*. **1999**, *105*(1), 9–16.

Henderson, K.L.; Belden, J.B.; Coats, J.R. Mass balance of metolachlor in a grassed phytoremediation system. *Environmental Science and Technology*. **2007**, *41*(11), 4084–4089.

Horne, J.; Sutherland, T.D.; Harcourt, R.L.; Russell, R.J.; Oakesthott, J.G. Identification of an opd (organophosphate degradation) gene in an Agrobacterium isolate. *Applied and Environmental Microbiology*. **2002**, *68*, 3371–3376.

Hussain, S.; Hartley, C.J.; Shettigar, M.; Pandey, G. Bacterial biodegradation of neonicotinoid pesticides in soil and water systems. FEMS *Microbiology Letters*. **2016**, *363*(23).

Hussain, S.; Siddique, T.; Arshad, M.; Saleem, M. Bioremediation and phytoremediation of pesticides: recent advances. *Critical Reviews in Environmental Science and Technology.* **2009**, *39*(10), 843–907.

Iqbal, M.; Zafar, S.I. Palm petiolar felt-sheath as a new and convenient material for the immobilization of microalgal cells. *Journal of Industrial Microbiology and Biotechnology.* **1997**, *19*(2), 139–141.

Jiang, J.D.; Zhang, R.F.; Li, R.; Gu, J.D.; Li, S.P. Simultaneous biodegradation of methyl parathion and carbofuran by a genetically engineered microorganism constructed by mini-Tn5 transposon, *Biodegradation.* **2007**, *18*, 403–412.

Johri, A.K.; Dua, M.; Tuteja, D.; Saxena, R.; Saxena, D.M.; Lal, R. Genetic manipulations of microorganisms for the degradation of hexachlorocyclo hexane. *FEMS Microbiology Reviews.* **1996**, *19*(2), 69–84.

Kadakol, J.C.; Kamanavalli, C.M.; Shouche, Y. Biodegradation of Carbofuran phenol by free and immobilized cells of *Klebsiella pneumoniae* ATCC13883T. *World Journal of Microbiology and Biotechnology.* **2011**, *27*(1), 25–29.

Kamal, M.Z.; Nashwa, A.H.; Fetyan, A.; Ibrahim, A.M.; Sherif, E.N. Biodegradation and detoxification of malathion by of *Bacillus thuringiensis* MOS-5. *Australian Journal of Basic and Applied Sciences.* **2008**, *2*, 724–732.

Kang, D.G.; Kim, J.Y.H.; Cha, H.J. Enhanced detoxification of organophosphates using recombinant Escherichia coli with co-expression of organophosphorus hydrolase and bacterial hemoglobin. *Biotechnology Letters.* **2002**, *24*, 879–883.

Karigar, C.S.; Rao, S.S. Role of microbial enzymes in the bioremediation of pollutants: a review. *Enzyme Research.* **2011**, *2011*. https://doi.org/10.4061/2011/805187

Kaur, H.; Kapoor, S.; Kaur, G. Application of ligninolytic potentials of a white-rot fungus *Ganoderma lucidum* for degradation of lindane. *Environmental Monitoring and Assessment.* **2016**, *188*(10), 588.

Khan, S.; Zaffar, H.; Irshad, U.; Ahmad, R.; Khan, A.R.; Shah, M.M.; Bilal, M.; Iqbal, M.; Naqvi, T. Biodegradation of malathion by *Bacillus licheniformis* strain ML-1. *Archives of Biological Sciences.* **2016**, *68*(1), 51–59.

Kim, Y.K.; Kim, S.H.; Choi, S.C. Kinetics of endosulfan degradation by *Phanerochaete chrysosporium. Biotechnology Letters.* **2001**, *23*(2), 163–166.

Kulshrestha, G.; Kumari, A. Fungal degradation of chlorpyrifos by *Acremonium* sp. strain (GFRC-1) isolated from a laboratory-enriched red agricultural soil. *Biology and Fertility of Soils.* **2011**, *47*(2), 219–225.

Latifi, A.M.; Khodi, S.; Mirzaei, M.; Miresmaeili, M.; Babavalian, H. Isolation and characterization of five chlorpyrifos degrading bacteria. *African Journal of Biotechnology.* **2012**, *11*(13), 3140–3146

Ma, Y.; Zhai, S.; Mao, S.Y.; Sun, S.L.; Wang, Y.; Liu, Z.H.; Dai, Y.J.; Yuan, S. Co-metabolic transformation of the neonicotinoid insecticide imidacloprid by the new soil isolate *Pseudoxanthomonas indica* CGMCC 6648. *Journal of Environmental Science and Health, Part B.* **2014**, *49*(9), 661–70.

Marcacci, S. A phytoremediation approach to remove pesticides (atrazine and lindane) from contaminated environment. *EPFL.* **2004**.

Mishra, V.K.; Upadhyay, A.R.; Tripathi, B.D. Bioaccumulation of heavy metals and two organochlorine pesticides (DDT and BHC) in crops irrigated with secondary treated waste water. *Environmental Monitoring and Assessment.* **2009**, *156*(1–4), 99–107.

Moore, M.T.; Tyler, H.L.; Locke, M.A. Aqueous pesticide mitigation efficiency of *Typha latifolia* (L.), *Leersia oryzoides* (L.) Sw., and *Sparganium americanum* Nutt. *Chemosphere.* **2013**, *92*(10), 1307–1313

Mulla, S.I.; Talwar, M.P.; Bagewadi, Z.K.; Hoskeri, R.S.; Ninnekar, H.Z. Enhanced degradation of 2-nitrotoluene by immobilized cells of Micrococcus sp. strain SMN-1. *Chemosphere.* **2013**, *90*(6), 1920–1924.

Mulla, S.I.; Talwar, M.P.; Hoskeri, R.S.; Ninnekar, H.Z. Enhanced degradation of 3-nitrobenzoate by immobilized cells of *Bacillus flexus* strain XJU-4. *Biotechnology and Bioprocess Engineering.* **2012**, *17*(6), 1294–1299.

Nyakundi, W.O.; Magoma, G.; Ochora, J.; Nyende, A.B. Biodegradation of diazinon and methomyl pesticides by white rot fungi from selected horticultural farms in rift valley and central provinces, Kenya. In *Scientific Conference Proceedings.* **2012**.

Ortiz-Hernández, M.L.; Sanchez-Salinas, E.; Castrejon Godinez, M.L.; Dantan Gonzalez, E.; Popoca Ursino, E.C. Mechanisms and strategies for pesticide biodegradation: opportunity for waste, soils and water cleaning. *Revista Internacional de Contaminación Ambiental.* **2013**, *29.*

Ortiz-Hernández, M.L.; Sánchez-Salinas, E.; Dantán-González, E.; Castrejón-Godínez, M.L. Pesticide biodegradation: mechanisms, genetics and strategies to enhance the process. In *Biodegradation-life of Science.* **2013**, Intech Open.

Pal, R.; Chakrabarti, K.; Chakraborty, A.; Chowdhury, A. Degradation and effects of pesticides on soil microbiological parameters-A review. *International Journal of Agricultural Research.* **2006**, *1*(33), 240–258.

Pandey, G.; Dorrian, S.J.; Russell, R.J.; Oakeshott, J.G. Biotransformation of the neonicotinoid insecticides imidacloprid and thiamethoxam by *Pseudomonas* sp. 1G. *Biochemical and Biophysical Research Communications.* **2009**, *380*(3), 710–714.

Pascal-Lorber, S.; Laurent, F. Phytoremediation techniques for pesticide contaminations. In *Alternative Farming Systems, Biotechnology, Drought Stress and Ecological Fertilisation.* **2011**, 77–105. Springer, Dordrecht.

Ramírez-Sandoval, M.; Melchor-Partida, G.N.; Muñiz-Hernández, S.; Girón-Pérez, M.I.; Rojas-García, A.E.; Medina-Díaz, I.M.; Robledo-Marenco, M.L.; Velázquez-Fernández, J. Phytoremediatory effect and growth of two species of Ocimum in endosulfan polluted soil. *Journal of Hazardous Materials.* **2011**, *192*(1), 388–392.

Sarma, H. Metal hyperaccumulation in plants: a review focusing on phytoremediation technology. *Journal of Environmental Science and Technology.* **2011**, *4*(2), 118–138.

Scott, C.; Pandey, G.; Hartley, C.J.; Jackson, C.J.; Cheesman, M.J.; Taylor, M.C.; Pandey, R.; Khurana, J.L.; Teese, M.; Coppin, C.W.; Weir, K.M. The enzymatic basis for pesticide bioremediation. *Indian Journal of Microbiology.* **2008**, *1*, 48(1), 65.

Sethunathan, N.; Yoshida, T. A *Flavobacterium* sp. that degrades diazinon and parathion. *Canadian Journal of Microbiology.* **1973**, *19*, 873–875. DOI:10.1139/m73-138.

Siddique, T.; Okeke, B.C.; Arshad, M.; Frankenberger, W.T. Biodegradation kinetics of endosulfan by *Fusarium ventricosum* and a *Pandoraea* species. *Journal of Agricultural and Food Chemistry.* **2003**, *51*(27), 8015–8019.

Sonoki, T.; Kajita, S.; Ikeda, S.; Uesugi, M.; Tatsumi, K.; Katayama, Y.; Iimura, Y. Transgenic tobacco expressing fungal laccase promotes the detoxification of environmental pollutants. *Applied Microbiology and Biotechnology.* **2005**, *67*, 138–142.

Stomp, A.M.; Han, K.H.; Wilbert, S.; Gordon, M.P.; Cunningham, S.D. Genetic strategies for enhancing phytoremediation. *Annals of the New York Academy of Sciences.* **1994**, *721*(1), 481–491.

Su, C.; Jiang, L.; Zhang, W. A review on heavy metal contamination in the soil worldwide: Situation, impact and remediation techniques. *Environmental Skeptics and Critics.* **2014**, *3*, 24–38.

Sutherland, T.D.; Horne, I.; Weir, K.M.; Coppin, C.W.; Williams, M.R.; Selleck, M.; Russell, R.J.; Oakeshott, J.G. Enzymatic bioremediation: from enzyme discovery to applications. *Clinical and Experimental Pharmacology and Physiology.* **2004**, *31*(11), 817–821.

Tallur, P.N.; Mulla, S.I.; Megadi, V.B.; Talwar, M.P.; Ninnekar, H.Z. Biodegradation of cypermethrin by immobilized cells of Micrococcus sp. strain CPN 1. *Brazilian Journal of Microbiology.* **2015**, *46*(3), 667–672.

Talwar, M.P.; Ninnekar, H.Z. Biodegradation of pesticide profenofos by the free and immobilized cells of *Pseudoxanthomonas suwonensis* strain HNM. *Journal of Basic Microbiology.* **2015**, *55*(9), 1094–1103.

Velázquez-Fernández, J.B.; Martínez-Rizo, A.B.; Ramírez-Sandoval, M.; Domínguez-Ojeda, D. Biodegradation and bioremediation of organic pesticides. In *Pesticides-recent Trends in Pesticide Residue Assay.* **2012**, Intech Open.

Wang, B.E.; Hu, Y.Y. Comparison of four supports for adsorption of reactive dyes by immobilized *Aspergillus fumigates* beads. *Journal of Environmental Sciences.* **2007**, *19*,451–457

Wang, X.; Wu, N.; Guo, J.; Chu, X.; Tian, J.; Yao, B.; Fan, Y. Phytodegradation of organophosphorus compounds by transgenic plants expressing abacterial organophosphorus hydrolase. *Biochemical and Biophysical Research Communications.* **2008**, *365*(3), 453–458.

Xie, S.; Liu, J.; Li, L.; Qiao, C. Biodegradation of malathion by *Acinetobacter johnsonii* MA19 and optimization of cometabolism substrates. *Journal of Environmental Sciences.* **2009**, *21*, 76–82.

Xie, W.; Meng, Q.S.; Wu, Q.J.; Wang, S.L.; Yang, X.; Yang, N.N.; Li, R.M.; Jiao, X.G.; Pan, H.P.; Liu, B.M.; Su, Q. Pyrosequencing the *Bemisia tabaci* transcriptome reveals a highly diverse bacterial community and a robust system for insecticide resistance. *PLoS One.* **2012**, *7*(4):e35181.

Yu, Y.L.; Fang, H.; Wang, X.; Wu, X.M.; Shan, M.; Yu, J.Q. Characterization of a fungal strain capable of degrading chlorpyrifos and its use in detoxification of the insecticide on vegetables. *Biodegradation.* **2006**, *17*(5), 487–494.

Zhang, B.; Bai, Z.; Hoefel, D.; Tang, L.; Wang, X.; Li, B.; Li, Z.; Zhuang, G. The impacts of cypermethrin pesticide application on the non-target microbial community of the pepper plant phyllosphere. *Science of the Total Environment.* **2009**, *407*(6), 1915–1922.

Zhang, C.; Wang, S.; Yan, Y. Isomerization and biodegradation of beta-cypermethrin by Pseudomonas aeruginosa CH7 with biosurfactant production. *Bioresource Technology.* **2011**, *102*(14), 7139–7146.

Zhang, Y.; An, J.; Ye, W.; Yang, G.; Qian, Z.G.; Chen, H.F.; Cui, L.; Feng, Y. Enhancing the promiscuous phosphotriesterase activity of a thermostable lactonase (GkaP) for the efficient degradation of organophosphate pesticides. *Applied and Environmental Microbiology.* **2012**, *78*(18), 6647–6655.

Zheng, C.; Zhou, J.; Wang, J.; Qu, B.; Lu, H.; Zhao, H. Aerobic degradation of nitrobenzene by immobilization of *Rhodotorula mucilaginosa* in polyurethane foam. *Journal of Hazardous Materials.* **2009**, *168*(1), 298–303.

Zhou, L.Y.; Zhang, L.J.; Sun, S.L.; Ge, F.; Mao, S.Y.; Ma, Y.; Liu, Z.H.; Dai, Y.J.; Yuan, S. Degradation of the neonicotinoid insecticide acetamiprid via the N-carbamoylimine derivate (IM-1-2) mediated by the nitrile hydratase of the nitrogen-fixing bacterium *Ensifer meliloti* CGMCC 7333. *Journal of Agricultural and Food Chemistry.* **2014**, *62*(41), 9957–9964.

CHAPTER 8

Microbiological Aspects of Pesticide Remediation in Freshwater and Soil Environs

REZWANA ASSAD,* IRSHAD AHMAD SOFI, IQRA BASHIR, IFLAH RAFIQ, ZAFAR AHMAD RESHI, and IRFAN RASHID

Department of Botany, University of Kashmir, Srinagar 190006, Jammu and Kashmir, India

Corresponding author. E-mail: rezumir@gmail.com

ABSTRACT

Pesticides usage has gradually augmented and evolved ever since their initial application in the middle of the 20th century. Currently, usage of pesticides is an indispensable tool for escalating the crop yield as well as protection of economically important crops. Conversely, these pesticides are a major cause of pollution in aquatic and soil ecosystems and most of the pesticides belong to groups that are generally persistent and toxic. Persistent pesticide residues that accumulate and magnify over time, have adverse impacts on environmental and human health. Subsequently, systematic surveillance and management of pesticide utilization is crucial. Conventional methods aimed to eradicate pesticide residues are inefficient and presently bioremediation of these pollutants is a theme of attention with reference to ecological cleaning. This chapter gives an overview of potential "microbioremediation" strategies and use of potential microalgal–bacterial–fungal consortia for recuperation of pesticide polluted freshwater and soil environments. Information about potential microbial consortia that can be availed for remediation of range of pesticides can serve as baseline data for formulation of various remediation strategies and pesticide management projects.

8.1 INTRODUCTION

Occurrence of environmental pollution due to certain anthropogenic activities like increasing urbanization, technological advancement, rapid industrialization, precarious agricultural practices, and swift population expansion has been the focus of attention for environmentalists and conservationists worldwide. Among these activities, rapid population growth necessitated increase in the productivity of crops. Although green revolution increased crop productivity but simultaneously it led to excessive and unsystematic application of chemicals especially pesticides in public health and agricultural sectors. This undue and unsystematic use of pesticides instigated pollution of freshwater and soil environments.

Pesticides ("Pest": "harmful undesirable organism" and "cide": "kill") encompass algicides (eradicate algae), antifoliants, avicides (eradicate birds), bactericides (eradicate bacteria), disinfectants, fungicides (eradicate fungi), herbicides (eradicate weeds), insect and animal repellents, insecticides (eradicate insects), miticides (eradicate mites), molluscicides (eradicate snails), nematicides (eradicate nematodes), piscicides (eradicate fishes), rodenticides (eradicate rodents), and viricides (eradicate viruses). As stated by Food and Agriculture Organization, pesticide is any substance premeditated to manage disease vectors for animals and humans, in addition to pests menacing agrarian and/or industrial produce (Li and Jennings, 2017).

8.2 ECOLOGICAL AND HEALTH RISKS OF PESTICIDES

Even though, utilization of pesticides improves food security by increasing crop yield and reducing postharvest loses of economically important crops (Bonner and Alavanja, 2017) but Pimentel (1995) reported that out of all used pesticides, merely 0.1% arrive at their target life form and rest unused 99.9% enter the environment, which has resulted in making ecosystems unstable, resistance development in insects, elimination of parasites, predators, pollinators and decomposers, resurgence of minor pests, and destruction of useful insects (Tewari et al., 2012; Singh et al., 2014). Additionally, long term pesticide contamination destructs soil and water microbial diversity (Megharaj et al., 2000). Due to the dearth of a comprehensible pesticides management approach, large amounts of pesticide residues have stockpiled over the time (Dasgupta et al., 2010).

Soil and water ecosystems act as important sinks for unnecessary and excessive pesticides. Even after proper application, several pesticides penetrate the ecosystem through runoff incidents which subsequently leads to the pollution of both aquatic environments and soil systems. Subsequently, the fate of these chemicals is often vague since these can pollute other areas that are far away from the zones where these were actually applied. Therefore, restoration of areas contaminated with pesticides is an extremely difficult task (Gavrilescu, 2005).

Disproportionate use and high persistence rates of pesticides result in environmental pollution which often leads to public health crises. The lipophilic attribute of various pesticides can result in their accumulation in living organisms and their consequent rampant upright transport via food chain leads to bioaccumulation and biomagnification (Nayak et al., 2018; FAO, 2019). Exposure to widely used pesticides can cause cancer, neurological disorders, endocrine disruption, cardiopulmonary disorders, geno and cytotoxicity, dysfunctioning of immune system, skin diseases, kidney failures, fatal deformities, reproductive disorders, miscarriages, and even mortality (Kamel and Hoppin, 2004; Ortiz-Hernández et al., 2013; McLellan et al., 2019). Despite the known fact that the persistence of pesticides in water and soil systems has caused global environmental security, global public health, and socio-economic concerns (Chaussonnerie et al., 2016) still the use of these chemicals cannot be stopped entirely.

8.3 SUSTAINABLE REMEDIATION APPROACHES

With the ever-increasing environmental concerns, decontamination and recovery of pesticide-polluted sites in a safe, efficient, and economical way has become universal precedence. Conventional remediation approaches are pricey, environmentally perilous, mostly inefficient, and inefficacious. Bioremediation or biological remediation practices have been investigated for decades as an alternative sustainable environmental clean-up technology. In bioremediation technique microorganisms, fungi, plants, or only their certain enzymes are employed to detoxify and disintegrate the environmental contaminants (Tewari et al., 2012). It has emerged as an economical, eco-friendly, efficient, and most promising approach for remediating and cleaning pesticides-polluted ecosystems (Tewari et al., 2012). Consequently, this technology has evolved as a powerful practical technology for the remediation of diverse pollutants (Ortiz-Hernández et al., 2013).

Bioremediation can be achieved in three ways: (1) Bioattenuation, wherein the usual practice of degradation is employed; (2) Biostimulation, wherein electron donors or acceptors, water, and nutrients are used for intentional stimulation of pesticide degradation; and (3) Bioaugmentation, where microbes with confirmed potential of degrading pesticides are added (Cycoń et al., 2017).

8.4 MICROBIOLOGICAL ASPECTS OF PESTICIDE REMEDIATION

8.4.1 APPLICATION OF ALGAL–BACTERIAL–FUNGAL CONSORTIA

In the sphere of ecological restoration science, it is crucial to investigate degradation of pesticides by employing microbes in freshwater and soil environments. Microbioremediation, an effective tool for pollution control and sustainable development, engrosses bacterial bioremediation, mycoremediation, and phycoremediation. These involve the usage of bacteria, fungi, and microalgae, respectively, for cleaning of contaminated environment. This method relies on the potential of microbes to convert pollutants into harmless/less toxic compounds (Rani and Dhania, 2014). This chapter draws light on microorganisms possessing the pesticide degradation potential, either alone or in consortia and presents an outline of microbioremediation strategies and recuperation of pesticide polluted freshwater and soil environments by means of potential microalgal–bacterial–fungal consortia (Figure 8.1). The correlation between the three kingdoms "algae, bacteria and fungi" provides a unique prospect to utilize this mutualism as polymicrobial association for pesticide bioremediation. However, further in-depth study is required to elucidate these associations. Benefit that is acquired by bacteria and fungi by an algal mutualism is the addition of substitute sources of carbon by the microalgae, hence promoting the remediation ability of bacteria and fungi by escalating their quantity (number) and biomass.

Traditionally microalgae, bacteria, and fungi were used independently for remediation purposes but overtime use of consortia was reported to be mega effectual in several bioaugmentation research projects (Fu and Secundo, 2016). In consortia, microalgae may be proficient in accumulating and removing one kind of contaminant, while bacteria and fungi can target other types of contaminants from aquatic and soil ecosystems, thereby enhancing the overall efficacy of bioremediation. Consortial approach with broader biological spectrum strengthens the sustenance and overall competitive ability of consortia. Consequently, the combination of catabolic pathways of different microorganisms can facilitate their synergistic and harmonious growth under unfavorable conditions and also prove more effective in treating the polluted sites (Pino and Peñuela, 2011).

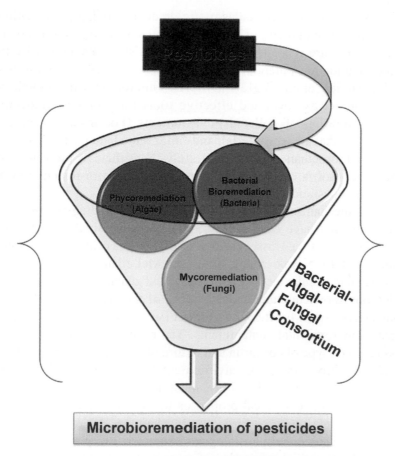

FIGURE 8.1   Microbioremediation of pesticides using bacterial–algal–fungal consortium.

Subsequently, there are numerous studies wherein algae–bacteria or algae–fungi consortia have been used for pesticide bioremediation purposes (Fu and Secundo, 2016) but, this is the first attempt of its kind wherein potential algae–bacteria–fungal consortia for remediation of a range of pesticide in freshwater and soil environs has been highlighted. The permutation of microalgae, bacteria, and fungi has immense potential in the pesticide remediation technology. Indeed, the biodegradation by means of multitaxon (algal–bacterial–fungal) consortia is a perfect self-sustaining approach, in comparison to traditional engineering technologies, that have numerous drawbacks like the formation of secondary contaminants, uneconomical, and practical inapplicability under a few circumstances (Subashchandrabose et al., 2013).

Microorganisms, recognized as pesticide degraders have been cultured and segregated from range of sites polluted by several types of pesticides. Currently, there are repositories of microorganisms categorized by their pesticide degradation potential in different laboratories across the world (Ortiz-Hernández et al., 2013). Different microorganisms together in consortia can act as novel and effective tools for pollutant biodegradation and restoration of polluted environments. For instance, there are several reports that algae, bacterial, and fungi remediate pesticide dichloro diphenyl trichloroethane (DDT) separately. Nonetheless, these microbes together in consortia can prove to be more effective in remediating this deadly banned pesticide.

Detailed mechanism of microbioremediation including pathways and enzymes involved have been discussed in detail by Singh (2008), Ortiz-Hernández et al. (2013), Hussein et al. (2016); Sharma et al. (2016), Hultberg and Bodin (2018), Nayak et al. (2018), and McLellan et al. (2019). Novel analytical and molecular techniques like the next generation sequencing have extended our knowledge and understanding of the mechanisms (how) and the characterization of functional performer (who) that cause biodegradation of organic environmental contaminants. The pace of pesticide degradation process relies on type of consortia and nature of pesticide and can be further affected by a broad range of environmental factors like light intensity, oxygen tension, nutrients, pH, salinity, substrate bioavailability, temperature, humidity, and physicochemical properties like chemical structure, concentration, molecular weight, and toxicity. Furthermore, several biochemical, ecological, molecular, and physiological characteristics play an imperative function in the microbial conversion of these pollutants (Ortiz-Hernández et al., 2013).

Various potential microbial consortia that can be employed as an effectual means for pesticide remediation in freshwater and soil environs are discussed in Table 8.1. This information can be used for remediation of range of pesticides and can serve as baseline data for various remediation strategies and pesticide management projects.

TABLE 8.1 Potential Microbial Consortiums for Pesticide Remediation in Freshwater and Soil Environs

S. No.	Pesticide	Algae	References	Bacteria	References	Fungi	References
1.	Acephate	*Chlamydomonas mexicana*	Kumar et al. (2011)				
2.	Acibenzolar-S-methyl			*Bacillus pumilus*	Myresiotis et al. (2012)		
3.	Alachlor	*Chlorella* sp. *Scenedesmus* sp.	Matamoros and Rodriguez (2016)			*Phanerochaete chrysosporium*	McFarland et al. (1996)
4.	Aldicarb			*Achromobacter* sp. *Arthrobacter* sp. *Nocardia* sp. *Pseudornonas* sp.	Karns et al. (1986) Read (1987)		
				Pseudomonas sp.	Chaudhry and Ali (1988)		
5.	Aldrin			*Bacillus* sp. *Micrococcus* sp. *Pseudomonas* sp.	Patil et al. (1970)	*Aspergillus niger*	Korte and Porter (1970)
						Trichoderma viridae	Patil et al. (1970)
				Bacillus sp. *Micromonospora* sp. *Mycobacterium* sp. *Nocardia* sp. *Pseudomonas* sp. *Streptomyces* sp. Thermoactinomyces	Ferguson and Korte (1977)		

TABLE 8.1 *(Continued)*

S. No.	Pesticide	Algae	References	Bacteria	References	Fungi	References
				Pseudomonas fluorescens	Erick et al. (2006)		
6.	Aniline			*Pseudomonas multivorans*	Helm and Reber (1979)		
				Alcaligenes faecalis	Surovtseva and Vol'nova (1980)		
				Rhodococcus erythropolis	Aoki et al. (1983)		
				Pseudomonas sp.	Kaminski et al. (1983) Anson and Mackinnon (1984)		
				Moraxela sp.	Zeyer et al. (1985)		
7.	Anthracene					*Acremonium* sp.	Ma et al. (2014)
8.	Atrazine	*Scenedesmu obliquus*	Mofeed and Mosleh (2013)	*Arthrobacter nicotinovorans*	Aislabie et al. (2005)	*Cerrena maxima Coriolus hirsutus Coriolopsis fulvocinerea*	Koroleva et al. (2001)
		Chlamydomonas mexicana	Kabra et al. (2014)	*Cryptococcus laurentii*	Evy et al. (2012)	*Phanerochaete chrysosporium*	Reddy and Mathew (2001)

TABLE 8.1 *(Continued)*

S. No.	Pesticide	Algae	References	Bacteria	References	Fungi	References
		Chlorella vulgaris	Hussein et al. (2016)	Pichia kudriavzevii	Evy et al. (2013)	Agrocybe semiorbicularis Auricularia auricola Coriolus versicolor Dichotomitus squalens Flammulina velupites Hypholoma fasciculare Phanerochaete velutina Pleurotus ostreatus Stereum hirsutum	Bending et al. (2002)
				Raoultella planticola	Swissa et al. (2014)	Pleurotus ostreatus	Pereira et al. (2013)
				Bacillus sp. Burkholderia sp. Pseudomonas alcaligenes Pseudomonas sp.	Dutta et al. (2016)		

TABLE 8.1 *(Continued)*

S. No.	Pesticide	Algae	References	Bacteria	References	Fungi	References
9.	Barban				Marty et al. (1986)		
10.	Bensulfuron-methyl					*Penicillium pinophilum*	Peng et al. (2012)
11.	Benzo[k]fluoranthene					*Fusarium flocciferum* *Pleurotus ostreatus* *Trametes versicolor* *Trichoderma* sp.	Baldrian et al. (2008)
12.	Benzo[a]anthracene					*Fusarium flocciferum* *Pleurotus ostreatus* *Trametes versicolor* *Trichoderma* sp.	Baldrian et al. (2008)
						Pleurotus ostreatus	Bhattacharya et al. (2014)

TABLE 8.1 *(Continued)*

S. No.	Pesticide	Algae	References	Bacteria	References	Fungi	References
13.	Benzo[a]pyrene					*Fusarium flocciferum* *Pleurotus ostreatus* *Trametes versicolor* *Trichoderma* sp.	Baldrian et al. (2008)
14.	Benzo[ghi]perylene					*Fusarium flocciferum* *Pleurotus ostreatus* *Trametes versicolor* *Trichoderma* sp.	Baldrian et al. (2008)
15.	1-Methyl-2-propynyl-*m*-chlorocarbonilate			*Pseudomonas cepacia*	Vega et al. (1985)		
16.	Bisphenol	*Agmenellum quadruplicatum*	Cerniglia et al. (1979)				
		Monoraphidium braunii	Gattullo et al. (2012)				
17.	Butylate			*Flavobacterium* sp.	Mueller et al. (1988)		

TABLE 8.1 *(Continued)*

S. No.	Pesticide	Algae	References	Bacteria	References	Fungi	References
18.	Carbamate					*Sphingomona syanoikayae*	Cases et al. (2005)
19.	Carbaryl			*Pseudomonas aeruginosa* *Bacillus* sp. *Micrococcus* sp. *Pseudomonas* sp. *Rhodococcus* sp.	Chang and Sun (1981) Rajagopal et al. (1984) Larkin and Day (1986)		
20.	Carbofuran	*Chlorella vulgaris*	Hultberg et al. (2016) Hussein et al. (2016) Hultberg and Bodin (2018)	*Achromobacter* sp. *Pseudomonas* sp. *Arthrobacter* sp. *Bacillus* sp. *Micrococcus* sp. *Azospirillum lipoferum* *Pseudomonas cepacia* *Streptomyces* sp.	Felsot et al. (1981) Felsot et al. (1982) Rajagopal et al. (1984) Venkateswarlu and Sethunathan (1984)	*Aspergillus niger*	Hultberg and Bodin (2018)

TABLE 8.1 (Continued)

S. No.	Pesticide	Algae	References	Bacteria	References	Fungi	References
				Nocardia sp.	Venkateswarlu and Sethunathan (1985)		
				Achromobacter sp.	Karns et al. (1986)		
				Flavobacterium sp. Pseudomonas sp.	Chaudhry and Ali (1988)		
				Citrobacter sp.	Chaussonnerie et al. (2016)		
21.	Carfentrazone-ethyl	Chlorella vulgaris	Hultberg et al. (2016) Hultberg and Bodin (2018)	Pseudomonas cepacia	Vega et al. (1985)	Aspergillus niger	Hultberg and Bodin (2018)
22.	Chlordecone			Moraxela sp.	Zeyer et al. (1985)		
23.	3-Chloroaniline			Micrococcus sp.	Mallick et al. (1999)		
				Pseudomonas sp.	Singh et al. (2003a)		
24.	Chlorpyrifos	Chlorella sp. Scenedesmus sp.	Matamoros and Rodriguez (2016)	Enterobacter sp.	Singh et al. (2004)	Cladosporium cladosporioides Ganoderma sp.	Chen et al. (2015)

TABLE 8.1 *(Continued)*

S. No.	Pesticide	Algae	References	Bacteria	References	Fungi	References
				Alcaligenes faecalis	Yang et al., (2005)	*Streptomyces* sp.	Fuentes et al. (2013)
				Klebsiella sp.	Ghanem et al. (2007)	*Aspergillus terreus*	Silambarasan and Abraham (2013)
				Sphingomonas sp.	Li et al. (2007)	*Aspergillus flavus*	Kurniati et al. (2014)
				Serratia sp.	Xu et al. (2007)	*Pleurotus coccineus Pleurotus gigantea Trametes versicolor*	Gouma et al. (2019)
				Verticillium sp.	Fang et al. (2008)		
				Pseudomonas aeruginosa	Lakshmi et al. (2008)		
				Paracoccus sp.	Xu et al. (2008)		
				Bacillus pumilus	Anwar et al. (2009)		

TABLE 8.1 *(Continued)*

S. No.	Pesticide	Algae	References	Bacteria	References	Fungi	References
				Pseudomonas nitroreducens	Korade and Fulekar (2009)		
				Bacillus licheniformis	Zhu et al. (2010)		
				Pseudomonas stutzeri	Awad et al. (2011)		
				Bacillus subtilis *Micrococcus luteus* *Pseudomonas aeruginosa*	Bhuimbar et al. (2011)		
				Acremonium sp.	Kulshrestha and Kumari (2011)		
				Synechocystis sp. (cyanobacterium)	Singh et al. (2011)		
				Cellulomonas fimi *Phanerochaete chrysosporium*	Barathidasan et al. (2014)		
				Bacillus sp. *Penicillium* sp. *Streptomyces thermocarboxydus*	Fawzy et al. (2014)		

TABLE 8.1 *(Continued)*

S. No.	Pesticide	Algae	References	Bacteria	References	Fungi	References
				Azospirillum lipoferum *Paenibacillus polymyxa*	Romeh and Hendawi (2014)		
25.	Chlortoluron			*Pseudomonas alcaligenes*	Marty et al. (1986)	*Mortierella* sp.	Badawi et al. (2009)
26.	Chrysene			*Pseudomonas cepacia*	Vega et al. (1985)	*Fusarium flocciferum* *Pleurotus ostreatus* *Trametes versicolor* *Trichoderma* sp.	Baldrian et al. (2008)
27.	Chlorpropham [isopropyl-*N*-(3-chlorophenyl) carbamate]			*Pseudomonas cepacia*	Vega et al. (1985)		
				Pseudomonas alcaligenes	Marty et al. (1986)		
28.	Citalopram					*Bjerkandera adusta* *Bjerkandera* sp. *Phanerochaete chrysosporium*	Rodarte-Morales et al. (2011)

TABLE 8.1 *(Continued)*

S. No.	Pesticide	Algae	References	Bacteria	References	Fungi	References
29.	Clomazone	*Chlorella vulgaris*	Hultberg et al. (2016)				
30.	Chlorophenyl ether	*Aulacoseira granulata Microcystis aeruginosa Scenedesmus quadricauda*	Guanzon et al. (1996)				
31.	Cyanazine	*Chlorella vulgaris*	Hultberg et al. (2016) Hultberg and Bodin (2018)			*Aspergillus niger*	Hultberg and Bodin (2018)
32.	Cypermethrin					*Pseudomonas auroginosa*	Bhosle and Nasreen (2013)
33.	Cyprodinil	*Chlorella vulgaris*	Ardal (2014)				
34.	Dichlorodiph-enyldichloroethane					*Trichoderma* sp.	Ortega et al. (2011)
35.	Dichlorodiph-enyldichloroethylene			*Phanerochaete chrysosporium*	Bumpus et al. (1993)		

TABLE 8.1 *(Continued)*

S. No.	Pesticide	Algae	References	Bacteria	References	Fungi	References
36.	Desmedipham			*Pseudomonas putida*	Knowles and Benezet (1981)		
37.	Diazinon			*Serratia marcescens*	Abo-Amer (2011)		
				Serratia liquefaciens *Serratia marcescens*	Hussaini et al. (2013)		
				Bacillus amyloliquefaciens *Bacillus licheniformis* *Bacillus pseudomycoides* *Pseudomonas aeruginosa*	Thabit and El-Naggar (2013)		
				Brevundimonas diminuta *Burkholderia caryophylli* *Pseudomonas peli*	Mahiudddin et al. (2014)		

TABLE 8.1 *(Continued)*

S. No.	Pesticide	Algae	References	Bacteria	References	Fungi	References
38.	Dibenzo[a,h] anthracene					*Fusarium flocciferum Pleurotus ostreatus Trametes versicolor Trichoderma* sp.	Baldrian et al. (2008)
39.	3,4-Dichloroaniline			*Pseudomonas putida*	You and Bartha (1982)		
40.	1,4-Dichlorobenzene			*Pseudomonas* sp.	Spain and Nishino (1987)		
41.	DDT	*Scenedesmus obliquus*	Semple et al. (1999)	*Aerobacter aerogenes*	Wedemeyer (1966)	*Trichoderma viride*	Matsumura and Boush (1968) Patil et al. (1970)
		Anabaena sp. *Chlorococcum* sp. *Nostoc* sp.	Megharaj et al. (2000)	*Arthrobacter* sp. *Bacillus* sp. *Micrococcus* sp. *Pseudomonas* sp.	Patil et al. (1970)	*Phanerochaete chrysosporium*	Reddy and Mathew (2001)
		Euglena gracilis	DeLorenzo et al. (2001)	*Pseudomonas acidovorans*	Hay and Foch (1998)	*Boletus edulis Gymnopilus viscidus Laccaria bicolor Laccaria scaburm*	Huang et al. (2007)

TABLE 8.1 *(Continued)*

S. No.	Pesticide	Algae	References	Bacteria	References	Fungi	References
		Cylindrotheca sp. *Euglena gracilis Scenedesmus obliquus*	Priyadarshani et al. (2011)	*Terrabacter* sp.	Aislabie et al. (1999)	*Pleurotus ostreatusto*	Purnomo et al. (2010)
		Anabaena sp. *Aulosira fertilissima Chlorococcum* sp. *Nostoc* sp.	Subashchandrabose et al. (2013)	*Ralstonia eutropha*	Hay and Focht (2000)		
		Dunaliella sp. *Cylindrotheca* sp.	Biswas et al. (2015)	*Basea thiooxidans*	Pesce and Wunderlin (2004)		
				Pseudomonas sp.	Kamanavalli and Ninnekar (2005)		
				Eubacterium limosum	Yim et al. (2008)		
				Sphingobacterium sp.	Fang et al. (2010)		
				Acinetobacter sp. *Moraxella* sp. *Neisseria* sp. *Pseudomonas* sp.	Velázquez-Fernández et al. (2012)		

TABLE 8.1 *(Continued)*

S. No.	Pesticide	Algae	References	Bacteria	References	Fungi	References
42.	2,4,Dichlorophenoxy acetic acid			*Azoarcus* sp.	Ortiz et al. (2013)		
				Cupriavidus pinatubonensis	Kraiser et al. (2013)	*Aspergillus niger*	Faulkner and Woodcock (1964)
43.	Dichlorovos			*Flavobacterium* sp.	Ning et al. (2012)		
				Acinetobacter sp. *Proteus vulgaris* *Serratia* sp. *Vibrio* sp.	Agarry et al. (2013)		
44.	Dieldrin and endrin	*Agmenellum quadraplicatum* *Dunaliella* sp.	Patil et al. (1972) Matsumoto et al. (2009)	*Bacillus* sp.	Matsumura and Boush (1967)	*Phlebia brevispora*	Kamei et al. (2010)
				Pseudomonas sp.	Matsumura et al. (1968)	*Mucor racemosus*	Kataoka et al. (2010)
				Arthrobacter sp. *Bacillus* sp. *Micrococcus* sp.	Patil et al. (1970)	*Pachycephala aurea Phlebia acanthocystis Phlebia brevispora*	Xiao et al. (2011a)

TABLE 8.1 *(Continued)*

S. No.	Pesticide	Algae	References	Bacteria	References	Fungi	References
				Arthobacter sp. *Bacillus* sp.	Jagnow and Halder (1972)	*Mucor* sp. *Trichoderma* sp.	Velázquez-Fernández et al. (2012)
				Aerobacter sp. *Bacillus* sp. *Burkholderia* sp. *Micrococcus* sp. *Pseudomonas* sp.	Velázquez-Fernández et al. (2012)	*Cordyceps brongniartii* *Cordyceps militaris*	Xiao and Kondo (2013)
45.	Difenoconazole	*Chlorella vulgaris*	Hultberg et al. (2016) Hultberg and Bodin (2018)			*Aspergillus niger*	Hultberg and Bodin (2018)
						Fusarium oxysporum *Lecanicillium saksenae* *Lentinula edodes* *Penicillium brevicompactum*	Shi et al. (2012)
46.	Difluorobenzens			*Rhodococcus opacus*	Zaitsev et al. (1995)		

TABLE 8.1 *(Continued)*

S. No.	Pesticide	Algae	References	Bacteria	References	Fungi	References
				Rhodococcus sp.	Rapp and Gabriel-Jürgens (2003)		
				Labrys portucalensi	Moreira et al. (2012)		
47.	Dimethoate	*Chlorella vulgaris*	Hussein et al. (2016)				
48.	Dimethomorph/ pyrimethanil	*Scenedesmus obliquus* *Scenedesmus quadricauda*	Dosnon-Olette et al. (2010)				
49.	Diuron					*Phanerochaete chrysosporium*	Fratila-Apachitei et al. (1999)
						Beauveria bassiana *Cunninghamella elegans* *Mortirella isabellina*	Tixier et al. (2000)

TABLE 8.1 *(Continued)*

S. No.	Pesticide	Algae	References	Bacteria	References	Fungi	References
						Agrocybe semiorbicularis Auricularia auricola Coriolus versicolor Dichotomitus squalens Flammulina velupites Hypholoma fasciculare Phanerochaete velutina Pleurotus ostreatus Stereum hirsutum	Bending et al. (2002)
						Aspergillus niger Beauveria bassina Cunninghamella elegans Mortierella isabellina	Tixier et al. (2002)
						Mortierella sp.	Badawi et al. (2009)

TABLE 8.1 (Continued)

S. No.	Pesticide	Algae	References	Bacteria	References	Fungi	References
						Coriolus versicolor Dichotomitus squalens Flammulina velupites Pleurotus ostreatus Stereum hirsutum	Bending et al. (2010)
						Ganoderma lucidum	Da Silva Coelho et al. (2010)
						Aspergillus niger	Marco-Urrea et al. (2015)
50.	Dromecarb					*Aspergillus versicolor*	Knowles and Benezet (1981)
51.	Endosulfan	*Chlorococcum* sp. *Scenedesmus* sp.	Sethunathan et al. (2004)	*Bacillus subtilis*	Awasthi et al. (1999)	*Aspergillus niger*	Bhalerao and Puranik (2007)
				Bacillus sp. *Staphylococcus* sp.	Kumara and Philipa (2006)	*Trametes hirsute*	Kamaei et al. (2011)

TABLE 8.1 *(Continued)*

S. No.	Pesticide	Algae	References	Bacteria	References	Fungi	References
				Pseudomonas aeruginosa	Jayashree and Vasudevan (2007)	*Aspergillus fumigates* *Candida* sp. *Mucor* sp. *Pencillium* sp.	Mohanasrinivasan et al. (2013)
				Anabaena sp.	Shivaramaiah (2010)		
				Nocardiopsis sp.	Dudhagara et al. (2012)		
				Burkholderia sp. *Pseudomonas aeruginosa* *Pseudomonas spinosa*	Velázquez-Fernández et al. (2012)		
				Alcaligenes faecalis	Kong et al. (2013)		
				Bacillus sp. *Micrococcus* sp. *Pseudomonas* sp. *Staphylococcus* sp.	Mohanasrinivasan et al. (2013)		
				Pseudomonas fluorescens	Giri et al. (2014)	*Pleurotus eryngii Coprinus comatus*	Wang et al. (2017)

TABLE 8.1 *(Continued)*

S. No.	Pesticide	Algae	References	Bacteria	References	Fungi	References
52.	Endosulfansulphage			*Klebsiella* sp.	Singh and Singh (2014)	*Trametes hirsute*	Kamaei et al. (2011)
53.	Endrin			*Arthrobacter* sp. *Bacillus* sp. *Micrococcus* sp. *Pseudomonas* sp. *Trichoderma viride*	Patil et al. (1970)		
54.	S-Ethyl dipropylthio-carbamate			*Alcaligenes* sp. *Bacillus* sp. *Diheterospora* sp. *Epicoccum purpurascens Fusarium oxysporum Micrococcus* sp. *Paecilomyces lilacinus Penicillium* sp. *Pseudomonas* sp.	Lee (1984)		
				Arthrobacter sp.	Tam et al. (1987)		
				Flavobacterium sp.	Mueller et al. (1988)		
				Rhodococcus sp.	Dick et al. (1990)		

TABLE 8.1 *(Continued)*

S. No.	Pesticide	Algae	References	Bacteria	References	Fungi	References
55.	Fenamiphos	*Chlamydomonas* sp. *Chlorella* sp. *Nostoc* sp. *Scenedesmus* sp. *Stichococcus* sp.	Singh et al. (2003b)	*Brevibacterium* sp.	Cáceres et al. (2008b)		
		Chlorococcum sp. *Pseudokirchner-iella subcapitata*	Cáceres et al. (2008b)	*Brevundimonas* sp. *Cupriavidus* sp. *Microbacterium* sp. *Ralstonia* sp. *Sinorhizobium* sp.	Cabrera et al. (2010)		
56.	Fenhexamid	*Scenedesmu obliquus*	Mofeed and Mosleh, (2013)				
57.	Fenitrothion			*Flavobacterium* sp.	Adhya et al. (1981)		
				Burkholderia sp.	Hayatsu et al. (2000)		
58.	Fludioxonil	*Chlorella vulgaris*	Hultberg and Bodin (2018)			*Aspergillus niger*	Hultberg and Bodin (2018)
59.	Fluoranthene					*Acremonium* sp.	Ma et al. (2014)

TABLE 8.1 *(Continued)*

S. No.	Pesticide	Algae	References	Bacteria	References	Fungi	References
60.	Fluorene					Acremonium sp.	Ma et al. (2014)
61.	Fluoxetine					Bjerkandera adusta Bjerkandera sp. Phanerochaete chrysosporium	Rodarte-Morales et al. (2011)
62.	Fluroxypyr	Chlamydomonas reinhardtii	Zhang et al. (2011) Jin et al. (2012)	Unknown consortium of soil bacteria	Tao and Yang (2011)		
63.	Folpet					Gongronella sp. Rhizopus stolonifer	Martins et al. (2014)
64.	Glyphosate	Leptolyngbya boryana Microcystis aeruginosa Nostoc punctiforme	Hove-Jensen et al. (2014)	Penicillium citrium	Pothuluri et al. (1998)	Trametes versicolor	Pizzul et al. (2009)
		Oscillatoria limnetica	Salman and Abdul-Adel (2015)	Bacillus megaterium Pseudomonas aeruginosa	Al-Arfaj et al. (2013)		

TABLE 8.1 *(Continued)*

S. No.	Pesticide	Algae	References	Bacteria	References	Fungi	References
		Emiliania huxleyi Isochrysis galbana Skeletonema costatum Phaeodactylum tricornutum	Wang et al. (2016)	*Agrobacterium radiobacter Burkholderia pseudomallei Ochrobactrum anthropi Sinorhizobium meliloti*	Hove-Jensen et al. (2014)		
				Pseudomonas sp.	Zhao et al. (2015)		
65.	Lindane	*Anabaena* sp. *Nostoc ellipsosporum*	Kuritz and Wolk (1995)	*Paucimobilis* sp. *Sphingomonas* sp.	Pesce and Wunderlin (2004)	*Phanerochaete chrysosporium*	Reddy and Mathew (2001)
		Nodularia sp. *Cyanothece* sp. *Nostoc* sp. *Synechococcus* sp. *Oscilatoria* sp.	El-Bestawy et al. (2007)				
		Scenedesmus intermedius	González et al. (2012)	*Burkholderia* sp. *Flavobacterium* sp. *Pseudomonas* sp. *Vibrio* sp.	Velázquez-Fernández et al. (2012)	*Pleurotus ostreatus*	Rigas et al. (2005)

TABLE 8.1 *(Continued)*

S. No.	Pesticide	Algae	References	Bacteria	References	Fungi	References
		Anabaena azotica	Zhang et al. (2012)	*Streptomyces* sp.	Benimeli et al. (2008)	*Ganoderma australe*	Rigas et al. (2007)
		Chlorella sp. *Scenedesmus* sp.	Matamoros and Rodríguez (2016)				
66.	HCN			*Pseudomonas aeruginosa*	Sharma et al. (2009)		
				Sphingomonas sp.	Manickam et al. (2012)		
67.	Heptachlor			*Phanerochaete chrysosporium*	Arisoy and Kolankaya (1998)	*Pleyrotus acanthocystis*	Xiao et al. (2011b)
				Phlebia sp.	Xiao et al. (2011b)	*Aspergillus niger*	Bhalerao (2012)
						Phanerochaete ostreatus	Purnomo et al. (2013)
68.	Heptachlor epoxide			*Phlebia* sp.	Xiao et al. (2011b)		
69.	Hexazinone	*Chlorella vulgaris*	Hultberg et al. (2016)				
70.	Imidacloprid	*Chlamydomonas mexicana*	Kumar et al. (2011)				
71.	Isopropyl-N-phenylcarbamate			*Pseudomonas alcaligenes*	Marty et al. (1986)		

TABLE 8.1 *(Continued)*

S. No.	Pesticide	Algae	References	Bacteria	References	Fungi	References
72.	Isoproturon	*Scenedesmus quadricauda*	Dosnon-Olette et al. (2010)			*Alternaria* sp. *Basidiomycete* strain Gr177 *Mortierella* sp. *Mucor* sp. *Phoma eupyrena*	Rønhede et al. (2005)
		Chlamydomonas reinhardtii	Bi et al. (2012)			*Cunninghamella elegans*	Hangler et al. (2007)
		Chlorella vulgaris	Hussein et al. (2016)			*Mortierella* sp.	Badawi et al. (2009)
73.	Isoprothiolane	*Aulacoseira granulata Microcystis aeruginosa Scenedesmus quadricauda*	Guanzon et al. (1996)				
74.	Linuron					*Mortierella* sp.	Badawi et al. (2009)
						Pleurotus coccineus Pleurotus gigantean Trametes versicolor	Gouma et al. (2019)

TABLE 8.1 *(Continued)*

S. No.	Pesticide	Algae	References	Bacteria	References	Fungi	References
75.	Malathion	*Aspergillus oryzae* *Nostoc muscorum* *Streptomyces platensis*	Ibrahim et al. (2014)			*Fusarium oxysporum*	Peter et al. (2015)
		Chlorella vulgaris *Scenedesmus quadricuda* *Spirulina platensis*	Abdel-Razek et al. (2019)				
76.	Mandipropamid	*Chlorella vulgaris*	Ardal (2014)				
77.	Methyl-4-nitrophenyl thiophosphate	*Aulacoseira granulata* *Microcystis aeruginosa* *Scenedesmus quadricauda*	Guanzon et al. (1996)				
78.	Mesotrione	*Amphora coffeaeformis* *Pediastrum tetras* *Ankistrodesmus fusiformis*	Moro et al. (2012)				

TABLE 8.1 *(Continued)*

S. No.	Pesticide	Algae	References	Bacteria	References	Fungi	References
		Microcystis sp. *Scenedesmus quadricauda*	Ni et al. (2014)				
79.	Metalaxyl	*Chlorella vulgaris*	Ardal (2014)			*Agrocybe semiorbicularis Auricularia auricola Coriolus versicolor Dichotomitus squalens Flammulina velupites Hypholoma fasciculare Phanerochaete velutina Pleurotus ostreatus Stereum hirsutum*	Bending et al. (2002)
80.	Metfluorazon	*Chlorella fusca*	Thies et al. (1996)			*Gongronella* sp. *Rhizopus stolonifer*	Martins et al. (2014)

TABLE 8.1 *(Continued)*

S. No.	Pesticide	Algae	References	Bacteria	References	Fungi	References
81.	Methomyl			Stenotrophomonas maltophilia	Mohamed (2009)		
82.	Methoxychlor	Chlorella sp.	Semple et al. (1999)				
83.	Methyl parathion	Bifidobacterium animalis Chlorella vulgaris Nostoc linckia Nostoc muscorum Phormidium foveolarum Scenedesmus bijugatus	Megharaj et al. (1994)	Pseudomonas sp.	Chaudhry et al. (1988)	Aspergillus sydowii Penicillium decaturense	Alvarenga et al. (2014)
		Scenedesmus obliquus	Semple et al. (1999)	Bacillus sp.	Sharmila et al. (1989)		
		Nostoc muscorumwas	Megharaj et al. (2000)	Pseudomonas fluorescens	Zboinska et al. (1992)		
		Euglena gracilis	DeLorenzo et al. (2001)	Plesiomonas sp.	Cui et al. (2001)		
				Stenotrophomonas maltophilia	Mohamed (2009)		

TABLE 8.1 *(Continued)*

S. No.	Pesticide	Algae	References	Bacteria	References	Fungi	References
		Euglena gracilis *Scenedesmus obliquus*	Priyadarshani et al. (2011)	*Bacillus* sp. *Stenotrophomonas* sp. *Proteus* sp. *Proteus vulgaris* *Acinetobacter* sp. *Flavobacterium* sp. *Pseudomonas putida* *Citrobacter freundii* *Pseudomonas aeruginosa* *Pseudomonas* sp.	Pino and Peñuela (2011)		
				Cyanobacteria	Ibrahim et al. (2014)		
				Pseudomonas sp.	Wang et al. (2014)		
84.	Metoalcholar	*Chlorella vulgaris*	Hussein et al. (2016)				
85.	Metribuzin					*Pleurotus coccineus* *Pleurotus gigantean* *Trametes versicolor*	Gouma et al. (2019)

TABLE 8.1 *(Continued)*

S. No.	Pesticide	Algae	References	Bacteria	References	Fungi	References
86.	Mirex	*Chlorococcum* sp.	Semple et al. (1999)				
		Chlamydomonas sp. *Chlorococcum* sp. *Dunaliella* sp.	Priyadarshani et al. (2011)				
87.	Molinate	*Chlorella vulgaris*	Hussein et al. (2016)				
88.	Monocrotophos	*Chlorella vulgaris* *Nostoc linckia* *Plasmodium tenue* *Scenedesmus bijugatus* *Scenedesmus elongates*	Megharaj et al. (1987)	*Arthrobacter atrocyaneus* *Bacillus megaterium*	Bhadbhade et al. (2002)	*Aspergillus flavus Fusarium pallidoroseum Macrophomina* sp.	Jain et al. (2014)
				Paracoccus sp. *Aspergillus niger* *Pseudomonas stutzeri*	Jia et al. (2007) Jain et al. (2012) Barathidasan and Reetha (2013)		
89.	Naphthalene	*Chlorella* sp.	Semple et al. (1999)			*Acremonium* sp.	Ma et al. (2014)

TABLE 8.1 *(Continued)*

S. No.	Pesticide	Algae	References	Bacteria	References	Fungi	References
		Selenastrum capricornutum	Gavrilescu (2010)				
		Dunaliella sp., *Cylindrotheca* sp.	Biswas et al. (2015)				
90.	1-Naphthol			*Pseudomonas aeruginosa*	Chapalamadugu and Chaudhry (1991)		
91.	Organopollutants					*Phanerochaete chrysosporium*	Sasek (2003)
92.	Parathion			*Flavobacterium* sp.	Sethunathan and Yoshida (1973)	*Bjerkandera adusta Phanerochaete chrysosporium Pleurotus ostreatus*	Jauregui et al. (2003)
				Bacillus sp. *Pseudomonas* sp.	Siddaramappa et al. (1973)		
				Pseudomonas stutzeri	Daughton and Hsieh (1977)		

TABLE 8.1 *(Continued)*

S. No.	Pesticide	Algae	References	Bacteria	References	Fungi	References
93.	Pendimethalin	*Chlorella vulgaris*	Hussein et al. (2016)	*Arthrobacter* sp. *Bacillus* sp.	Nelson et al. (1982)	*Fusarium oxysporum Lecanicillium saksenae Lentinula edodes Penicillium brevicompactum*	Shi et al. (2012)
94.	Pentachlorobenzene	*Chlorella* sp. *Scenedesmus* sp.	Matamoros and Rodríguez (2016)				
95.	Pentachlorophenol			*Arthrobacter* sp.	Stanlake and Finn (1982)	*Lentinus edodes*	Pletsch et al. (1999)
				Flavobacterium sp.	Crawford and Mohn (1985)	*Trametes versicolor*	Tuomela et al. (1999)
						Anthrocophyllum discolor	Rubilar et al. (2007)
96.	Phenanthrene	*Chlorella sorokiniana Pseudomonas migulae*	Muñoz et al. (2003)			*Acremonium* sp.	Ma et al. (2014)
		Selenastrum capricornutum	Gavrilescu (2010)			*Phanerochaetes ordida*	Turlo (2014)

TABLE 8.1 *(Continued)*

S. No.	Pesticide	Algae	References	Bacteria	References	Fungi	References
97.	Phenmedipham	*Chlorella vulgaris*	Hultberg and Bodin (2018)	*Flavobacterium* sp.	Knowles and Benezet (1981)	*Aspergillus niger*	Hultberg and Bodin (2018)
98.	Phenol	*Euglena gracilis*	DeLorenzo et al. (2001)				
99.	Pirimicarb	*Chlorella vulgaris*	Hultberg et al. (2016)				
100.	Polychlorinated biphenyls			*Dehalobacter restrictus* *Desulfito bacterium* *Desulfomonile tiedjei*	Borja et al. (2005)	*Doratomyces nanus* *Doratomyces purpureofuscus* *Doratomyces verrucis-porus* *Myceliophthor-athermophila* *Phoma eupyren* *Therm-oascus crustaceus*	Mouhamadou et al. (2013)
101.	Polychlorinated dibenzofurans					*Phanerochaete chrysosporium*	Wu et al. (2013)
102.	Prometryne	*Chlamydomonas reinhardtii*	Jin et al. (2012)	*Bacillus* sp. *Ochrobactrum* sp.	Zhou et al. (2012)		

TABLE 8.1 *(Continued)*

S. No.	Pesticide	Algae	References	Bacteria	References	Fungi	References
103.	Propamocarb	*Chlorella vulgaris*	Ardal (2014) Hultberg et al. (2016)				
104.	Propamocarb hydrochloride			*Bacillus amyloliquefaciens Bacillus pumilus*	Myresiotis et al. (2012)		
105.	Propanil	*Chlorella vulgaris*	Hussein et al. (2016)				
106.	Propoxur			*Achromobacter* sp.	Karns et al. (1886)		
107.	Pyrene					*Fusarium flocciferum Pleurotus ostreatus Trametes versicolor Trichoderma* sp.	Baldrian et al. (2008)
						Pseudo trametes gibbosa	Wen et al. (2011)
108.	Pyrethrin					*Sphingomonas yanoikayae*	Cases et al. (2005)
109.	Pyriproxin	*Chlorella vulgaris*	Hussein et al. (2016)				

TABLE 8.1 *(Continued)*

S. No.	Pesticide	Algae	References	Bacteria	References	Fungi	References
110.	Quinalphos	*Chlorella vulgaris* *Nostoc linckia* *Plasmodium tenue* *Scenedesmus bijugatus* *Scenedesmus elongates*	Megharaj et al. (1987)				
111.	Simazine	*Chlorella vulgaris*	Hussein et al. (2016)			*Phanerochaete chrysosporium* *Penicillium steckii* *Fusarium verticillioides* *Bjerkandera adusta* *Bjerkandera* sp. *Phanerochaete chrysosporium*	Mougin et al. (1997) Kodama et al. (2001) Urlacher et al. (2004) Alzahrani (2009) Rodarte-Morales et al. (2011)
112.	Sulfamethoxazole			*Pseudomonas alcaligenes*	Marty et al. (1986)		
113.	Methyl 3,4-dich-lorophenylcarbamate						

TABLE 8.1 *(Continued)*

S. No.	Pesticide	Algae	References	Bacteria	References	Fungi	References
114.	Terbuthylazine	*Chlorella vulgaris*	Hultberg et al. (2016) Hultberg and Bodin (2018)			*Agrocybe semiorbicularis Auricularia auricola Coriolus versicolor Dichotomitus squalens Flammulina velupites Hypholoma fasciculare Phanerochaete velutina Pleurotus ostreatus Stereum hirsutum*	Bending et al. (2002)
						Fusarium oxysporum Lecanicillium saksenae Lentinula edodes Penicillium brevicompactum	Shi et al. (2012)

TABLE 8.1 *(Continued)*

S. No.	Pesticide	Algae	References	Bacteria	References	Fungi	References
115.	Thiamethoxam			*Bacillus amyloliquefaciens Bacillus pumilus Bacillus subtilis*	Myresiotis et al. (2012)		
116.	Toxaphene	*Chlorella* sp.	Semple et al. (1999)	*Trichoderma viridae*	Patil et al. (1970)		
117.	2,4,5-Trichlorophenol			*Bjerkandera* sp.	Lacayo et al. (2006)	*Clitocybe maxima*	Zhou et al. (2015)
118.	Trinexapac ethyl	*Chlorella vulgaris*	Hultberg et al. (2016) Hultberg and Bodin (2018)			*Aspergillus niger*	Hultberg and Bodin (2018)
119.	Vernolate			*Flavobacterium* sp.	Mueller et al. (1988)		
120.	Vydate					*Trichoderma harzianum Trichoderma viride*	Helal and Abo-El-Seoud (2015)

8.5 CONCLUSION AND FUTURE PROSPECTS

From the advent of green revolution, the escalation of pesticide applications has brought their long-term detrimental impacts on the natural environment and public health into attention. Pesticide polluted aquatic ecosystems and soils can altogether be remediated by means of microalgal–bacterial–fungal consortia. As against conventional methods, microbioremediation is an economical, environment-friendly, and efficient technique for remediating pesticide-contaminated ecosystems. The usage of consortia proves more effective in treating the polluted sites. However, advanced research is required to identify potential multitaxon (algal–bacterial–fungal) consortia that can be together employed for remediation of diverse pesticides in a particular system.

8.6 GAPS AND FUTURE PROSPECTS

There are several gaps in the arena of microbioremediation that further necessitate in-depth study, namely:

1. There are several limitations of using bacterial–microalgal–fungal consortia like it is difficult to estimate outcomes of remediation derived from laboratory and miniature pilot size studies. Therefore, such types of studies must be carried out in field under natural conditions. Another constraint of using microalgal–bacterial–fungal consortia is the enduring perpetuation of homeostasis among species. The future prospects of employing a microalgae–bacterial–fungal consortium gyrate around extenuating these constraints and can be improved in many aspects.

2. *Unavailability in market:* The unavailability of microbial consortia hinders their popularization and use for clean-up of pesticides. Additionally, all kinds of pesticides cannot be degraded by a single microbial strain, which necessitates advanced research.

3. *Use of omics approach:* Majority of microbes in environment are nonculturable and the functional aspect of many microbial genes and proteins is still not known. Metagenomics and functional genomics can help in understanding the overall structure and function of microbial community, and thus presents a potential method to comprehend the molecular mechanism of bioremediation, development of

proficient consortia, and in achieving effective pesticide bioremediation under natural conditions.

4. Although numerous studies exist on identification of microbial genes and enzymes engaged in pesticide degradation, but their broad practical applications are comparatively less reported. So, there is an imperative requirement to use this data to formulate triumphant enzymatic bioremediation schemes in natural polluted environments.

5. Moreover, the use of cell-free enzyme preparations from algae/bacteria/fungi to degrade and detoxify pollutants efficiently can be a novel alternative, as it overcomes several of the limitations related to microbial growth under natural state.

6. *Use of biopesticides:* As against conventional agrochemicals, use of biological-derived products recognized as biopesticides especially microbial biopesticides like Plant Growth Promoting Bacteria/Rhizobacteria (PGPB or PGPR), having biofertilizer and biopesticide properties, for sustainable crop production and disease management is an effective approach. The biodegradability and specificity are the key benefits of biopesticides, thus avoiding pollution caused by traditional pesticides. In spite of two decades of exhaustive research, and the accumulation of innumerable potentially valuable microbes across world, their successful commercialization has not yet been achieved. Use of biopesticides can help in increasing the crop production without harming the environment.

KEYWORDS

- **microalgae**
- **microbioremediation**
- **mycoremediation**
- **pesticide pollution**
- **phycoremediation**

REFERENCES

Abdel-Razek, M. A.; Abozeid, A. M.; Eltholth, M. M.; Abouelenien, F. A.; El-Midany, S. A.; Moustafa, N. Y.; Mohamed, R. A. Bioremediation of a pesticide and selected heavy metals in wastewater from various sources using a consortium of microalgae and cyanobacteria. *Slov. Vet. Res.* **2019**.

Abo-Amer, A. Biodegradation of diazinon by *Serratia marcescens* DI101 and its use in bioremediation of contaminated environment. *J. Microbiol. Biotechnol.* **2011**, *21*, 71–80.

Adhya, T. K.; Barik, S.; Sethunathan, N. Hydrolysis of selected organophosphorus insecticides by two bacterial isolates from flooded soil. *J. Appl. Bacteriol.* **1981**, *50*, 167–172.

Agarry, S. E.; Olu-arotiowa, O. A.; Aremu, M. O.; Jimoda, L. A. Biodegradation of Dichlorovos (Organophosphate Pesticide) in soil by bacterial isolates. *J. Natural. Sci. Res.* **2013**, *3*, 12–16.

Aislabie, J.; Bej, A. K.; Ryburn, J.; Lloyd, N.; Wilkins, A. Characterization of *Arthrobacter nicotinovorans* HIM, an atrazine-degrading bacterium, from agricultural soil, New Zealand. *FEMS Microbiol. Ecol.* **2005**, *52*, 279–286.

Aislabie, J.; Davison, A. D.; Boul, H. L.; Franzmann, P. D.; Jardine, D. R.; Karuso, P. Isolation of *Terrabacter* sp. strain DDE-1, which metabolizes 1,1-dichloro-2,2-bis(4 chlorophenyl) ethylene when induced with biphenyl. *Appl. Environ. Microbiol.* **1999**, *65*, 5607–5611.

Al-Arfaj, A.; Abdel-Megeed, A.; Ali, H. M.; Al-Shahrani, O. Phyto-microbial degradation of glyphosate in Riyadh area. *J. Pure. Appl. Microbiol.* **2013**, *7*, 1351–1365.

Alzahrani, A. M. Insects cytochrome P450 enzymes: Evolution, functions and methods of analysis. *Glob. J. Mol. Sci.* **2009**, *4*, 167–179.

Anson, J. G.; Mackinnon, G. Novel *Pseudomonas* plasmid involved in aniline degradation, *Appl. Environ. Microbiol.* **1984**, *48*, 868.

Anwar, S.; Liaquat, F.; Khan, Q. M.; Khalid, Z. M.; Iqbal, S. Biodegradation of chlorpyrifos and its hydrolysis product 3,5,6-trichloro-2-pyridinol by *Bacillus pumilus* strain C2A1. *J. Hazard. Mater.* **2009**, *168*, 400–405.

Aoki, K.; Ohtsuka, K.; Shinke, R.; Nishira, H. Isolation of aniline assimilation bacteria an physiological characterisation of aniline biodegradation in *Rhodococcus erythropolis* AN-13. *Agric. Biol. Chem.,* **1983**, *47*, 2569.

Ardal, E. Phycoremediation of pesticides using microalgae. Master's Thesis **2014**.

Arisoy, M.; Kolankaya, N. Biodegradation of Heptachlor by *Phanerochaete chrysosporium* ME 446: The toxic effects of Heptachlor and its Metabolites on Mice. *Turk. J. Biol.* **1998**, *22*, 427–434.

Awad, N. S.; Sabit, H. H.; Abo-Aba, S. E. M.; Bayoumi, R. A. Isolation, characterization and fingerprinting of some chlorpyrifos-degrading bacterial strains isolated from Egyptian pesticides-polluted soils. *Afr. J. Microbiol. Res.* **2011**, *5*, 2855–2862.

Awasthi, N.; Kumar, A.; Makkar, R.; Cameotra, S. S. Biodegradation of soil-applied endosulfan in the presence of a biosurfactant. *J. Environ. Sci. Health.* **1999**, *34*, 793–803.

Badawi, N.; Ronhede, S.; Olsson, S.; Kragelund, B. B.; Johnsen, A. H.; Jacobsen, O. S. Metabolites of the phenylurea herbicides chlorotoluron, diuron, isoproturon and linuron produced by the soil fungus *Mortierella* sp. *Environ. Pollut.* **2009**, *157*, 2806–2812.

Baldrian, P. Wood-inhabiting ligninolytic basidiomycetes in soils: Ecology and constraints for applicability in bioremediation. *Fungal. Ecol.* **2008**, *1*, 4–12.

Barathidasan, K.; Reetha, D. Microbial degradation of monocrotophos by *Pseudomonas stutzeri*. *Indian Streams. Res.* **2013**, *3*, 1.

Barathidasan, K.; Reetha, D.; John-Milton, D.; Sriram, N.; Govin-Dammal, M. Biodegradation of chlorpyrifos by co-culture of *Cellulomonas fimi* and *Phanerochaete chrysosporium*. *Afr. J. Microbiol. Res.* **2014**, *8*, 961–966.

Bending, G. D.; Friloux, M.; Walker, A. Degradation of contrasting pesticides by white rot fungi and its relationship with ligninolytic potential. *FEMS Microbiol. Lett.* **2002**, *212*, 9–63.

Benimeli, C. S.; Fuentes, M. S.; Abate, C. M.; Amoroso, M. J. Bioremediation of lindane-contaminated Soil by *Streptomyces* sp. M7 and its effects on *Zea mays* growth. *Int. Biodeterior. Biodegrad.* **2008**, *61*, 233–239.

Bhadbhade, B. J.; Sarnaik, S. S.; Kanekar, P. P. Biomineralization of an organophosphorus pesticide, Monocrotophos, by soil bacteria. *J. Appl. Microbiol.* **2002**, *93*, 224–234.

Bhalerao, T. S. Bioremediation of endosulfan-contaminated soil by using bioaugmentation treatment of fungal inoculant *Aspergillus niger. Turk. J. Biol.* **2012**, *36*, 561–567.

Bhalerao, T. S.; Puranik, P. R. Biodegradation of organochlorine pesticide, endosulfan, by a fungal soil isolate, *Aspergillus niger. Int. Biodeter. Biodegrad.* **2007**, *59*, 315–321.

Bhattacharya, S.; Das, A.; Prashanthi, K.; Palaniswamy, M.; Angayarkanni, J. Mycoremediation of benzo[a]pyrene by *Pleurotus ostreatus* in the presence of heavy metals and mediators. *3 Biotech.* **2014**, *4*, 205–211.

Bhosle, N. P.; Nasreen, S. Remediation of cypermethrin-25 EC by microorganisms. *Eur. J. Exp. Biol.* **2013**, *3*, 144–152.

Bhuimbar, M. V.; Kulkarni A. N.; Ghosh, J. S. Detoxification of chlorpyriphos by *Micrococcus luteus* NCIM 2103, *Bacillus subtilis* NCIM 2010 and *Pseudomonas aeruginosa* NCIM 2036. *Res. J. Envir. Earth. Sci.* **2011**, *3*, 614–619.

Bi, Y. F.; Miao, S. S.; Lu, Y. C.; Qiu, C. B.; Zhou, Y.; Yang, H. Phytotoxicity, bioaccumulation and degradation of isoproturon in green algae. *J. Hazard. Mater.* **2012**, *243*, 242–249.

Biswas, K.; Paul, D.; Sinha, S. N. Biological agents of bioremediation: A concise review. *Front. Environ. Microbiol.* **2015**, *1*, 39–43.

Bonner, M. R.; Alavanja, M. C. R. Pesticides, human health, and food security. *Food Energy Secur.* **2017**, *6*, 89–93.

Borja, J., Taleon, D. M.; Auresenia, J.; Gallardo, S. Polychlorinated biphenyls and their biodegradation. *Proc. Biochem. J.* **2005**, *40*, 1999–2013.

Bumpus, J. A.; Powers, R. H.; Sun, T. Biodegradation of DDE (1,1-Dichloro-2,2-bis(4-hlorophenyl)ethane) by *Phanerochaete chrysosporium. Mycol. Res.* **1993**, *97*, 95–98.

Cabrera, J. A.; Kurtz, A.; Sikora, R. A.; Schouten, A. Isolation and characterization of fenamiphos degrading bacteria. *Biodegradation* **2010**, *21*, 1017–1027.

Cáceres, T.; Megharaj, M.; Naidu, R. Toxicity and transformation of fenamiphos and its metabolites by two micro algae *Pseudokirchneriella subcapitata* and *Chlorococcum* sp. *Sci. Total. Environ.* **2008a**, *398*, 53–59.

Cáceres, T. P.; Megharaj, M.; Naidu, R. Biodegradation of the pesticide fenamiphos by ten different species of green algae and cyanobacteria. *Curr. Microbiol.* **2008b**, *57*, 643–646.

Cases, I.; de Lorenzo, V. Promoters in the environment: Transcriptional regulation in its natural context. *Nat. Rev. Microbiol.* **2005**, *3*, 105–118.

Cerniglia, C. E.; Gibson, D. T.; Van Baalen, C. Algal oxidation of aromatic hydrocarbons: Formation of l-naphthol from naphthalene by *Agmenellum quudruplicatum*, strain PR-6. Biochem. *Biophys. Res. Commun.* **1979**, *88*, 50–58.

Chang, Y.; Tan, Y.; Sun, M. Biodegradation mechanism of organophosphate pesticides in the aquatic ecosystem. *Huanjing Kexue xuebao.* **1981**, *1*, 115.

Chapalamadugu, S.; Chaudhry, G. R. Hydrolysis of carbaryl by a *Pseudomonas* sp. and construction of a microbial consortium that completely metabolizes carbaryl. *Appl. Environ. Microbiol.* **1991**, *57*, 744–750.

Chaudhry, G. R.; Ali, A. N. Bacterial metabolism of carbofuran. *Appl. Environ. Microbiol.* **1988,** *54*, 1414–1419.

Chaudhry, G. R.; Ali, A. N.; Wheeler, W. B. Isolation of a methyl parathion-degrading *Pseudomonas* sp. that possesses DNA homologous to the opd gene from a *Flavobacterium* sp. *Appl. Environ. Microbiol.* **1988,** *54*, 288–293.

Chaussonnerie, S.; Saaidi, P. L.; Ugarte, E.; Barbance, A.; Fossey, A.; Barbe, V.; Fouteau, S. Microbial degradation of a recalcitrant pesticide: Chlordecone. *Front. Microbiol.* **2016,** *7*, 2025.

Chen, M.; Xu, P.; Zeng, G.; Yang, C.; Huang, D.; Zhang, J. Bioremediation of soils contaminated with polycyclic aromatic hydrocarbons, petroleum, pesticides, chlorophenols and heavy metals by composting: Applications, microbes and future research needs. *Biotechnol. Adv.* **2015,** *33*, 745–755.

Crawford, R. L.; Mohn, W. W. Microbiological removal of pentachlorophenol from soil using a *Flavobacterium. Enzyme Microb. Technol.* **1985,** *7*, 617–620.

Cui, Z.; Li, S.; Fu, G. Isolation of methyl parathion- degrading strain M6 and cloning of the methyl parathion hydrolase gene. *Appl. Environ. Microbiol.* **2001,** *67*:4922–4925.

Cycoń, M.; Mrozik, A.; Piotrowska-Seget, Z. Bioaugmentation as a strategy for the remediation of pesticide-polluted soil: A review. *Chemosphere.* **2017,** *172*, 52–71.

Da Silva Coelho, J.; de Oliveira, A. L.; de Souza, C. G. M.; Bracht, A.; Peralta, R. M. Effect of the herbicides bentazon and diuron on the production of ligninolytic enzymes by *Ganoderma lucidum. Int. Biodeter. Biodegrad.* **2010,** *64*, 156–161.

Dasgupta, S.; Meisner, C.; Wheeler, D. Stockpiles of obsolete pesticides and cleanup priorities: A methodology and application for Tunisia. *J. Environ. Manage.* **2010,** *91*, 824–830.

Daughton, C. G.; Hsieh, D. P. Parathion utilization by bacterial symbionts in a chemostat. *Appl. Environ. Microbiol.* **1977,** *34*, 175–184.

DeLorenzo, M. E.; Geoffrey, I. S.; Philippe, E. R. Toxicity of pesticides to aquatic microorganisms: A review. *J. Environ. Toxicol. Chem.* **2001,** *20*, 84–98.

Dick, W. A.; Ankumah, R. A.; McClung, G.; Abou-Assaf, N. Enhanced degradation of s-ethyl N,N-dipropyl carbamothioate in soil and by an isolated soil microorganism. In Enhanced Biodegradation of Pesticides in the Environment, Racke, K. D. and Coats, J. R., Eds., American Chemical Society, Washington, DC, USA. **1990,** *98*.

Dosnon-Olette, R.; Trotel-Aziz, P.; Couderchet, M.; Eullaffroy, P. Fungicides and herbicide removal in *Scenedesmus* cell suspensions. *Chemosphere.* **2010,** *79*, 117–123.

Dudhagara, P.; Bhalani, S.; Bhatt, S.; Ghelani, A. degradation of organophosphate and organochlorine pesticides in liquid culture by marine isolate *Nocardiopsis* species and its bioprospectives. *J. Environ. Res. Dev.* **2012,** *7*, 995–1001.

Dutta, A.; Vasudevan, V.; Nain, L.; Singh, N. Characterization of bacterial diversity in an atrazine degrading enrichment culture and degradation of atrazine, cyanuric acid and biuret in industrial wastewater. *J. Environ. Sci. Health B.* **2016,** *51*, 24–34.

El-Bestawy, E.; El-Salam, Z.; Mansy, E. R. H. Potential use of environmental cyanobacterial species in bioremediation of lindane-contaminated effluents. *Int. Biodeterior. Biodegrad.* **2007,** *59*, 180–92.

Erick, R. B., Juan, A. O.; Paulino, P.; Torres, L. G. Removal of aldrin, dieldrin, heptachlor, and heptachlor epoxide using activated carbon and/or *pseudomonas fluorescens* free cell cultures. *J. Environ. Sci. Health B.* **2006,** *41*, 553–569.

Evy, A. A.; Jaseetha, A. S., Das, N. Atrazine degradation in liquid culture and soil by a novel yeast *Pichia kudriavzevii* strain Atz-EN-01 and its potential application for bioremediation. *J. Appl. Pharma. Sci.* **2013,** *3,* 035–043.

Evy, A. A. M.; Lakshmi, V.; Das, N. Biodegradation of atrazine by *Cryptococcus laurentii* isolated from contaminated agricultural soil. *J. Microbiol. Biotech. Res.* **2012,** *2,* 450–457.

Fang, H.; Dong, B.; Yan, H.; Tang, F.; Yu, Y. Characterization of a bacterial strain capable of degrading DDT congeners and its use in bioremediation of contaminated soil. *J. Hazard. Mater.* **2010,** *184,* 281–289.

Fang, H.; Xiang, Y. Q.; Hao, Y. J.; Chu, X. Q.; Pan, X. D.; Yu, J. Q.; Yu, Y. L. Fungal degradation of chlorpyrifos by *Verticilium* sp. DSP in pure cultures and its use in bioremediation of contaminated soil and pakchoi. *Int. Biodeter. Biodegr.* **2008,** *61,* 294–303.

Faulkner, J. K.; Woodcock, D. Metabolism of 2, 4-dichlorophenoxyacetic acid ('2, 4-D') by *Aspergillus niger* van Tiegh. *Nature.* **1964,** *203,* 865.

Fawzy.; I. E.; Hend, A. M.; Osama, N. M.; Khaled, M. G.; Ibrahim, M. G. Biodegradation of chlorpyrifos by microbial strains isolated from agricultural wastewater. *J. Am. Sci.* **2014,** *10,* 98–108.

Felsot, A.; Maddox, J. V.; Bruce, W. Enhanced microbial degradation of carbofuran in soils with histories of furadan use. *Bull. Environ. Contamin. Toxicol.* **1981,** *26,* 781–788.

Ferguson, J. A.; Korte, F. Epoxidation of aldrin to exo-dieldrin by soil bacteria. *Appl. Environ. Microbiol.* **1977,** *34,* 7–13.

Fratila-Apachitei, L. E.; Hirst, J. A.; Siebel, M. A.; Gijzen, H. J. Diuron degradation by *Phanerochaete chrysosporium* BKM-F-1767 in synthetic and natural media. *Biotechnol. Lett.* **1999,** *21,* 147–154.

Fu, P.; Secundo, F. Algae and their bacterial consortia for soil bioremediation. *Chem. Eng. Trans.* **2016,** *49,* 427–432.

Fuentes, M. S.; Briceño, G. E.; Saez, J. M.; Benimeli, C. S.; Diez, M. C.; Amoroso, M. J. Enhanced removal of a pesticides mixture by single cultures and consortia of free and immobilized *Streptomyces* strains. *Bio Med.* **2013,** *1,* 1–9.

Gattullo, C. E.; Bährs, H.; Steinberg, C. E. W.; Loffredo, E. Removal of bisphenol A by the freshwater green alga *Monoraphidium braunii* and the role of natural organic matter. *Sci. Total Environ.* **2012,** *416,* 501–506.

Gavrilescu, M. Fate of pesticides in the environment and its bioremediation. *Eng. Life Sci.* **2005,** *5,* 497–526.

Gavrilescu, M. Environmental biotechnology: Achievements, opportunities and challenges. *Dynamic Biochem Process Biotech. Mol. Biol.* **2010,** *4,* 1–36.

Ghanem, I.; Orfi, M.; Shamma, M. Biodegradation of chlorpyrifos by *Klebsiella* sp. isolated from an activated sludge sample of waste water treatment plant in Damascus Folia. *Microbiol.* **2007,** *52,* 423–427.

Giri, K.; Rawat, A. P.; Rawat, M.; Rai, J. P. N. Biodegradation of hexachlorocyclohexane by two species of *bacillus* isolated from contaminated soil. *Chem. Ecol.* **2014,** *30,* 97–109.

González, R.; García-Balboa, C.; Rouco, M.; Lopez-Rodas, V.; Costas, E. Adaptation of micro-algae to lindane: A new approach for bioremediation. *Aquat. Toxicol.* **2012,** *109,* 25–32.

Gouma, S.; Anastasia, A.; Papadaki, A.A.; Markakis, G.; Magan, N.; Goumas, D. Studies on pesticides mixture degradation by white rot fungi. *JEE.* **2019,** *20,* 16–26.

Guanzon Jr, N. G.; Fukuda, M.; Nakahara, H. Accumulation of agricultural pesticides by three freshwater microalgae. *Fish. Sci.,* **1996,** *62,* 690–697.

Hangler, M.; Jensen, B.; Rønhede, S. R. Inducible hydroxylation and demethylation of the herbicide isoproturon by *Cunninghamella elegans*. *FEMS Microbiol. Lett.* **2007**, *268*, 254–260.

Hay, A. G.; Foch, D. D. Cometabolism of 1,1-dichloro-2,2-bis (4-chlorophenyl) ethylene by *Pseudomonas acidovorans* M3GY grown on biphenyl. *Appl. Environ. Microbiol.* **1998**, *64*, 2141–2146.

Hay, A. G.; Focht, D. D. Transformation of 1,1-dichloro-2, 2-(4-chlorophenyl) ethane (DDD) by *Ralstonia eutropha*-strain A5. *FEMS Microbiol. Ecol.* **2000**, *31*, 249–253.

Hayatsu, M.; Hirano, M.; Tokuda, S. Involvement of two plasmids in fenitrothion degradation by *Burkholderia* sp. strain NF100. *Appl. Environ. Microbiol.* **2000**, *66*, 1737–1740.

Helal, I. M.; Abo-El-Seoud, M. A. Fungal biodegradation of pesticide vydate in soil and aquatic system. In 4th International Conference on Radiation Sciences and Applications. **2015**, 13–17.

Helm, V.; Reber, H. Investigation on the regulation of aniline utilization in *Pseudomonas multivorans* strain AN- 1, J. *Appl. Microbiol. Biotechnol.* **1979**, *7*, 191–199.

Hove-Jensen, B.; Zechel, D. L.; Jochimsen, B. Utilization of glyphosate as phosphate source: Biochemistry and genetics of bacterial carbon-phosphorus lyase. *Microbiol. Mol. Biol. Rev.* **2014**, *78*, 176–197.

Huang, Y.; Zhao, X.; Luan, S. Uptake and biodegradation of DDT by 4 ectomycorrhizal fungi. *Sci. Total. Environ.* **2007**, *385*, 235–241.

Hultberg, M.; Bodin, H. Effects of fungal-assisted algal harvesting through biopellet formation on pesticides in water. *Biodegradation* **2018**, *29*, 557–565.

Hultberg, M.; Bodin, H.; Ardal, E.; Asp, H. Effect of microalgal treatments on pesticides in water. *Environ. Technol.* **2016**, *37*, 893–898.

Hussaini, S. Z.; Shaker, M.; Iqbal, M. A. Isolation of bacterial for degradation of selected pesticides. *Adv. Biores.* **2013**, *4*, 82–85.

Hussein, M. H.; Abdullah, A. M.; Eladal, E. G.; El-Din, N. I. B. Phycoremediation of some pesticides by microchlorophyte alga, *Chlorella* sp. *J. Fertil. Pestic.* **2016**, *7*, 2.

Ibrahim, W. M.; Karam, M. A.; El-Shahat, R. M.; Adway, A. A. Biodegradation and utilization of organophosphorus pesticide malathion by cyanobacteria. *BioMed Res. Int.* **2014**.

Jagnow, G.; Halder, K. Evolution of CO_2 from soil incubated with dieldrin-14C. *Soil. Biolog. Biochem.* **1972**, *4*, 43.

Jain, R.; Garg, V.; Singh, K. P.; Gupta, S. Isolation and characterization of monocrotophos degrading activity of soil fungal isolate *Aspergillus Niger* MCP1 (ITCC7782.10). *Int. J. Environ. Sci.* **2012**, *3*, 841–850.

Jain, R.; Garg, V.; Yadav, D. In vitro comparative analysis of monocrotophos degrading potential of *Aspergillus flavus*, *Fusarium pallidoroseum* and *Macrophomina* sp. *Biodegradation* **2014**, *25*, 437–446.

Jauregui, J.; Valderrama, B.; Albores, A.; Vazquez-Duhalt, R. Microsomal transformation of organophosphorus pesticides by white rot fungi. *Biodegradation* **2003**, *14*, 397–406.

Jayashree, R.; Vasudevan, N. Effect of tween 80 added to the soil on the degradation of endosulfan by *Pseudomonas aeruginosa*. *Int. J. Environ. Sci. Tech.* **2007**, *4*, 203–210.

Jia, K. Z.; Li, X. H.; He, J.; Gu, L. F.; Ma, J. P.; Li, S. P. Isolation of a monocrotophos-degrading bacterial strain and characterization of enzymatic degradation. *Huan Jing Ke Xue* **2007**, *28*, 908–912.

Jin, Z. P.; Luo, K.; Zhang, S.; Zheng, Q.; Yang, H. Bioaccumulation and catabolism of prometryne in green algae. *Chemosphere.* **2012**, *87*, 278–284.

Kabra, A. N.; Ji, M. K.; Choi, J.; Kim, J. R.; Govindwar, S. P.; Jeon, B. H. Toxicity of atrazine and its bioac cumulation and biodegradation in a green microalga, *Chlamydomonas mexicana*. *Environ. Sci. Pollut. Res. Int.* **2014**, *21*, 12270–12278.

Kamaei, I.; Takagi, K.; Kondo, R. Degradation of endosulfan and endosulfansulphate by white-rot fungus *Trametes hirsuta*. *J. Wood Sci.* **2011**, *57*, 317.

Kamanavalli, C. M.; Ninnekar, H. Z. Biodegradation of DDT by a *Pseudomonas* Species. *Curr. Microbiol.* **2005**, *48*, 10–13.

Kamei, I.; Takagi, K.; Kondo, R. Bioconversion of dieldrin by wood-rotting fungi and metabolite detection. *Pest. Manag. Sci.* **2010**, *66*, 888–891.

Kamel, F.; Hoppin, J. A. Association of pesticide exposure with neurologic dysfunction and disease. *Environ. Health Perspect.* **2004**, *112*, 950–958.

Kaminski, U.; Janke, D.; Prauser, H.; Fritsche, W. Degradation of anilines and monochloroanilines by *Rhodococcus* sp. AN 177 and a pseudomonad: A comparative study. *Z. Allg. Mikrobiol.* **1983**, *4*, 235–246.

Karns, J. S.; Mulbry, W. W.; Nelson, J. O.; Kearney, P. C. Metabolism of carbofuran by a pure bacterial culture. *Pest. Biochem. Physiol.* **1986**, *25*, 211–217.

Kataoka, R.; Takagi, K.; Kamei, I.; Kiyota, H.; Sato, Y. Bio-degradation of dieldrin by a soil fungus isolated from a soil with annual endosulfan applications. *Environ. Sci. Technol.* **2010**, *44*, 6343–6349.

Knowles, C. O.; Benezet, H. J. Microbial degradation of the carbamate pesticides desmedipham, phenmedipham, promecarb, and propamocarb. *Bull. Environ. Contam. Toxicol.* **1981**, *27*, 529–533.

Kodama, T.; Ding, L.; Yoshida, M.; Yajima, M. Biodegradation of anstriazine herbicide, simazine. *J. Mol. Catal. B Enzym.* **2001**, *11*, 1073–1078.

Kong, L.; Zhu, S.; Zhu, L.; Xie, H.; Su, K.; Yan, T.; Wang, J.; Wang, J.; Wang, F.; Sun, F. Biodegradation of organochlorine pesticide endosulfan by bacterial strain *Alcaligenes faecalis* JBW4. *J. Environ. Sci.* **2013**, *25*, 2257–2264.

Korade, D. L.; Fulekar, M. H. Rhizosphere remediation of chlorpyrifos in mycorrhizospheric soil using ryegrass. *J. Hazard. Mater.* **2009**, *172*, 1344–1350.

Koroleva, O. V.; Stepanova, E. V.; Landesman, E. O.; Vasilchenko, L. G.; Khromonygina, V. V.; Zherdev, A. V.; Rabinovich, M. L. In vitro degradation of the herbicide atrazine by soil and wood decay fungi controlled through ELISA technique. *Toxicol. Environ. Chem.* **2001**, *80*, 175–188.

Korte, F.; Porter, P. E. Minutes of the Fifth Meeting of the IUPAC Terminal Pesticide Residues. Erbach, West Germany, **1970**.

Kraiser, T.; Stuardo, M.; Manzano, M.; Ledger, T.; González, B. Simultaneous assessment of the effects of an herbicide on the triad: Rhizobacterial community, an herbicide degrading soil bacterium and their plant host. *Plant Soil.* **2013**, *366*, 377–388.

Kulshrestha, G.; Kumari, A. Fungal degradation of chlorpyrifos by *Acremonium* sp. strain (GFRC-1) isolated from a laboratory-enriched red agricultural soils. *Biol. Fertil. Soils.* **2011**, *47*, 219–225.

Kumar, K.; Dasgupta, C. N.; Nayak, B.; Lindblad, P.; Das, D. Development of suitable photobioreactors for CO_2 sequestration addressing global warming using green algae and cyanobacteria. *Bioresour. Technol.* **2011**, *102*, 4945–4953.

Kumara, M.; Philipa, L. Endosulfan mineralization by bacterial isolates and possible degradation pathway identification. *Bioreme. J.* **2006**, *10*, 179–190.

Kuritz, T.; Wolk, C. P. Use of filamentous cyanobacteria for bio degradation of organic pollutants. *Appl. Environ. Microbiol.* **1995**, *161*, 234–236.

Kurniati, E.; Arfarita, N.; Imai, T.; Higuchi, T.; Kanno, A.; Yamamoto, K.; Sekine, M. Potential bioremediation of mercury-contaminated substrate using filamentous fungi isolated from forest soil. *J. Environ. Sci.* **2014**, *26*, 1223–1231.

Lacayo, R. M.; Terrazas, E.; van Bavel, B.; Mattiasson, B. Degradation of toxaphene by *Bjerkandera* sp. strain BOL13 using waste biomass as a co-substrate. *Appl. Microbiol. Biotechnol.* **2006**, *71*, 549–554.

Lakshmi, C. V.; Kumar, M.; Khanna, S. Biotransformation of chlorpyrifos and bioremediation of contaminated soil. *Int. Biodeter. Biodegr.* **2008**, *62*, 204–209.

Larkin, M. J.; Day, M. J. The metabolism of carbaryl by three bacterial isolates, *Pseudomom* spp. (NCIB 12042 and 12043) and *Rhodococcus* sp. (NCIB 12038) from garden soil. *J. Appl. Bacteriol.* **1986**, *60*, 233–242.

Lee, A. EPTC degrading microorganisms isolatedfrom a soil previously exposed to EPTC. *Soil Biol. Biochem.* **1984**, *16*, 529–531.

Li, X.; He, J.; Li, S. Isolation of a chlorpyrifos-degrading bacterium, *Sphingomonas* sp. strain Dsp-2, and cloning of the mpd gene. *Res. Microbiol.* **2007**, *158*, 143–149.

Li, Z.; Jennings, A. Worldwide regulations of standard values of pesticides for human health risk control: A review. *Int. J. Environ. Res. Public. Health.* **2017**, *14*, 826.

Ma, X.; LingWu, L.; Fam, H. Heavy metal ions affecting the removal of polycyclic aromatic hydrocarbons by fungi with heavy-metal resistance. *Appl. Microbiol. Biotechnol.* **2014**, *98*, 9817–9827.

Mahiudddin, M.; Fakhruddin, A. N. M.; Chowdhury, M. A. Z.; Rahman, M. A.; Alam, M. K. Degradation of the organophosphorus insecticide diazinon by soil bacterial isolate. *Int. J. Biotechnol.* **2014**, *3*, 12–23.

Mallick, K.; Bharati, K.; Banerji, A.; Shakil, N. A.; Sethunathan, N. Bacterial degradation of chlorpyrifos in pure cultures and in soil. *Bull. Environ. Contam. Toxicol.* **1999**, *62*, 48–54.

Manickam, N.; Bajaj, A.; Saini, H. S.; Shanker, R. Surfactant mediated enhanced biodegradation of hexachlorocyclohexane (HCH) isomers by *Sphingomonas* sp. NM05. *Biodegrad.* **2012**, *23*, 673–682.

Marco-Urrea, E.; Garcia-Romera, I.; Aranda, E. Potential of non-ligninolytic fungi in bioremediation of chlorinated and polycyclic aromatic hydrocarbons. *New Biotechnol.* **2015**, *32*, 620–628.

Martins, T. M.; Núñez, O.; Gallart-Ayala, H.; Leitão, M. C.; Galceran, M. T.; Pereira, C. S. New branches in the degradation pathway of monochlorocatechols by *Aspergillus nidulans*: a metabolomics analysis. *J. Hazard. Mater.* **2014**, *268*, 264–272.

Marty, J. L.; Khafif, T.; Vega, D.; Bastide, J. Degradation of phenyl carbamate herbicides by *Pseudomonas alcaligens* isolated from soil. *Soil Biol. Biochem.* **1986**, *18*, 649–653.

Matamoros, V.; Rodriguez, Y. Batch vs continuous-feeding operational mode for the removal of pesticides from agricultural run-off by microalgae systems: A laboratory scale study. *J. Hazard. Mater.* **2016**, *309*, 126–132.

Matsumoto, E.; Kawanaka, Y.; Yun, S. J.; Oyaizu, H. Bioremediation of the organochlorine pesticides, dieldrin and endrin, and their occurrence in the environment. *Appl. Microbiol. Biotechnol.* **2009**, *84*, 205–216.

Matsumura, F.; Boush, G. M. Dieldrin degradation by soil microorganisms. *Science.* **1967**, *156*, 959–961.

Matsumura, F.; Boush, G. M. Degradation of insecticides by a soil fungus *Trichoderma viride*. *J. Econom. Entomol.* **1968**, *61*, 610–612.

Matsumura, F.; Boush, G. M.; Tai, A. Breakdown of dieldrin in the soil by a microorganism. *Nature.* **1968**, *219*, 965–967.

McFarland, M.; Salladay, D.; Ash, D.; Baiden, E. Composting treatment of alachlor impacted soil amended with the white rot fungus *Phanerochaete chrysosporium. Hazard. Waste. Hazard. Mater.* **1996**, *13*, 363–373.

McLellan, J.; Gupta, S. K.; Kumar, M. Feasibility of using bacterial-microalgal consortium for the bioremediation of organic pesticides: application constraints and future prospects. In Application of Microalgae in Wastewater Treatment, pp. 341–362. Springer, Cham. **2019**.

Megharaj, M.; Kantachote, D.; Singleton, I.; Naidu, R. Effects of long-term contamination of DDT on soil microflora with special reference to soil algae and algal transformation of DDT. *Environ. Pollut.* **2000**, *109*, 35–42.

Megharaj, M.; Madhavi, D. R.; Sreenivasulu, C.; Umamaheswari, A.; Venkateswarlu, K. Biodegradation of methyl parathion by soil isolates of microalgae and cyanobacteria. *Bull. Environ. Contam. Toxicol.* **1994**, *53*, 292–297.

Megharaj, M.; Venkateswarlu, K.; Rao, A. S. Metabolism of monocrotophos and quinalphos by algae isolated from soil. *Bull. Environ. Contam. Toxicol.* **1987**, *39*, 251–256.

Mofeed, J.; Mosleh, Y. Y. Toxic responses and antioxidative enzymes activity of *Scenedesmus obliquus* exposed to fenhexamid and atrazine, alone and in mixture. *Ecotoxicol. Environ. Saf.* **2013**, *95*, 234–240.

Mohamed, M. S. Degradation of methomyl by the novel bacterial strain *Stenotrophomonas maltophilia* M1. *Electron. J. Biotechnol.* **2009**, *12*, 1–6.

Mohanasrinivasan, V.; Suganthi, V.; Selvarajan, E.; Subathra-Devi C.; Ajith, E.; Muhammed, F. N. P.; Sreeram, G. Bioremediation of endosulfan contaminated soil. *Res. J. Chem. Environ.* **2013**, *17*, 93–101.

Moreira, I. S.; Amorim, C. L.; Carvalho, M. F.; Castro, P. M. L. Degradation of difluorobenzenes by the wild strain *Labrys portucalensis. Biodegradation* **2012**, *23*, 653–662.

Moro, C. V.; Bricheux, G.; Portelli, C.; Bohatier, J. Comparative effects of the herbicides chlortoluron and mesotrione on freshwater microalgae. *Environ. Toxicol. Chem.* **2012**, *31*, 778–786.

Mougin, C.; Laugero, C.; Asther, M.; Chaplain, V. Biotransformation of striazine herbicides and related degradation products in liquid culture by the white rot fungus *Phanerochaete chrysosporium. Pest. Sci.* **1997**, *49*, 169–177.

Mouhamadou, B.; Faure, M.; Sage, L.; Marcais, J.; Souard, F.; Geremia, R. A. Potential of autochthonous fungal strains isolated from contaminated soils for degradation of polychlorinated biphenyls. *Fungal Biol.* **2013**, *117*, 268–274.

Mueller, J. G.; Skipper, H. D.; Wine, E. L. Loss of butylate-utilizing ability by a *Flavobacterium* sp. *Pest. Biochem. Physiol.* **1988**, *32*, 189–196.

Muñoz, R.; Guieysse, B.; Mattiasson, B. Phenanthrene biodegradation by an algal-bacterial consortium in two-phase partitioning bioreactors. *Appl. Microbiol. Biotechnol.* **2003**, *61*, 261–267.

Myresiotis, C. K.; Vryzas, Z.; Papadopoulou-Mourkidou, E. Biodegradation of soil-applied pesticides by selected strains of plant growth-promoting rhizobacteria (PGPR) and their effects on bacterial growth. *Biodegradation* **2012**, *23*, 297–310.

Nayak, S. K.; Dash, B.; Baliyarsingh, B. Microbial remediation of persistent agro-chemicals by soil bacteria: An overview. In Microbial Biotechnology, pp. 275–301. Springer, Singapore, **2018**.

Nelson, L. M. Biologically induced hydrolysis of parathion in soil: Isolation of hydrolyzing bacteria. *Soil Biol. Biochem.* **1982**, *14*, 219–222.

Nelson, M. L.; Yaron, B.; Nye, P. H. Biologically induced hydrolysis of parathion in soil: Kinetics and modelling. *Soil Biol. Biochem.* **1982**, *14*, 223–228.

Ni, Y.; Lai, J.; Wan, J.; Chen, L. Photosynthetic responses and accumulation of mesotrione in two freshwater algae. *Environ. Sci. Processes. Impacts.* **2014**, *16*, 2288–2294.

Ning, J.; Gang, G.; Bai, Z.; Hu, Q.; Qi, H.; Ma, A.; Zhuan, X.; Zhaung, G. In situ enhanced bioremediation of dichlorvos by a phyllosphere *Flavobacterium* strain. *Front. Environ. Sci. Eng.* **2012**, *6*, 231–237.

Ortega, N. O.; Nitschke, M.; Mouad, A. M.; Landgraf, M. D.; Rezende, M. O. O.; Seleghim, M. H. R.; Sette, L. D.; Porto, A. L. M. Isolation of Brazilian marine fungi capable of growing on DDD Pesticide. *Biodegradation* **2011**, *22*, 43–50.

Ortiz, I.; Velasco, A.; Borgne, S. L.; Revah, S. Biodegradation of DDT by stimulation of indigenous microbial populations in soil with co substrates. *Biodegrad.* **2013**, *10532*, 9578–9581.

Ortiz-Hernández, M. L.; Sánchez-Salinas, E.; Dantán-González, E.; Castrejón-Godínez, M. L. Pesticide biodegradation: Mechanisms, genetics and strategies to enhance the process. In Biodegradation-Life of Science. IntechOpen. **2013**.

Patil, K. C.; Matsumura, F.; Boush, G. M. Degradation of endrin, aldrin, and DDT by soil microorganisms. *J. App. Microbiol,* **1970**, *19*, 879–881.

Patil, K.; Matsumura, F.; Boush, G. Metabolic transformation of DDT, dieldrin, aldrin, and endrin by marine microorganisms. *Environ. Sci. Technol.* **1972**, *6*, 629–632.

Peng, X.; Huang, J.; Liu, C.; Xiang, Z.; Zhou, J.; Zhong, G. Biodegradation of bensulphuron-methyl by a novel *Penicillium pinophilum* strain BP-H-02. *J. Hazard. Mater.* **2012**, *213*, 216–22.

Pereira, P. M.; Sobral Teixeira, R. S.; de Oliveira, M. A. L.; da Silva, M.; Ferreira, V. S. Optimized atrazine degradation by *Pleurotus ostreatus* INCQS 40310: An alternative for impact reduction of herbicides used in sugarcane crops. *J. Microb. Biochem. Technol. S.* **2013**, *12*, 006.

Pesce, S. F.; Wunderlin, D. A. Biodegradation of lindane by a native bacterial consortium isolated from contaminated river sediment. *Int. Biodeter. Biodegrad.* **2004**, *54*, 255–260.

Peter, L.; Gajendiran, A.; Mani, D.; Nagaraj, S.; Abraham, J. Mineralization of malathion by *Fusarium oxysporum* strain JASA1 isolated from sugarcane fields. *Environ. Prog. Sustain. Energy.* **2015**, *34*, 112–116.

Pimentel, D. Amounts of pesticides reaching target pests: environmental impacts and ethics. *J. Agric. Environ. Ethics.* **1995**, *8*, 17–29.

Pino, N.; Peñuela, G. Simultaneous degradation of the pesticides methyl parathion and chlorpyrifos by an isolated bacterial consortium from a contaminated site. *Int. Biodeter. Biodegr.* **2011**, *65*, 827–831.

Pizzul, L.; Castillo, M. D. P.; Stenström, J. Degradation of glyphosate and other pesticides by ligninolytic enzymes. *Biodegrad.* **2009**, *20*, 751–759.

Pletsch, M.; de Araujo, B.; Charlwood, B. Novel biotechnological approaches in environmental remediation research. *Biotechnol. Adv.* **1999**, *17*, 679–687.

Pothuluri, J. V.; Chung, Y. C.; Xiong, Y. Biotransformation of 6-nitrochrysene. *Appl. Environ. Microbiol.* **1998**, *64*, 3106–3109.

Priyadarshani, I.; Sahu, D.; Rath, B. Microalgal bioremediation: current practices and perspectives. *J. Biochem. Tech.* **2011**, *3*, 299–304.

Purnomo, A. S.; Mori, T.; Kamei, I., Nishii, T.; Kondo, R. Application of mushroom waste medium from *Pleurotus ostreatus* for bioremediation of DDT-contaminated soil. *Int. Biodeterior. Biodegrad.* **2010**, *64*, 397–402.

Purnomo, A. S.; Mori, T.; Putra, S. R.; Kondo, R. Biotransformation of heptachlor and heptachlor epoxide by white-rot fungus *Pleurotus ostreatus. Int. Biodeter. Biodegrad.* **2013**, *82*, 40–44.

Rajagopal, B. S.; Panda, S.; Sethunathan, N. Accelerated degradation of carbaryl and carbofuran in a flooded soil pretreated with hydrolysis products, I-naphthol and carbofuran phenol. *Bull. Environ. Contam. Toxicol.* **1984**, *36*, 827–832.

Rani, K.; Dhania, G. Bioremediation and biodegradation of pesticide from contaminated soil and water-a noval approach. *Int. J. Curr. Microbiol. Appl. Sci.* **2014**, *3*, 23–33.

Rapp, P.; Gabriel-Jürgens, L. H. Degradation of alkanes and highly chlorinated benzenes, and production of biosurfactants, by a psychrophilic *Rhodococcus* sp. and genetic characterization of its chlorobenzene dioxygenase. *Microbiology.* **2003**, *149*, 2879–2890.

Read, D. C. Greatly accelerated microbial degradation of aldicarb in re-treated field soil, in flooded soil, and in water. *J. Econ. Entomol.* **1987**, *80*, 156–163.

Reddy, C.; Mathew, Z. Bioremediation potential of white rot fungi. In. Gadd, G. (Ed.), Fungi in Bioremediation. Cambridge University Press. Cambridge, U.K. **2001**.

Rigas, F.; Dritsa, V.; Marchant, R.; Papadopoulou, K.; Avramides, E. J.; Hatzianestis, I. Biodegradation of Lindane by *Pelourotus ostreatus* via Central Composite Design. *Environ. Int.* **2005**, *31*, 191–196.

Rigas, F.; Papadopoulou, K.; Dritsa, V.; Doulia, D. Bioremediation of a soil contaminated by lindane utilizing the fungus *Ganoderma australe* via response surface methodology. *J. Hazard. Mater.* **2007**, *140*, 325–332.

Rodarte-Morales, A. I.; Feijoo, G.; Moreira, M. T., Lema, J. M. Degradation of selected pharmaceutical and personal care products (PPCPs) by white-rot fungi. *World. J. Microbiol. Biotechnol.* **2011**, *27*, 1839–1846.

Romeh, A. A.; Hendawi, M. Y. Bioremediation of certain organophosphorus pesticides by two biofertilizers, Paeni-bacillus (*Bacillus*) *polymyxa* (Prazmowski) and *Azospirillum lipoferum* (Beijerinck). *J. Agric. Sci. Tech.* **2014**, *16*, 265–276.

Rønhede, S.; Jensen, B.; Rosendahl, S.; Kragelund, B. B.; Juhler, R. K.; Aamand, J. Hydroxylation of the herbicide isoproturon by fungi isolated from agricultural soil. *Appl. Environ. Microbiol.* **2005**, *71*, 7927–7932.

Rubilar, O.; Feijoo, G.; Diez, C.; Lu-Chau, T. A.; Moreira, M. T.; Lema, J. M. Biodegradation of pentachlorophenol in soil slurry cultures by *Bjerkandera adusta* and *Anthracophyllum discolor. Ind. Eng. Chem. Res.* **2007**, *46*, 6744–6751.

Salman, J. M.; Abdul-Adel, E. Potential use of cyanophyta species *Oscillatoria limnetica* in bioremediation of organophosphorus herbicide glyphosate. *Mesop. Environ. J.* **2015**, *1*, 15–26.

Sasek, V. Why mycoremediations have not yet come to practice. In Sasek, V. et al. (Eds.), In: The Utilization of Bioremediation to Reduce Soil Contamination: Problems and Solutions, Kluwer Academic Publishers. **2003**, 247–276.

Semple, K. T.; Ronald, B. C.; Stefan, S. Biodegradation of aromatic compounds by microalgae. Mini review. *FEMS. Microbiol. Lett.* **1999,** *170,* 291–300.

Sethunathan, N.; Megharaj, M.; Chen, Z. L.; Williams, B. D.; Lewis, G.; Naidu, R. Algal degradation of a known endocrine disrupting insecticide, α-endosulfan, and its metabolite, endosulfan sulfate, in liquid medium and soil. *J. Agric. Food Chem.* **2004,** *52,* 3030–3035.

Sethunathan, N.; Yoshida, T. A *Flavobacterium* that degrades diazinon and parathion. *Can. J. Microbiol.* **1973,** *19,* 873–875.

Sharma, A.; Pankaj, P. K.; Gangola, S.; Kumar, G. Microbial degradation of pesticides for environmental cleanup. *Bioremed. Ind. Pollut.* **2016.**

Sharma, S.; Singh, P.; Raj, M.; Chadha, B. S.; Saini, H. S. Aqueous phase partitioning of hexachlorocyclohexane (HCH) isomers by biosurfactant produced by *Pseudomonas aeruginosa* WH-2. *J. Hazard. Mater.* **2009,** *171,* 1178–1182.

Sharmila, M.; Ramanand, K.; Sethunathan, N. Effect of yeast extract on the degradation of organophosphorus insecticides by soil enrichment and bacterial cultures. *Can. J. Microbiol.* **1989.** *35,* 1105–1110.

Shi, H.; Pei, L.; Gu, S.; Zhu, S.; Wang, Y.; Zhang, Y.; Li, B. Glutathione S-transferase (GST) genes in the red flour beetle, *Tribolium castaneum,* and comparative analysis with five additional insects. *Genomics.* **2012,** *100,* 327–335.

Shivaramaiah, H. M. Biodegradation of endosulfan by *Anabaena* pesticide. *Res. J.* **2010,** *22,* 125–128.

Siddaramappa, R.; Rajaram, K. P.; Sethunathan, N. Degradation of parathion by bacteria isolated from flooded soil. *Appl. Microbiol.* **1973,** *26,* 846–849.

Silambarasan, S.; Abraham, J. Ecofriendly method for bioremediation of chlorpyrifos from agricultural soil by novel fungus *Aspergillus terreus* JAS1. *Water Air Soil Pollut.* **2013,** *224,* 1369.

Singh, B. K.; Walker, A.; Morgan, J. A.; Wright, D. J. Effects of soil pH on the biodegradation of chlorpyrifos and isolation of a chlorpyrifos-degrading bacterium. *Appl. Environ. Microbiol.* **2003a,** *69,* 5198–206.

Singh, B. K.; Walker, A.; Morgan, J. A.; Wright, D. J. Role of soil pH in the development of enhanced biodegradation of fenamiphos. *Appl. Environ. Microbiol.* **2003b,** *69,* 7035–43.

Singh, B. K.; Walker, A.; Morgan, J. A.; Wright, D. J. Biodegradation of chloropyrifos by *Enterobacter* strain B-14 and its use in bioremediation of contaminated soil. *Appl. Environ. Microbiol.* **2004,** *70,* 4855–4863.

Singh, D. K. Biodegradation and bioremediation of pesticide in soil: Concept, method and recent developments. *Indian. J. Microbiol.* **2008,** *48,* 35–40.

Singh, D. P.; Khattar, J. I. S.; Nadda, J.; Singh, Y.; Garg, A.; Kaur, N.; Gulati, A. Chlorpyrifos degradation by the *Cyanobacterium synechocystis* sp. strain PUPCCC 64. *Environ. Sci. Pollut. Res.* **2011,** *18,* 1351–1359.

Singh, M.; Singh, D. K. Biodegradation of endosulfan in broth medium and in soil microcosm by *Klebsiella* sp. M3. *Bull. Environ. Contam. Toxicol.* **2014,** *92,* 237–242.

Singh, R.; Singh, P.; Sharma, R. Microorganism as a tool of bioremediation technology or cleaning environment: A review. *Proc. Int. Acad. Ecol. Environ. Sci.* **2014,** *4,* 1–6.

Spain, J. C.; Nishino, S. F. Degradation of 1,4-Dichlorobenzene by a *Pseudomonas* sp. *Appl. Environ. Microbiol.* **1987,** *53,* 1010–1019.

Stanlake, G. J.; Finn, R. K. Isolation and characterization of a pentachlorophenol-degrading bacterium. *Appl. Environ. Microbiol.* **1982,** *44,* 1421–1427.

Subashchandrabose, S. R.; Ramakrishnan, B.; Megharaj, M.; Venkateswarlu, K.; Naidu, R. Mixotrophic cyanobacteria and microalgae as distinctive biological agents for organic pollutant degradation. *Environ. Int.* **2013,** *51,* 59–72.

Surovfseva, E. G.; Vol'nova, A. I. Aniline as the sole source of carbon, nitrogen and energy for *Alcaligenes faecalis. Microbiologiya.* **1980,** *49,* 49.

Swissa, N.; Nitzan, Y.; Langzam, Y.; Cahan, R. Atrazine biodegradation by a monoculture of *Raoultella planticola* isolated from a herbicides wastewater treatment facility. *Int. Biodeter. Biodegrad.* **2014,** *92,* 6–11.

Tam, A. C.; Behki, R. M.; Khan, S. U. Isolation and characterization of an EPTC-degrading *Arthrobacter* strain and evidence for plasmid-associated EPTC degradation, *Appl. Environ. Microbiol.* **1987,** *53,* 1088.

Tao, L.; Yang, H. Fluroxypyr biodegradation in soils by multiple factors. *Environ. Monit. Assess.* **2011,** *175,* 227–238.

Tewari, L.; Saini, J.; Arti. Bioremediation of pesticides by microorganisms: General aspects and recent advances. In Bioremediation of Pollutants. I.K. International Publishing House Pvt. Ltd. New Delhi. **2012.**

Thabit T. M.; El-Naggar, M. A. Diazinon decomposition by soil bacteria and identification of degradation products by GC-MS. *Soil. Environ.* **2013,** *32,* 96–102.

Thies, F.; Backhaus, T.; Bossmann, B.; Grimme, L. H. Xenobiotic biotransformation in unicellular green algae. Involvement of cytochrome P450 in the activation and selectivity of the pyridazinone pro-herbicide metflurazon. *Plant Physiol.* **1996,** *112,* 361–70.

Tixier, C.; Bogaerts, P.; Sancelme, M.; Bonnemoy, F.; Twagilimana, L.; Cuer, A. Fungal biodegradation of a phenylurea herbicide, diuron: Structure and toxicity of metabolites. *Pest Manag. Sci.* **2000,** *56,* 455–462.

Tixier, C.; Sancelme, M.; Aït-Aïssa, S.; Widehem, P.; Bonnemoy, F.; Cuer, A. Biotransformation of phenylurea herbicides by a soil bacterial strain, *Arthrobacter* sp. N2: Structure, ecotoxicity and fate of diuron metabolite with soil fungi. *Chemosphere.* **2002,** *46,* 519–526.

Tuomela, M.; Lyytikainen, M.; Oivanen, P.; Hatakka, A. Mineralization and conversion of pentachlorophenol (PCP) in soil inoculated with the white rot fungus *Trametes versicolor. Soil Biol. Biochem.* **1999,** *31,* 65–74.

Turlo, J. The biotechnology of higher fungi-current state and perspectives. *Folia. Biol. Oecol.* **2014,** *10,* 49–65.

Urlacher, V. B.; Lutz-Wahl, S.; Schmid, R. D. Microbial P450 enzymes in biotechnology. *Appl. Microbiol. Biotechnol.* **2004,** *64,* 317–325.

Vega, D.; Bastide, J.; Coste, C. Isolation from soil and growth characteristics of a CIPC-degrading strain of *Pseudomoms cepacia. Soil Biol. Biochem.,* **1985,** *17,* 541.

Velázquez-Fernández, J. B.; Martínez-Rizo, A. B.; Ramírez-Sandoval, M.; Domínguez-Ojeda, D. Biodegradation and bioremediation of organic pesticides. In Pesticides-Recent Trends in Pesticide Residue Assay. IntechOpen. **2012.**

Venkateswarlu, K.; Sethunathan, N. Degradation of carbofuran by *Azospirillum lipoferum* and *Streptomyces* spp. isolated from flooded soil. *Bull. Environ. Contam. Toxicol.* **1984,** *33,* 556–560.

Venkateswarlu, K.; Sethunathan, N. Enhanced degradation of carbofuran by *Pseudomoms cepacia* and *Nocardia* sp. in the presence of growth factors. *Plant Soil.* **1985,** *84,* 445.

Wang, C.; Lin, X.; Li, L.; Lin, S. Differential growth responses of marine phytoplankton to herbicide glyphosate. *PLoS One.* **2016,** *11,* e0151633.

Wang, L.; Chi, X. Q.; Zhang, J. J.; Sun, D. L.; Zhou, N. Y. Bioaugmentation of a methyl parathion contaminated soil with *Pseudomonas* sp. strain WBC 3. *Int. Biodeter. Biodegrad.* **2014,** *87,* 116–121.

Wang, Y.; Zhang, B.; Chen, N.; Wang, C.; Feng, S.; Xu, H. Combined bioremediation of soil co-contaminated with cadmium and endosulfan by *Pleurotus eryngii* and *Coprinus comatus. J. Soils. Sediments.* **2017,** *18,* 2136–2147.

Wedemeyer, G. Dechlorination of DDT by *Aerobacter aerogenes. Science.* **1966,** *152,* 647–647.

Wen, J.; Gao, D.; Zhang, B.; Liang, H. Co-metabolic degradation of pyrene by indigenous white-rot fungus *Pseudotrametes gibbosa* from the northeast China. *Int. Biodeter. Biodegradation* **2011,** *65,* 600–604.

Wu, J.; Zhao, Y.; Liu, L.; Fan, B.; Li, M. Remediation of soil contaminated with decarbrominated diphenyl ether using white rot fungi. *J. Environ. Eng. Landsc. Manag.* **2013,** *21,* 171–179.

Xiao, P. F.; Kondo, R. Biodegradation of dieldrin by *Cordyceps* fungi and detection of metabolites. *Appl. Mech. Mater.* **2013,** *295,* 30–34.

Xiao, P.; Mori, T.; Kamei, I.; Kondo, R. A novel metabolic pathway for biodegradation of DDT by the white rot fungi, *Phlebia lindtneri* and *Phlebia brevispora. Biodegrad.* **2011a,** *22,* 859–867.

Xiao, P.; Mori, T.; KameiI, I.; Kondo, R. Metabolism of organochlorine pesticide heptachlor and its metabolite heptachlor epoxide by white rot fungi, belonging to genus *Phlebia. FEMS Microbiol. Lett.* **2011b,** *314,* 140–146.

Xu, G. M.; Li, Y. Y.; Zheng, W.; Peng, X.; Li, W.; Yan, Y. C. Mineralization of chlorpyrifos by co-culture of *Serratia* and *Trichosporon* sp. *Biotechnol. Lett.* **2007,** *29,* 1469–1473.

Xu, G.; Zheng, W.; Li, Y.; Wang, S.; Zhang, J.; Yan, Y. Biodegradation of chlorpyrifos and 3,5,6-trichloro-2-pyridinol by a newly isolated *Paracoccus* sp. strain TRP. *Int. Biodeter. Biodegrad.* **2008,** *62,* 51–56.

Yang, L.; Zhao, Y. H.; Zhang, B. X.; Yang, C. H.; Zhang, X. Isolation and characterization of a chlorpyrifos and 3,5,6-trichloro-2-pyridinol degrading bacterium. *FEMS Microbiol. Lett.* **2005,** *251,* 67–73.

Yim, Y. J.; Seo, J.; Kang, S. I.; Ahn, J. H.; Hur, H. G. Reductive dechlorination of methoxychlor and DDT by human intestinal bacterium *Eubacterium limosum* under anaerobic conditions. *Arch. Environ. Contam. Toxicol.* **2008,** *54,* 406–411.

You, I. S.; Bartha, R. Stimulation of 3,4-dichloroaniline mineralization by aniline. *Appl. Environ. Microbiol.* **1982,** *44,* 678.

Zaitsev, G.; Uotila, J. S.; Tsitko, I. V.; Lobanok, A. G.; Salkinoja-Salo-nen, M. S. Utilization of halogenated benzenes, phenols, and benzoates by *Rhodococcus opacus* GM-14. *Appl. Environ. Microbiol.* **1995,** *61,* 4191–4201.

Zboinska, E.; Lejczak, B.; Kafarski, P. Organophosphonate utilization by the wild-type strain of Pseudomonas fluorescens. *Appl. Environ. Microbiol.* **1992,** *58,* 2993–2999.

Zeyer, J.; Wasserfallen, A.; Timmis, K. N. Microbial mineralization of ring-substituted anilines through an ortho-cleavage pathway, *Appl. Environ. Microbiol.* **1985,** *50,* 447.

Zhang, H.; Hu, C.; Jia, X.; Xu, Y.; Wu, C.; Chen, L.; Wang, F. Characteristics of γ-hexachlorocyclohexane biodegradation by a nitrogen-fixing cyanobacterium, *Anabaena azotica. J. Appl. Phycol.* **2012,** *24,* 221–225.

Zhang, S.; Qiu, C. B.; Zhou, Y.; Jin, Z. P.; Yang, H. Bioaccumulation and degradation of pesticide fluroxypyr are associated with toxic tolerance in green alga *Chlamydomonas reinhardtii. Ecotoxicology.* **2011,** *20,* 337–347.

Zhao, H.; Zhu, J.; Liu, S.; Gao, H.; Zhou, X.; Tao, K. Bioremediation potential of glyphosate-degrading *pseudomonas* spp. strains isolated from contaminated soil. *J. Gen. Appl. Microbiol.* **2015,** *61*, 165–170.

Zhou, W.; Li, Y.; Min, M.; Hu, B.; Zhang, H.; Ma, X.; Li, L.; Cheng, Y.; Ruan, R. Growing wastewater-borne microalga *Auxenochlorella prototothecoides* UMN280 on concentrated municipal wastewater for simultaneous nutrient removal and energy feedstock production. *Appl. Energy* **2012,** *98*, 433–440.

Zhou, Z.; Chen, Y.; Liu, X.; Zhang, K.; Xu, H. Interaction of copper and 2,4,5-trichlorophenol on bioremediation potential and biochemical properties in co-contaminated soil incubated with *Clitocybe maxima. RSC Adv.* **2015,** *5*, 42768–42776.

Zhu, J.; Zhao, Y.; Qiu, J. Isolation and application of a chlorpyrifos-degrading *Bacillus licheniformis* ZHU-1. *Afr. J. Microbiol. Res.* **2010,** *4*, 2410–2413.

WEB-LINKS

FAO. Prevention and disposal of obsolete pesticides. Retrieved from http://www.fao.org/agriculture/crops/obsolete-pesticides/what-dealing/obs-pes/en/ (Accessed on May 6, 2019) **2019.**

CHAPTER 9

Application of Beneficial Microorganisms with High Efficient Biosorption Potential for the Bioremediation of Pesticide Contamination of Freshwater and Soil Environment

CHARLES OLUWASEUN ADETUNJI,[1*] OSIKEMEKHA ANTHONY ANANI,[2] and CHUKWUEBUKA EGBUNA[3,4]

[1]*Applied Microbiology, Biotechnology and Nanotechnology Laboratory, Department of Microbiology, Edo State University Uzairue, PMB 04, Auchi, Edo State, Nigeria*

[2]*Laboratory of Ecotoxicology and Forensic Biology, Department of Biological Science, Faculty of Science, Edo State University Uzairue, Edo State, Nigeria*

[3]*Africa Centre of Excellence in Public Health and Toxicological Research (ACE-PUTOR), University of Port-Harcourt, Rivers State, Nigeria.*

[4]*Department of Biochemistry, Faculty of Natural Sciences, Chukwuemeka Odumegwu Ojukwu University, Anambra State 431124, Nigeria*

Corresponding author. E-mail: adetunjicharles@gmail.com

ABSTRACT

The highest rate of industrialization recorded globally has been documented as the major reason for the high rate of environmental pollution majorly through the utilization of pesticides. The constant application of pesticides for the management of agricultural pests had led to high rates of bioaccumulation, high pesticide residue in preharvest and postharvest food commodities, disruption of the ecosystem, climate changes, and impairment of human health as a result of exposure to various hazards. Therefore, there is a need to search for a sustainable solution that will mitigate all the highlighted problems. The application of biosorption has been discovered as a typical example of biotechnological techniques to several conventional techniques. The uniqueness of their wide application for the eco-restoration of heavily polluted soil and water could be linked to their eco-friendly, cost-effectiveness, sustainable, significant reduction of heavily polluted soil with several pesticides. Therefore, this chapter provides a comprehensive review of the application of biosorption as a next-generation biotechnology technique for the bioremediation of heavily polluted soil. Moreover, the modes of action and the structure of beneficial microorganisms are highlighted in detail. The advantages and demerits of biosorption as a sustainable biotechnology tool were also highlighted.

9.1 INTRODUCTION

In every agricultural activity that requires planting and harvesting, pesticides are used. Pesticides are used to eradicate uninvited pests and insects that hinder the productivity and yields of growing crops. The majorly used pesticides in agricultural activities globally are "glyphosate" and "atrazine" (PANA, 2016a, b).

The residues of pesticides set-off in the environment have a tendency to remain in the biosphere for a very long time, and persist in the biological food chain; thus making the entire ecosystem unbalanced. The Geological Survey of United State of American led a study on the impact(s) of pesticides in the ecosystem and reported that about 90% of the water and aquatic resources were adulterated with pesticides (Chhunthang and Katoch, 2017). These were consequents from the huge agricultural activities influenced by pesticide usage on urban lands (Aktar et al., 2009).

It is estimated that about 99% of the deaths linked to pesticides ensue in emerging nations such as Nigeria, and have also been recorded that about 25% of global production of pesticides are used therein (Ojo and Joshua, 2016). In Nigeria, about $400m yearly is spent on pesticides importation—Nigerian Stored Products Research Institute (Punch Newspaper, 2017). The application of these pesticides has resulted in an inordinate hazard to plants storage safety and human health in this region and the turnover gained by the use of these pesticides to protect plants is negated in view of the related health risk involved. However, the Government of Nigeria has set a unity threshold (0.01 mg/L) for the discharge of all types of agrochemicals (pesticides) into a national watercourse (Kola and Lawal, 1999). Nevertheless, most of these chemicals have their different shelf-life, biodegradable patterns, noxiousness, persistence nature, and accumulation tendency which may not be eco-friendly to most aquatic biota.

Glyphosate (round-up) is one of the widely used pesticides in Nigeria. It belongs to the class of pesticides called "organophosphates" precisely a phosphonate. Glyphosate is widely used for killing unwanted plant (weeds) that compete with its nutrients by the following mode of action: inhibition of the flora enzyme 5-enolpyruvylshikimate-3-phosphate synthase (United States EPA, 2007). The WHO in March 2015 categorized glyphosate as "possibly cancer-causing in humans" (group 2A). This was founded in the in vitro epidemiological studies on fauna (Cressey, 2015; Guyton et al., 2015; IARC, 2015). A study gap conducted by the European Food Safety Authority in November 2015 revealed that glyphosate is improbable DNA damaging substance (genotoxic) that can cause cancer in humans (European Food Safety Authority, 2016). On the other hand, the European Chemicals Agency categorized glyphosate as triggering severe eye injury and poisonous to water biota. However, it did not indicate if it is one of the main causes of cancer or has impact on reproductory structures (ECHA, 2017).

Nanotech (Nanotechnology) is a scientific operation of how ions, atoms, and molecules are manipulated in a larger supra-molecular scale (Drexler and Eric, 1986, 1992). Numerous researchers have focused their hard work on the growth of more cheaper, natural, eco-friendly, and sustainable way of using microbes in combating flora disease-causing organisms (Adetunji and Sarin, 2017). The application of beneficial microorganisms with high efficient biosorption potential for the bioremediation of pesticide contamination of freshwater and soil environment is a bio-nanotechnology approach which is the focus of this study.

The use of microbes as biosorption nanoprocess to biodegradation pesticides in different media have been used by Mishra et al. (2014), Abdelmalek and Salaheldin (2016), Patra et al. (2012), Kanhed et al. (2012), Bramhanwade et al. (2016), and Adetunji et al. (2017a, b). However, there is paucity of literature on the use of it in the decontamination of pesticides in water and soil.

9.2   BIO-SORPTION POTENTIAL FOR THE BIOREMEDIATION OF PESTICIDE, HEAVY METALS, AND OTHER XENOBIOTICS CONTAMINANTS IN FRESHWATER AND SOIL ENVIRONMENT

Mustapha and Halimoon (2015) reviewed the biosorption potential of microbes on heavy metals in the surroundings. The main focus was on industrial effluent and sediment comprising heavy metals. They opined that the use of conservative techniques in decontaminating pollutants is very expensive and ineffective. However, they proposed the use of microbes in the biosorption of environmental contaminants in waste stream. This is based on the fact that they have the ability to bind metallic ions and metabolize it to create a nontoxic residue that will not recontaminate the aquatic system.

Gupta et al. (2015) wrote a comprehensive review of the bioadsorbents remediation potentials of heavy metals by microbes. The authors stated that biosorption procedure has been recognized as physiognomies of dead dry weight of both cellulose and microbes bind to repair metallic ion contaminants from semi-liquid suspension. They also suggested that biosorption process has high efficiency of decontaminating pollutants such as heavy metals to a very low concentration. In conclusion, they proposed biosorption as one of the best greener alternatives in the bioremediation of contaminants compared to other techniques because of its multimechanistic role in the ecosystem.

Ahalya et al. (2003) evaluated the biosorption of heavy metals by microbes. The authors reported that pollutants such as lead (Pb), chromium (Cr), mercury (Hg), uranium (Ur), selenium (Se), zinc (Zn), arsenic (As), cadmium (Cd), gold (Au), silver (Ag), copper (Cu), and nickel (Ni) introduced into the watercourse via industrial and farming activities have significantly impacted the biota therein and consequently man along the food chain. In conclusion, they proposed a sustainable biotech approach using microorganisms as a biosorption remedy.

Lebeth et al. (2018) tested the biosorption potential of microorganisms (filamentous fungi isolates) in decontaminating cadmium in coastal water and sediment. The study focused on the biosorption capabilities of isolated fungi on cadmium (Cd) from coastal waters and sediments. Water and sediment samples collected from Ibo, Lapu-Lapu City, Cebu, Philippines were subjected to various potato dextrose agar plates containing serial dilution separation of Cd concentration; 25, 50, 75, and 100 ppm, to undergo bioremediation and biosorption processes. The results of the biological controlled experiment revealed distinct colonies that propagated on the maximum Cd concentration (100 ppm). This was further isolated into clean cultures. The findings of the in vitro biosorption assay showed that the fungi isolates of genus *Aspergillus* and *Penicillium* were the most dominant colonies and were able to decontaminate the Cd in the coastal water and sediment collected.

Abdia and Kazemi (2015) reviewed the biosorption potential of heavy metals comparing them with different biosorbents. The authors accounted that the key benefits of biosorption are its excellent metal recovery capability, eco-friendly ability, and it is economical. They also proposed natural dry mass for biosorbents purpose, sourced from by-products of flora and agricultural wastes.

Kanamarlapudi et al. (2018) evaluated the application of biosorption in the abstraction of heavy metals from waste-water. The authors reported that since water is very vital to humans in all ramifications, the attendant issues faced; pollution via effluents from industrial, agricultural, and other point sources activities, the conventional approach for the decontamination of metallic ions have not only really helped in toxicant removal, instead, it has complicated the bioremediation process by releasing eco-unfriendly xenobiotics into the remediating system. Moreover, they are not cheap and require technical expertise to execute.

The authors proposed a biosorption and bio-accumulation procedure using microbial isolates because they are cheap and eco-friendly. Moreover, these alternate methods have some pros or advantages above the conformist approaches. In conclusion, natural resources like industrial by-products, microbes' dry weight, and agricultural wastes were recommended as probable bio-sorbents for contaminants like heavy metal abstraction because of the existence of metal-binding capacity. Nonetheless, the future possibility for biosorption as a wastewater treatment is highly recommended.

Abatenh et al. (2017) reviewed the utilization of microbes in bioremediation of wastes. They reported that the dietary ability of microbes varied totally, that is why they are commonly used as biosorption amalgam for bioremediation procedures: decontamination, extermination, check, or reclamation of varied biochemical wastes in the environment via the all-encompassing activities of microbes. In addition, all these procedures are carried out in an enzymatic mode via bio-metabolizing. This will ensure that any toxin is not set-off after the reclamation process into the environment. The authors also reported that extant approaches and tactics used globally are biopiles, bio-augmentation, bio-attenuation, bio-venting, and bio-stimulation. In conclusion, they opined that each of these techniques has sole advantage(s) over each other.

Ahemad and Kibret (2013) reviewed the current developments in microbial biosorption of heavy metals. The authors stated that there are many bioremediation approaches as well as biosorption method to remediate heavy metal in contaminated regions. They reported that presently, the utilization of microbes for biosorption of heavy metals have been well thought-out to be one of the best dependable substitutes above the conservative technique (bioremediation). Finally, they proposed the utilization of microbes: fungi, bacteria, the best bio-sorbents for heavy metal.

Azubuike et al. (2016) evaluated the pros, prospects, and limitations of some classified bioremediation techniques. They reported that contaminants such as hydrocarbons, heavy metals, pesticides, and nuclear wastes have posed several health challenges in the ecosystem, as a result of noxiousness from them. The authors proposed bioremediation techniques using microbes for remediating contaminated sites for the reason that they are less economical, well-known to be effective to degrade xenobiotics, and environment friendly. In situ or ex situ approaches have been employed in degrading xenobiotics. However, ex situ procedures seem to be more cost-effective related to the in situ methods based on the facts on the materials needed on-site and incapability to successfully envisage and control the subsuperficial of contaminated regions. The authors conclusively recommend two main methods to improve bioremediation: bio-augmentation and bio-stimulation on condition that certain militating factors in the environment are well considered and check.

Gadd reviewed the significance and implication of using biosorption methods in decontaminating contaminants. He opined that biosorption is a physical and chemical process that needs mechanisms such as superficial complexation, precipitation, adsorption, absorption, and ion exchange. He further explained that biosorption is a characteristic of both existing and deceased organisms. The author further presaged biosorption as a favorable biotechnology for noxious waste abstraction as well as contaminant reclamation based on its efficacy, uncomplicatedness, equivalent process to conservative ion-interchange expertise, and accessibility of dry weight. He concluded that there is need to exploit the industrial framework in large scale utilization of microbes for bioremediation in waste treatment.

Prakash (2017) wrote a comprehensive review of the utilization of different fungal species and their capability to biodegrade several xenobiotic substances and their eventual reduction into harmless substances. Some of these fungal species also have the potential to absorb toxic pesticides and metabolized them into nontoxic compounds through the use of their unique enzymes. The stated that white rot fungi have the capability to degrade some environmental pollutants like pesticides and polyaromatic compounds. The authors also gave special insight into the mechanism utilized by which various fungal species utilized for the biodegradation of polyaromatic hydrocarbon and heavy metal. Moreover, it was stated that white rot fungi specially secrete various array of enzymes for the biodegradation of various contaminants most especially pesticide. Their review shows that micoremediation is a sustainable solution that could be used as a permanent replacement for various chemical treatments of hazardous compounds.

Wołejko et al. (2016) wrote a comprehensive review of the likelihood of utilizing several substrates that have potential that could enhance the bioremediation of contaminated soils with pesticides and heavy metals. The authors stated that bioengineering could be used as a sustainable solution that could facilitate adequate bioremediation process mainly through bio-augmentation and bio-stimulation. They utilized enzymatic, microbial, organic substrates, natural sorbents, and nanoparticles which could absorb pesticides and other pollutants from the contaminated environment. They also emphasized that the application of genetic engineering could be a sustainable biotechnological tool used in the selection of plant and microorganism that have a capability to effectively biodegrade pollutants like pesticides and heavy metal. The author also suggested that several trials need to be performed at the laboratory and at large scale using a field trial. Also, they also encourage the combination of interdisciplinary approaches from biochemistry, microbiology, environmental engineering, ecology, and processing engineering.

Karpouzas et al. (2005) performed an experiment using an enrichment culture techniques for the isolation of beneficial microorganisms that are capable of facilitating the process of biodegradation of organo-phosphorus pesticides containing ethoprophos and cadusafos, respectively, using soil from a potato monoculture area obtained from Northern Greece. These microorganisms were cultured using mineral salts medium amended with nitrogen-containing cadusafos (10 mg L^{-1}) as the only carbon source while the soil extract medium was utilized for the isolation of bacterial that are capable of cadusafos-degrading bacteria. The microorganisms capable of degrading these pesticides were coded CadII and CadI, respectively. They later subjected to molecular characterization using 16S rRNA genes. It was observed that isolate CadI exhibited 97.4% similarity to the 16S rRNA gene from a strain of Flavobacterium while CadII exhibited 99.7% similarity to the16S rRNA gene of a *Sphingomonas paucimobilis*. The author later observed that the two bacteria isolates possess the capability to absorb and biodegrade cadusafos within 48 h when cultured with the following media containing mineral salts medium and soil extract medium, respectively, with concurrent population growth. Their study showed that these bacteria portend the capability to utilize cadusafos as a carbon source. Moreover, it was observed that the degradation of cadusafos was enhanced when the mineral salts medium was amended with glucose when compared to the addition of succinate to mineral salts medium which led to the reduction in the degradation rate of cadusafos. Also, it was discovered that the two isolated were able to degrade ethoprophos which was a chemical analog of cadusafos but there was no significant biodegradation when tested with isofenphos and isazofos. Furthermore, the addition of *Sphingomonas paucimobilis* with inoculum densities (4.3×108 cells g^{-1}) in addition with soil freshly treated with ethoprophos and cadusafos resulted in enhanced biodegradation of the two nematocides used during this study. Their study showed that the bacteria used during their study possess the capability to absorb and biodegrade pesticides present in a contaminated environment and they can be used for sanitation of such heavily polluted environment.

Biofilms are structure of microbial community containing different microbial cells attached to the surface of each after which they are surrounded in a matrix of extracellular polymeric substances secreted by cells. Biofilm has been recognized as a biological tool produced from bacterial that could be utilized for the bioremediation of polluted environment such as pesticides, heavy metal, and other xenobiotic substances. Their biodegradation activity might be linked to the fact that could immobilize pollutants because of their high microbial biomass. The application of biofilm as a bioremediation tool has been documented in the treatment of polluted water, soil contaminated with several pollutants. Das et al. (2012) wrote a comprehensive on the utilization of biofilm as a biotechnological tool for the bioremediation of heavily polluted environment with several pollutants. The review work elaborates on the general application of biofilm for the eco-restoration of polluted environment which will serve as a baseline study for some other frontier research.

Tewari et al. (2017) wrote a comprehensive review of the application of beneficial microorganisms for the bioremediation of contaminated soil. The high rate of the industrialization and the introduction of green revolution have led to an increase in the utilization of synthetic pesticides for the control of various agricultural pests and diseases; thus mitigating against an increase in agricultural production. These chemical pesticides have been documented to play a crucial role by increasing the rate of food production but there are several adverse effects surrounding their utilization. Some of these disadvantages include the destruction of useful insects, elimination of parasites, predators and pollinators, resurgence of minor pests, and highly unstable ecosystem. Also, health hazard has been documented due to the continuous utilization of these noxious xenobiotic pesticides. Therefor there is a need to detoxify them. Some of the traditional methods that have been applied over years include incineration and landfills that might lead to the development of secondary contamination which might be linked to the leaching of pesticides into groundwater and surrounding soil and generation of emissions which are very toxic in nature. The authors also stated that the application of bioremediation technology could be utilized as a more eco-friendly and effective biotechnology solution for adequate clean-up of the polluted environment. This might be linked to the fact that the microorganism is capable of converting organic pollutants into partial or completely too stable nontoxic end-products and microbial biomass. Furthermore, it has been observed that most of these beneficial microorganisms that portend that capability to biodegrade several pollutants are found in the rhizosphere of a plant. Also, it was stated that microorganisms pose several biochemical modes of action utilized for the biodegradation of pesticides including aromatic nitroreductase processes, pesticide conjugation reactions, and formation, hydrolytic transformations, oxidative transformations by synthesizing various enzymes, and carbon–phosphorus bond cleavage reactions.

Javaid et al. (2016) wrote a comprehensive review of the application of biological control, agent for the removal of the pollutant from various agricultural soils. The author stated that the constant application of pesticides for the management of pests and disease and weeds has been a source of environmental concerns. They observed that the application of bioremediation techniques as a sustainable, cost-effective, environment friendly, and efficient methodology when compared to chemical and physical techniques which have been documented to be very costly together with several adverse effects on human health and environment. It has been documented that the process of bioremediation exhibits some level of sensitivity to the level of concentration of nitrogen and hydrogen peroxide within a particular microbial community as well as pH changes and temperature. The authors also suggested that there is a need to carry out modes of action involved in the process of absorption of pesticides and their effectual biotransformation.

Ding et al. (2013) demonstrated the biodegradation and biosorption of pyrene and phenanthrene using heat-killed and live *Phanerochaete chrysosporium* to establish the mechanism and factors that control the biodegradation of polycyclic aromatic hydrocarbons in aqueous solution. They also screened the best nutrient conditions (nitrogen and carbon source concentration), repeated-batch feed of polycyclic aromatic hydrocarbons on the rate of biodegradation and biosorption as well as their reaction to Cu^{2+}.

The result obtained shows that dead bodies of *P. chrysosporium* possess the capability to remove polycyclic aromatic hydrocarbons that might be linked to their high biosorption capability. The partition coefficients obtained for pyrene and phenanthrene were 17,500 L kg^{-1} and 4040 L kg^{-1}, respectively. Moreover, it was observed that the process of biodegradation and biosorption, constitute the major factor that led to the dissipation of polycyclic aromatic hydrocarbons by living *P. chrysosporium* in water. Also, it was observed that the biosorption of pyrene and phenanthrene was 52.21% and 19.71%, respectively. Furthermore, it was noticed that increase in the time of incubation between 3 to 40 days enhances the rate of biodegradation from 15.55% to 49.21% for pyrene and 20.40% to 60.62% for phenanthrene while there was a reduction in the level of polycyclic aromatic hydrocarbons in the fungal bodies. Also, the effect of carbon and nitrogen content of the nutrient condition also contributed to the high biodegradation of the pollutant. There was a significant increase in the bio-degradation of pyrene and phenanthrene, respectively, with 99.47% and 83.97% for pyrene and 99.55% and 92.77% for phenanthrene after 60 days of incubation. This might be linked to the improved biosorption capability of the fungal biomass when performed under carbon-rich nutrient and to stimulated-biodegradation when performed in the presence of limited nitrogen. It was also observed that the living *P. chrysosporium* exhibit the capability to remove the tested pollutant under the repeated-batch feed of phenanthrene and through a continuous and repeated process of biodegradation. The author stated that three cycles when incubated for 6 days will give biodegradation percentage up to 90%.

Trinder et al. (2016) hypothesized that dietary supplementation with *Lactobacillus* poses the capability to absorb and remove the level of toxicity in some organophosphate pesticides including chlorpyrifos and parathion. The author screened numerous species of *Lactobacillus* species for their capability to biodegrade 100 ppm of chlorpyrifos and parathion (CP) or parathion in De Man, Rogosa and Sharpe agar broth when cultured for 24 h in the growth medium.

The result shows that the *Lactobacillus* strains could not grow up to stationary-phase and exhibit irregular culture morphology when treated with pesticides. Moreover, it was observed that strain GG (LGG) of *Lactobacillus rhamnosus* which is pesticide-tolerant and strain GR-1 (LGR-1) of *L. rhamnosus* demonstrated that they possess the capability to absorb and biodegrade organophosphate pesticides available in the tested solution after 24 h of co-incubation. Also, it was observed that the metabolomics activity of strain *L. rhamnosus* GG was shown to be independent and no observable significant difference in the level of sequestration between live and heat-killed strains. Moreover, it was demonstrated that LGG and LGR-1 drastically decrease the absorption of CP or 100 M parathion when tested in an in vivo assay using Caco-2 Transwell model with the small intestine epithelium. The author also validates the potential of strain LGG and LGR-1 to influence the rate of sequestration on acute toxicity when tested on the newly enclosed *Drosophila melanogaster* flies that have been exposed to food without the addition of live LGG or with 10 M CP. The result obtained showed that LGG prevented CP-induced mortality. Their result shows that strain of *L. rhamnosus* used during this experiment could be used as a biotechnological tool for the reduction of toxic organophosphate pesticide which was exposed through passive binding. The study suggested that the experiment should be performed with livestock and clinical application should be carried out which might be linked to their affordability and their practical potential to be utilized as a supplement product together with food-grade bacteria. Their study showed that *Lactobacillus* mediated could reduce the level of toxicity in insects and diminish intestinal absorption when tested in applicable models. They could be of great application supplementing, apiary, livestock, and human foods containing probiotic microorganisms capable of decreasing organophosphate pesticide.

Basu (2014) screened the beneficial microorganism that is indigenous from various agricultural tracts available in Odisha which poses great capability to use as a bioremediation tool for eco-restoration of heavily polluted soil with chemical pesticides. These microorganisms are tested for their capability to degrade pesticides and their adsorption potential. Also, microbiological enrichment techniques were used to affirm their biosorption potential. Some other parameters such as pH and rpm were carried out to establish their adsorption efficiency.

The result obtained showed that 80% adsorption efficiency was obtained some parameters tested like agitation speed and pH was increased in the absence of heat treatment conditions of adsorbate in comparison to the result obtained in the presence of heat. Moreover, the application of mixed micro-bial consortia containing fungal and bacterial biomass plays a crucial role in the adsorption of toxic substances most especially the heavily contaminated sediments and soil with pesticides.

Diez (2010) wrote a comprehensive report on the practical application of microorganisms for the biodegradation of toxic pollutants in the environment including polychlorinated biphenyls, polycyclic aromatic hydrocarbons, and pesticides such as chlorophenols. The author highlighted some physico-chemical properties that enhance the rate of biosorption and biodegradation of pesticides in the soil. Moreover, detailed information about the rate of accumulation of pesticide residue in the details about residues of pesticides in superficial and groundwater water utilized in Chile and some foreign countries were provided in detail.

Furthermore, he also provided more details on the application of mycoremediation as a biotechnological tool for biodegradation of heavily polluted soil most especially with pesticides. Special emphasis was laid on the application of white rot fungus on their biodegradability potential on pesticides polluted soil. Some recent advances in the application of effective and simple systems like bi-lobed were utilized for the reduction of pesticide-contaminated environment most especially when filling the spraying equipment.

Adhikari et al. (2010) tested the application of *Bacillus* strain of S14 for its capability to adsorbed malathion effectively from the solution. This experiment was carried out to validate the capability of this strain to remove malathion from any aqueous solution. Malathion is a typical example of toxic organophosphate pesticide responsible for the high rate of environmental pollution. The result obtained from the experiment showed that the dry cells of Bacillus sp. S14 showed a high capability to remove malathion from solution. Moreover, it was observed that biosorption equilibrium was observed with 6 h. Also, the maximum biosorption value of 81.4% was observed from malathion when tested under the following environmental conditions such as dry biomass concentration 1 g L^{-1}, temperature 25 °C, and pH 6.5. The Freundlich and Langmuir isotherms tested showed that the latter showed an enhanced fit with data.

Furthermore, it was observed that the *Bacillus* sp. S14 can be stored up to 60 days most especially when stored in form of dried powdered cells most especially at room temperature while maintaining its biosorption efficiency. Their study shows that the dry cells of the strain used in this study (*Bacillus* sp. S14) can be utilized as a bio-sorbent for effective removal of malathion from aqueous solutions.

Gadd (1999) wrote a comprehensive review of the application of microorganism as a biosorption agent for the removal of a toxic substance which may be inorganic and organic in nature. It was also stated that biosorption is a biotechnological technique that involves both the dead and living microorganism as well as their component for effective removal of pollutants and from solution. It was also stated that biosorption is a physicochemical phenomenon that involves several mechanisms such as ion exchange, precipitation, absorption, surface complexation, and adsorption. It has been recognized as a sustainable technology because of numerous advantages which include the availability of biomass, efficiency, and analogous operation to conventional ion exchange technology. Examples of microorganisms and microsystems used in the process of biosorption, including fungi, bacteria, microalgae, have been documented for the removal of toxic substances like chemical pesticides, heavy metals, toxic metal, and radionuclides such as thorium and uranium. Their review work also highlighted several advantages, mesmerists, and the future potential of biosorption most especially in the industry as well as some biosorption that could be used for waste treatment in the environment.

Kumar et al. (2011) wrote a comprehensive review of the utilization of biological agents for the bioremediation of a heavily contaminated environment. It was stated in their review that bioremediation is one of the most effective biotechnological tools that could use for the management of toxic environment most especially heavily contaminated soil. It was as stated that bioremediation has been used for the cleaning of environment polluted worldwide most especially in Europe with large number of success. Moreover, the author stated that ex situ and in situ have been recognized as a special type of bioremediation used for the eco-restoration of heavily polluted soil with pesticides, landfill stabilization, biological carbon sequestration, endocrine disrupters, and mixed waste bio-treatment. Examples of such microorganism which have been identified in the bioremediation of contaminated sites include anaerobes, fungi, and aerobes, respectively.

Ahmad et al. (2018) wrote a comprehensive review on the application of microbial inoculation for the bioremediation of heavily polluted environment which includes pesticides and heavy metals. The world population has been observed to increase every day but there is a need to feed the ever-increasing population. The farmer most especially in the developing countries utilized pesticides including fertilizer as a typical example of agricultural input. Moreover, another scenario that led to an increase in the utilization of pesticides is the high rate of anthropogenic activities and industrialization has led to the high rate of accumulation of pollutants in the environment.

Some of the environmental strategies used previously have been observed with several adverse effects which include the development of a more resistant and toxic compound in the environment. Example of such techniques includes pyrolysis, recycling, incineration, and land-filling. Moreover, most of these techniques are very expensive and it becomes very difficult to apply them for the bioremediation of air, soil, and water. In view of the aforementioned, the application of green technologies has been discovered as a sustainable solution for the bioremediation of a heavily contaminated environment. Their review work discussed recent trends in the application of microbial inoculation for the bioremediation of heavily contaminated environment such as agrochemicals and the eco-restoration of heavily affected environment by pollution.

It has been discovered that the application of microbial-based inoculants will enhance quick uptake of nutrients, stimulate crop growth, and prevent plants from diseases and pests thereby serving as a permanent replacement to agrochemical used in food production. Their study also highlighted practical examples of microbial intervention that has been documented for the bioremediation of heavily polluted environment with heavy metals, poly-aromatic hydrocarbon, industrial effluents, and pesticides. It was stated that majority of such microorganism resides in the rhizosphere of plant roots which portends that capability to promote plant growth. Their review work showed that several microorganisms could be utilized for bioremediation of contaminated environment and as a biotechnological tool needed for the achievement of sustainable development.

Hai et al. (2012) performed a study to establish that combination of microorganisms and activated sludge cultures possessed a precise degradation pathway that could be utilized for the removal of recalcitrant organic substances and toxic pollutants from wastewater. Their study showed how most recalcitrant pesticides could be removed from the liquid mixture using white rote fungus and mixed culture of bacteria. The result obtained showed that the mixed culture containing fungus and bacteria was able to remove 62% 47%, 98% of alachlor, aldicarb, and atrazine available in the liquid phase, respectively, after incubating for a period of 14 days. The result obtained from their study showed the same trends with the previous result obtained for batches containing only acclimated fungus or formerly published elimination rates with nonacclimated bacterial cultures. Their study showed that biodegradation and biosorption were the main reasons responsible for the withdrawing of pesticides from the liquid phase.

Dhankhar and Hooda (2011) wrote a comprehensive review of the application of fungal biosorption for the removal of heavy metal from wastewater and other related organic and inorganic pollutant. Their study highlighted several practical experiments where biosorption has been utilized as a typical example of a biotechnological tool utilized for the bioremediation of polluted soil and their potential application for the removal of toxic substances from solutions. Much emphasis was laid on the application of fungal biomass for their application in numerous biosorption researches. The removal of heavy metal from the environment, especially wastewater, is now shifting from the use of conventional methods to the use of biosorption, which may be defined as the binding and concentration of selected heavy metal ions or other molecules on to certain biological material. Although most biosorption research concerns metal and related pollutants, including radionuclides, the term is now applied for particulates and all manner of organic pollutants as well. Such pollutants can be in gaseous, soluble, and insoluble forms. Biosorption is a physical process carried out through mechanisms such as ion exchange, surface complexation, and precipitation. It is a property of both living and dead organisms (and their components) and has been heralded as promising biotechnology for pollutant removal from solution. Various biomasses such as plant products (tree bark, peanut skin, sawdust, plant weeds, etc.) have been tested for metal biosorption with very encouraging results. In this comprehensive review, the biosorption ability of fungal biomass toward heavy metals is emphasized. Moreover, more emphasis was laid on the mechanism of action utilized by fungal bio-sorbents in the removal of pollutants such as pesticides and heavy metal. The authors stated that the fungal cell wall structure was assessed in terms of metal sequestration. The factors that influence the passive uptake of the pollutant were also stated. Also, the joining mechanism that regulates the absorption process was stated in detail which includes the major functional groups that enhance the process was also stated. Some recent trends on sorption kinetics, fungal bio-sorbent sorption, and sorption isotherms as well as the procedure involved in the quantification of metal–biomass relations was stated in detail. The authors also stated the current and future directions by which biosorption most especially on the metal-binding potential of fungal biomass on how they could be applied at the industrial level was highlighted.

It has been discovered that there are limited microbiological processes studies that have been proved not to be effective in analyzing the adverse influence of pesticides. Therefore, there is a need to search for more comprehensive techniques. It has also been documented that adverse effect of pesticides may induce alteration in the expression of certain genes as changes occur in the diversity. Feld et al. (2015) tested the influence of pesticides on the expression of a gene and focus specifically on the *amoA* gene which regulates the process involved in the oxidation of ammonia.

The author tested the effect of mancozeb, dazomet, and no pesticides on soil microcosms. The authors hypothesized that the quantity of *amoA* transcript and level of bacterial diversity reduces whenever pesticides are added to it and in order to test the hypothesis they utilized for reverse-transcription qPCR. This was verified using molecular techniques using diversity evaluation of the 16S ribosomal RNA and RNA genes, 454 sequencings which established the total number of total and active soil bacterial communities present in a particular environment. Their result shows that application of dazomet decreases archaeal *amoA* transcript numbers and bacterial number more than 2 log units while long-term influence for more than 28 days.

Moreover, the application of Mancozeb also reduced the *amoA* transcripts. The archaeal amoA transcripts and the bacterial showed a high level of sensitivity. Also, it was also showed that the level of nitrate production available in the Namended microcosms correlates with the number of bacterial *amoA* transcripts. Furthermore, it was discovered that Dazomet decreases the number of bacterial colony by 1 log unit but the population size was restored after 12 days. It was later observed that the diversity of the soil bacterial was re-established 12 h after application. The molecular analysis carried out established that *Proteobacteria* and *Firmicutes* were detected at day 12, while no adverse effect was recorded from mancozeb on the microbial diversity.

Rani et al. (2014) wrote a comprehensive review of the application of biodegradation and bioremediation of contaminated water and soil. It has been stipulated that there will be a drastic increase in the population of human being global, which implies that there is a need to feed the ever-increasing population. The authors also stated that there is an increase in the rate at which the application of hybrid seed, genetically modified crops, fertilizers, and pesticides are been applied. This might be linked to the urgent need of our time to increase food production that will cater for the ever-increasing population. The application of pesticides has been a known practice that could enhance the increase in food production and mitigates against several agricultural pests and diseases affecting agricultural crops. Some of them include nematode, insects, weeds, bacteria, and fungi.

The author highlighted that there are three major approaches that are important to bioremediation which include phytoremediation, bioremediation, bio-augmentation, and bio-stimulation. It was also stated that the biodegradation of pesticides using beneficial microorganisms like fungi, bacteria, and algae has been shown to be an economical, efficient, reliable, and economical methodology for the detoxification of unwanted chemicals. Moreover, they also stated that the utilization of genetically modified microorganisms using recombinant strain for degradation has been highlighted as a significant biotechnological solution that could enhance the process of soil bioremediation.

9.3 RECENT BIOTECHNOLOGICAL TECHNIQUES USED IN BIOREMEDIATION

Several researchers have documented numerous researches on the application of genetic engineering for improving the capability of microorganisms to biodegrade pesticides and hydrocarbon. The application of recombinant DNA techniques has been used to validate the several in vitro experimental assays that resolve several environmental challenges and eco-restoration of hazardous waste most especially under laboratory conditions. It has been established that genetically engineered microorganisms portend that capability degrades various pollutant under several defined conditions. The application of genetic modification technology has led to several recent advancements and sustainable technology that could resolve several bioremediation challenges. The process of bioremediation utilizes metabolic versatility and uses their gene diversity (Fulekar, 2009). It has been discovered that the genetic structure of these microorganisms equipped them as a sustainable biotechnological tool used several applications in resolving several environmental challenges. Some of these biotechnological techniques include biosorption, biodegradation, bioaccumulation, and biotransformation. Moreover, the most important indispensable blueprint of gene encrypting the enzymes that regulate the process of biodegradation could be found in the extra-chromosomal DNA and chromosomal of these microorganisms performing the process of bioremediation.

The application of recombinant DNA methods enhances the capability of microorganisms through the recognition of such degradative genes and their eventual transformation into appropriate host through necessary host using appropriate vector with the help of appropriate promoter. The application recombinant DNA technology utilized particle bombardment methodology, antisense RNA methodology, site-directed mutagenesis, antisense RNA technique, polymerase chain reaction, and electroporation. Moreover, the application of recombinant DNA technology has led to several tremendous progresses in the application of pollutant-degrading microorganisms through genetic alteration of specific regulatory and metabolic genes, strain improvement. All these are important biotechnological tools that led to all these highlights success toward the development of bioremediation tools which are economical, eco-friendly, safe, and sustainable. Their application also extends to their application in the soil, air, raw material from industrial, and water.

9.4 MECHANISM OF ACTION UTILIZED BY MICROORGANISM FOR THE BIOREMEDIATION OF CONTAMINATED ENVIRONMENT

The accomplishment of any bioremediation of heavily polluted environment leverages on the level of chemical structure of organic molecules available on the contaminated soil (Neilson and Allard, 2008; Beek, 2001). A typical example of such compounds that are difficult to metabolize by microorganisms during the process of biodegradation is xenophores (e.g., substitutions of H with Cl, NO_2, CN, and SO_3 groups). Therefore, most of these pollutants having such xenophores properties always constitute recalcitrant to microbial degradation (Talley, 2005; Alexander, 1999; Neilson and Allard, 2008; Beek, 2001). It has been discovered that most of these biodegradation process by these microorganisms utilize several hydrolytic enzymes for detoxification and eventual breaking down of most contaminants. These enzymes include *ligases, oxidoreductases, isomerases, hydrolases, transferases,* and *lyases.*

It has been discovered that toluene dioxygenase possesses the capability to degrade more than 100 different types of compounds which include chlorobenzene, nitrobenzene, and TCE. Other examples of other microbial biodegradative enzymes that facilitate the process of biodegradation include hydratases that regulate the addition of water to alkenes changing them into secondary alcohols, esterases, separate ester bonds by adding water; depolymerases, which hydrolyze polymers; glutathione *S*-transferase moves the thiol group to chlorinated compounds with affiliated dechlorination, racemases regulates the conversion of Land D-amino acid into CoA-ligase, which adds -S-CoA to fatty acids during the process of beta-oxidation, glutathione *S*-transferase regulates the movement of thiol group to chlorinated compounds with associated dechlorination (Neilson and Allard, 2008; Beek, 2001; Alvarez and Illman, 2006; Haggblom and Bossert, 2004; Rittmann and McCarty, 2001).

Moreover, it has been discovered that the properties of contaminants are crucial to contaminant–soil interactions, the movement of contaminants, and their capability to destroy and immobilize contaminants. Some of the significant properties of contaminant include molecular weight, Solubility in water, diffusion coefficient, density and aqueous solution chemistry, molecular weight, vapor pressure, and dielectric constant (Sara, 2003).

9.5   CONCLUSION, RECOMMENDATION, AND FUTURE DIRECTION

This study has highlighted various applications of biosorption techniques as a biotechnological tool for the bioremediation or eco-restoration of heavily polluted soils. Moreover, the modes of action through which beneficial microorganisms execute their bioremediation process most especially through the production of array of biotechnological important enzymes for breaking down of heavy pollutants present in soil and water. This study also listed array of beneficial microorganisms that have been utilized for the eco-restoration heavily polluted soil and water with various pesticides. This chapter also highlighted several in vitro, in vivo, ex situ, and in situ laboratory experiments where various beneficial microorganisms have been applied as a biotechnological tool for bioremediation of heavily polluted soil and water. There is a need to screening for more efficient strain from numerous utilized and different agro-ecology areas for high efficient strains that could perform detoxification of organic and inorganic pesticides and their effectual removal from heavily polluted soil and water.

KEYWORDS

- **beneficial microorganisms**
- **biosorption**
- **bioremediation**
- **pesticide contamination**
- **freshwater**

REFERENCES

Abatenh, E.; Gizaw, B.; Tsegaye, Z.; Misganaw Wassie Abatenh, E.; Gizaw, B.; Tsegaye, Z. et al. Application of microorganisms in bioremediation—review. *J. Environ. Microbiol.* **2017**, *1*(1), 02–09.

Abdelmalek, G.M.A.; Salaheldin, T.A. Silver nanoparticles as a potent fungicide for citrus phytopathogenic fungi. *J. Nanomed. Res.* **2016**, *3*(5), 00065.

Abdia, O.; Kazemi, M. A review study of biosorption of heavy metals and comparison between different biosorbents. *J. Mater. Environ. Sci.* **2015**, *6*(5), 1386–1399.

Adetunji, C.O.; Kumar, J.; Swaranjit, A.; Akpor, B. Synergetic effect of rhamnolipid from *Pseudomonas aeruginosa* C1501 and phytotoxic metabolite from *Lasiodiplodia pseudotheobromae* C1136 on *Amaranthus hybridus* L. and *Echinochloa crus-galli* weeds. *Environ. Sci. Pollut. Res.* **2017a**, *24*(15), 13700–13709. http://dx.doi.org/10.1007/s11356-017-8983-8.

Adetunji, C.O.; Neera, B.S. Impacts of biogenic nanoparticle on the biological control of plant pathogens: mini review. *Adv. Biotechnol. Microbiol.* **2017**, *7*, 3. doi: 10.19080/AIBM.2017.07.55571.

Adetunji, C.O.; Oloke, J.K.; Pradeep, M.; Jolly R.S.; Anil, K.S.; Swaranjit, S.C.; Bello, O.M. Characterization and optimization of a rhamnolipid from *Pseudomonas aeruginosa* C1501 with novel biosurfactant activities. *Sustain. Chem. Pharm.* **2017b**, *6*, 26–36.

Adhikari, S; Chattopadhyay, P; Ray, L. Biosorption of Malathion by dry cells of an isolated *Bacillus* sp. S14. *Chem. Speciation Bioavail.* **2010**, *22*(3), 207–213.

Ahalya, N.; Ramachandra, T.V.; Kanamadi, RD. Biosorption of heavy metals. *Res. J. Chem. Environ.* **2003**, *7*, 4.

Ahemad, M.; Kibret, M. Recent trends in microbial biosorption of heavy metals: a review. *Biochem. Mol. Biol.* **2013**, *1*(1), 19–26. doi:10.12966/bmb.06.02.2013.

Ahmad, M.; Pataczek, L.; Hilger.; T.H.; Zahir, Z.A.; Hussain.; A, Rasche, F.; Schafleitner, R.; Solberg, SØ. Perspectives of microbial inoculation for sustainable development and environmental management. *Front. Microbiol.* **2018**, *9*, 2992. doi:10.3389/fmicb.2018.02992.

Aktar, M.W.; Sengupta, D.; Chowdhury. Impact of pesticide use in agriculture: their benefits and hazards. *Interdiscip. Toxicol.* **2009**, *2*(1), 1–12.

Alexander, M. *Biodegradation and Bioremediation*, Academic Press, San Diego, CA, **1999**.

Azubuike, C.C.; Chikere, C.B.; Okpokwasili, G.C. Bioremediation techniques classification based on site of application: principles, advantages, limitations and prospects. *World J. Microbiol. Biotechnol.* **2016**, *32*, 180. doi.10.1007/s11274-016-2137-x.

Basu, A. Kinetic biosorption studies of chemical pesticides by microbial biomass isolated from agricultural soils. *Master Sci. Biotechnol.* **2014**, 1–56.

Beek, B. *The Handbook of Environmental Chemistry: Biodegradation and Persistence.* Springer-Verlag, Berlin, Heidelberg. **2001**, *2*, Part K.

Bramhanwade, K.; Shende, S.; Bonde.; S, Gade, A.; Rai, M. Fungicidal activity of Cu nanoparticles against *Fusarium* causing crop diseases. *Environ. Chem. Lett.* **2016**, *14*(2), 229–235.

Chhunthang, L.; Katoch S.S. Biosorption of Malathion pesticide using *Spirogyra* sp. *Inter. J. Environ. Agri. Res.*. **2017**, *3*, 3.

Cressey, D. Widely used herbicide linked to cancer. *Nature.* **2015**. doi:10.1038/nature.2015.17181.

Das, N.; Basak, L.V.G.; Abdul, S.J.; Abigail, M.E.A. Application of biofilms on remediation of pollutants—an overview. *J. Microbiol. Biotech. Res.* **2012**, *2*(5), 783–790.

Dhankhar, R.; Hooda, A. Fungal biosorption—an alternative to meet the challenges of heavy metal pollution in aqueous solutions. *Environ. Technol.* **2011**, *32*(5), 467–491. doi:10.1080/09593330.2011.572922.

Diez, M.C. Biological aspects involved in the degradation of organic pollutants. *J. Soil. Sci. Plant. Nutr.* **2010**, *10*(3), 244–267.

Ding, J.; Chen; B.L.; Zhu, L.Z. Biosorption and biodegradation of polycyclic aromatic hydrocarbons by *Phanerochaete chrysosporium* in aqueous solution. *Chin. Sci. Bull.* **2013**, *58*, 613–621. doi:10.1007/s11434-012-5411-9.

Drexler, K.; Eric. *Engines of Creation: The Coming Era of Nanotechnology.* Doubleday. **1986**.

Drexler, K.; Eric. *Nanosystems: Molecular Machinery, Manufacturing, and Computation.* John Wiley & Sons, New York, **1992**.

ECHA (European Chemicals Agency). *Glyphosate Not Classified as a Carcinogen-ECHA.* **2017**.

European Food Safety Authority. Effects in an Agricultural Soil Ecosystem as Measured by amoA Expression and Quantification. *Glyphosate Report* (PDF). EFSA. **2016a**.

European Food Safety Authority. *Glyphosate: EFSA Updates Toxicological Profile.* **2016b**. Efsa.europa.eu.

Feld, L.; Hjelmsø, M.H.; Nielsen, M.S.; Jacobsen, A.D.; Rønn, R.; Ekelund, F. et al. Pesticide side effects in an agricultural soil ecosystem as measured by amoA expression quantification and bacterial diversity changes. *PLoS One.* **2015**, *10*(5), 0126080. https://doi.org/10.1371/journal.pone.0126080.

Fulekar, M.H. Bioremediation of fenvalerate by *Pseudomonas aeruginosa* in a scale up bioreactor. *Romanian Biotechnol. Lett.* **2009**, *14*(6), 4900–4905

Gadd, GM. *Biosorption: Critical Review of Scientific Rationale, Environmental Importance and Significance for Pollution Treatment.* Society of Chemical Industry. **1999**. www.interscience.com, DOI 10.1002/jctb.1999.

Gupta, V.; Nayak, A.; Agarwal, S. Bioadsorbents for remediation of heavy metals: current status and their future prospects. *Environ. Eng. Res.* **2015**, *20*(1), 1–18. doi: https://doi.org/10.4491/eer.2015.018.

Guyton, K.Z.; Loomis, D.; Grosse, Y.; El Ghissassi, F.; Benbrahim-Tallaa, L.; Guha, N.; Scoccianti, C.; Mattock, H.; Straif, K. Carcinogenicity of tetrachlorvinphos, parathion, malathion, diazinon, and glyphosate. *Lancet Oncol.* **2015**, *16*(5), 490–491. doi:10.1016/S1470-2045(15)70134-8. PMID 25801782.

Haggblom, M.M.; Bossert, ID. *Dehalogenation: Microbial Processes and Environmental Applications,* Kluwer Academic Publishers, **2004**.

Hai, F.I.; Modin, O.; Yamamoto, K.; Fukushi, K.; Nakajima, F.; Nghiem, L.D. Pesticide removal by a mixed culture of bacteria and white rot fungi. *J. Taiwan Inst. Chem. Eng.* **2012**, (43), 459–462.

IARC (International Agency for Research on Cancer, World Health Organization). Press release: evaluation of five organophosphate insecticides and herbicides. *Monographs.* **2015**, *112*, 1–2.

Javaid, M.K.; Ashiq, M.; Tahir, M. Potential of biological agents in decontamination of agricultural soil. *Scientifica.* **2016**. http://dx.doi.org/10.1155/2016/1598325

Kanamarlapudi, S.L.R.K..; Chintalpudi, V.K.; Muddada, S. Application of biosorption for removal of heavy metals from wastewater. *Intechopen.* **2018**, http://dx.doi.org/10.5772/intechopen.77315.

Karpouzas, D.G.; Fotopoulou, A.; Menkissoglu-Spiroudi, U.; Singh, B.K. Non-specific biodegradation of the organophosphorus pesticides, cadusafos and ethoprophos, by two bacterial isolates. *FEMS Microbio. Ecol.* **2005**, *53*, 369–378.

Kola, R.I.; Lawal, S.L. Degradation of aquatic environment by agro-chemicals in the middle belt of Nigeria. In Osuntokun, A. (ed.), *Environmental Problems of Nigeria.* Ibadan, Davidson Press. **1999**, *78*–88.

Kumar, A.; Bisht, B.S.; Joshi, V.D.; Dhewa, T. Review on bioremediation of polluted environment: a management tool. *Inte. J. Environ. Sci.* **2011**, *1*(6), 1079–1093.

Lebeth, C.; Manguilimotan, Bitacura, J.C. Biosorption of cadmium by filamentous fungi isolated from coastal water and sediments. *Hindawi J. Toxicol.* **2018**, 7170510, 6. https://doi.org/10.1155/2018/7170510.

Mishra, S.; Singh, B.R.; Singh, A.; Keswani, C.; Naqvi, A.H. et al. Bio fabricated silver nanoparticles act as a strong fungicide against *Bipolaris sorokiniana* causing spot blotch disease in wheat. *PLoS One.* **2014**, *1*(9), 97881.

Mustapha, M.U.; Halimoon, N.B. Microorganisms and biosorption of heavy metals in the environment: a review paper. *Environ. Eng. Res.* **2015**, *20*(1), 5.

Neilson, A.H., Allard, A.S. Environmental degradation and transformation of organic chemicals, CRC Press/Taylor. **2008**.

NSPRI (Nigerian Stored Products Research Institute). *Nigeria Spends $400m Annually on Pesticides. Punch Newspaper.* **2017**. punching.com.

Ojo, J. Pesticides use and health in Nigeria. *IFE J. Sci.* **2016**, *8*, 4.

PANA (Pesticides Action Network, North America). *The Lynchpin of Industrial* Aga. **2016a**, http://www.panna.org/pesticides-big-picture/lynchpinindustrial-ag.

PANA (Pesticides Action Network, North America). *Pesticides Action Network, North America: Pesticides, The Big Picture.* **2016b**, http://www.panna.org/pesticides-bigpicture/.

Patra, P.; Mitra, S.; Debnath, N.; Goswami, A. Biochemical-, biophysical-, and microarray-based antifungal evaluation of the buffer mediated synthesized nano zinc oxide: an in vivo and in vitro toxicity study. *Langmuir.* **2012**, *28*(49), 16966–16978.

Prakash, V. Mycoremediation of environmental pollutants. *Inter. J. Chem. Tech. Res.* **2017**, *10*(3), 149–155.

Rani, K.; Dhania, G. Bioremediation and biodegradation of pesticide from contaminated soil and water—a novel approach. *Int. J. Curr. Microbiol. App. Sci.* **2014**, *3*(10), 23–33.

Rittmann, B.E.; McCarty, P.L. *Environmental Biotechnology: Principles and Applications.* McGraw Hill, **2001**.

Sara, M.N. *Site Assessment and Remediation Handbook,* Lewis Publishers. CRC Press, **2003**.

Talley, J.W. *Bioremediation of Recalcitrant Compounds.* In Talley, J. (ed), Taylor & Francis, **2005**.

Tewari, L.; Saini, J.; Arti. Bioremediation of pesticides by microorganisms: general aspects and recent advances. *Biomed. Pollut.* **2017**, 25–49.

Trinder, M.; McDowell, T.W.; Daisley, B.A.; Ali, S.N.; Leong, H.S.; Sumarah, M.W.; Reid, G. Probiotic *Lactobacillus rhamnosus* reduces organophosphate pesticide absorption and toxicity to *Drosophila melanogaster. Appl. Environ. Microbiol.* **2016**, *82*, 6204–6213. doi:10.1128/AEM.01510-16.

United States EPA. *Pesticide Market Estimates Agriculture, Home and Garden,* **2007**.

Wołejko, E., Wydro, U.; Łoboda, T. The ways to increase efficiency of soil bioremediation. *Ecol. Chem. Eng. S.* **2016**, *23*(1), 155–174. doi:10.1515/eces-2016-001.

CHAPTER 10

Biological Disease Control Agents in Organic Crop Production System

AJINATH DUKARE,[1*] SANGEETA PAUL,[2] PRIYANK MHATRE,[3] and PRATAP DIVEKAR[4]

[1]*Horticultural Crop Processing Division, ICAR-Central Institute of Post-Harvest Engineering and Technology, Abohar, Punjab 152116, India*

[2]*Division of Microbiology, ICAR-Indian Agricultural Research Institute, New Delhi 110012, India*

[3]*Division of Plant Protection, ICAR-Central Potato Research Station, Ooty, TN 643004, India*

[4]*ICAR-Indian Institute of Vegetable Research, Regional Research Station, Sargatia, Kushinagar 274406, UP, India*

Corresponding author. E-mail: ajinath111@gmail.com

ABSTRACT

Extreme and irrational use of synthetic chemicals for improved plant protection and plant productivity causes several adverse impacts on ecosystem. Considerable amount of economic losses has been caused to crop especially fungal as well as nematode pathogens. Thus, eco-safer methods of plant disease control are requisite for maintaining sustained crop productivity with eco-stability. Disease suppression using antagonistic microbes has been proposed as a feasible substitute for managing fungal diseases of crop plants. Microbe-mediated biological control reduces the severity of crop disease/damage, inflicted by numerous phytopathogens (fungi, nematode pests, etc.) via production of different antifungal compounds in natural conditions. In this approach, fungal pathogens are antagonized by microbial biocontrol agents through production secretion of mucolytic enzymes, antibiotics, volatile biocidal compounds, and/or competition for nutrients and space between biocontrol agents and pathogen. All these mechanisms are a powerful tool for lowering the harmful damages of pathogen through preventing their deleterious effects on crop growth. Microbes belonging to different taxonomic group of bacteria, viruses, and fungi are employed in the biological suppression of phytopathogens. Such bioagents can more efficiently grow, survive, and proliferate in several agro-horticultural ecosystems. In last few decades, such disease protection approach is considered a potential pest management strategy owing to accumulation of harmful chemical residues in biosphere. With this background, this chapter covers the aspects of crop disease control using biocontrol microbes, types of bioagents deployed in the biocontrol process, their modes of action, and commercialization aspect.

10.1 INTRODUCTION

Plant pathogens causing severe economic diseases in crop plants are the main peril to food production and environment stability globally. Among the pathogenic organisms, fungal pathogens are a chief disease-producing group imposing considerable economic losses in several agro-horticultural crops. They are important disease causative with more than 60% of the plant disease literature related to the fungal pathogens of crops (Hawksworth et al., 1995). Worldwide, about 10%–20% of agricultural production is influenced by pathogenic microbes that deny getting adequate food to almost 800 million peoples (Strange and Scott, 2005). Every year, crop diseases caused by fungi, parasitic nematodes, bacteria, and viruses result in the economic losses in terms of billions of dollars. According to the latest estimate, about 25% of crop wastage occurs globally primarily due to different pests and diseases (FAO, 2015). Such losses are more often in developing countries as agriculture is predominant sector in economic growth of such nations. Henceforth, such disease needs to be controlled for upholding the sufficiency and quality of produced foodstuff, feed, and fibers. Several strategies are being used to curb the plants diseases caused by several fungal pathogens, though, beyond the good cultivation practices, heavy dependency on synthetic fertilizers and pesticides for disease management have polluted our ecosystem and deteriorated the ecosystem (Dukare et al., 2013; Dukare and Paul, 2018). In addition, pathogens resistance against synthetic fungicides has developed. Due to all these reasons, present research is being focused more on developing alternative approaches to synthetic pesticides for disease management.

Among the alternatives approaches, pathogen biocontrol using antagonistic microorganism is capable and eco-safer method. Numerous kinds of plant- and soil-linked bacteria are exploited for alleviation of damaged in plants caused by several biotic as well as abiotic stresses (Paul et al., 2017). Biological suppression of crop diseases using plant growth-promoting rhizobacteria has been promoted for sustainable crop protection in organic agriculture system (Heydari and Pessarakli, 2010). In the biological control approach, antagonistic organisms prevent the activities, reproduction, and proliferation of phytopathogens occurring in their vicinity. A diversity of microbial biocontrol agents are accessible for utilizing; however, for auxiliary progress and successful acceptance, it will necessitate a better understanding of the intricate interactions among crops, people, and the ecosystem. Biocontrol methods could be well suited in a nonchemical and eco-friendly approach to sustainable crop production system. Consequently, biological control of plant diseases has now appeared as a broad concept, evident in the accounts, and encompasses several modes of actions. This chapter focuses on brief overview of concepts of microbial biological control agents (BCAs) used for sustainable disease management.

10.2   FUNDAMENTALS OF BIOLOGICAL DISEASE CONTROL

Biological control aims to reduce the population of disease causing pests or a pathogen through the use of naturally occurring organisms, manipulating the environment, or by mass introduction of bioagents in nature (Sterling, 1991). The organism responsible for suppressing pest or pathogen is termed as the BCAs. An initial trial for biological plant disease control was attempted against damping-off disease of conifer seedlings incited by *Pythium debaryanum* via soil introduction of fungal and bacterial antagonists. Weinding's (1936) findings of production of antifungal substance otoxin by *Trichoderma* species and parasitization of hyphae of *Rhizoctonia* spp. generated great enthusiasm in the biocontrol process. Yet, from the 1940s later chemical fungicides began to play an important way of crop protection from pathogen and biological control remained sidelined for considerable period of time.

In biocontrol phenomenon, various crop pathogens are antagonized by BCAs through secretion of lytic enzymes, antibiotics, siderophores, antifungal volatiles, or competition for minerals/space with pathogens. Such mechanisms of antagonistic organisms formulate them as a powerful weapon for lowering the adverse impacts of harmful pathogens on the crop.

Different species of organisms, belonging to several taxonomic groups such as bacteria, viruses, and fungi are being utilized for eco-safer disease control. In general, BCAs have the ability to grow, survive, and proliferate in numerous agro-horticultural ecosystems and are suitable for the best candidate for the biocontrol. An organism to be utilized as an ideal antagonist must possess certain attributes for successful biological control. Biological disease management is growing as a potential control strategy in recent years because of adverse impacts of excessive chemical usages associated with global ecological perspectives.

10.3   DIVERSITY OF MICROORGANISM EXPLOITED IN THE BIOLOGICAL DISEASE CONTROL APPROACH

10.3.1   BACTERIAL DISEASE CONTROL AGENTS

Bacteria are the predominant taxonomic group used for safer plant protection in the organic farming. Among the several genera of bacteria, different species from *Alcaligenes, Arthobacter, Bacillus, Enterobacter, Erwinia, Hafnia, Pseudomonas, Streptomyces, Serratia,* and *Xanthomonas* are identified as BCAs of crops (Weller, 1988). *Pseudomonas* spp. and *Bacillus* spp. are widely known bacteria for their ability of disease repression. The biocontrol actions of numerous *Pseudomonas* strains are ascribed to different mechanisms including mycoparasitism, antibiosis, root colonization, and competition. For example, *Pseudomonas fluorescens*, a ubiquitous soil bacterium, has been the most reported as potential candidate for pathogen inhibition. In bio-suppression, *Clavibacter michiganenis*, an etiological agent of tomato canker disease have been successfully managed using fluorescent pseudomonads (Amkraz et al., 2010). In a similar way, Bacillus species are also crucial in suppression of crop diseases induced by soil-borne fungal pathogens (Von Der Weid et al., 2002). The endospore-forming *Bacillus* spp. survives well under hostile environmental situations for extended period. Several species of *Bacillus* are recognized as efficient BCA for their use in greenhouse or open field environments (Kloepper et al., 2004). Likewise, certain species of other Gram-negative bacteria such as *Aeromonas, Arthrobacter, Enterobacter, Lysobacter,* and *Serratia* have also demonstrated potential suppression of pathogen activities. Furthermore, plant growth-enhancing actinomycetes also possesses broad-spectrum antibacterial or antifungal activities (El-Tarabily and Sivasithamparam,

2006; de Vasconcellos and Cardoso, 2009). Extracellular lytic enzymes such as chitinases, glucanase, protease, etc, producing bacterium holds a great promise as biological disease control due to their increased activities of mycoparasitism. For example, chitinolytic bacteria, *Bacillius subtilis,* and *P. fluorescens* and their chitinase enzymes exhibited antagonisim toward crop root rot pathogens (*Rhizoctonia solani* and *Fusarium solani*) in vitro assay (El-Mougy et al., 2009). Several authors have reported antifungal chitinase of *B. subtilis* demonstrating powerful growth inhibitory actions against several pathogenic fungi (Kobayashi et al., 2002; Yang et al., 2009).

10.3.2   FUNGI AS BIOLOGICAL DISEASE CONTROL AGENTS

Numerous species of fungi can also be utilized as crop BCAs that may provide useful fungicidal action against a variety of phytopathogens. Species in the fungal genera such as *Aspergillus, Ampelomyces, Chaetomium, Candida, Coniohyrium, Chaetomium, Cryptococcus, Gliocladium, Pythium, Rhodotorula,* and *Trichoderma* have been identified as antagonistic against plant pathogens (Soytong et al., 2005; Fravel, 2005; Ezziyyani et al., 2007). Among several fungi, use of *Trichoderma* species as BCAs against numerous plant pathogens including *Pythium* spp., *Fusarium* spp. *Phytophthora palmivora, Botrytis cinerea,* and *Rhizoctonia* spp. have been mentioned in several reports (Zeilinger and Omann, 2007). Some strains of *Trichoderma* like *T. harzianum, T. viride,* and *Trichoderma hamatum* are also reported as effective pathogen biocontrol agents (Harman, 2006). Another fungus, *Ampelomyces quisqualis* is mycoparasitic fungi that inhibits growth and kills powdery mildews by penetrating and killing hyphae (Kiss, 2003). Additionally, some species of *Chaetomium* and *Gliocladium* have shown biocidal property toward various soil-borne and seed-borne phytopathogens (Viterbo et al., 2007).

10.4   MODES OF ACTIONS OF BIOCONTROL AGENTS

BCAs use one or multiple mode of actions for suppression of growth, activities, or populations of invading plant pathogens of agricultural crops. Often more than single mechanism is operates in successfully preventing the pathogens growth (Dukare et al., 2018). Mechanisms of action of microbial antagonist responsible for disease control are discussed below.

10.4.1 MYCOPARASITISM

Mycoparasitism is the process that involves parasitism and death of pathogens due to direct action of specific BCA. During the process of mycoparasitism, growth of pathogen is suppressed and/or its propagules are killed. Fungi that are parasitic on other harmful fungi are usually called as mycoparasites. In general, process can be accomplished either through production of coiled structure around fungal hyphae or growing around it and/or producing cell wall breaking enzymes. Biological control of root-knot nematode by *Pasteuria penetrans* is a classical example of obligate bacterial hyper-parasitism. Disease suppression through parasitism is generally of common happening in kingdom fungi. Species of *Trichoderma* are well known for exhibiting mycoparasitism on pathogens such as *Rhizoctonia bataticola* and *Armillaria mellea* (Lo et al., 1998). In nature, wide ranges of fungi are parasitized by many mycoparasites and few of them are implicated in the fungal growth control occurring under natural ecosystem. Several mycoparasites of fungal pathogens such as *Coniothyrium minitans and Pythium oligandrum* are known to attack sclerotia and living fungal mycelium. Diverse hyper-parasites fungi such as *Acremonium alternatum, Ampelomyces quisqualis, Cladosporium oxysporum,* and *Gliocladium virens* causes parasitism in pathogen responsible for causing powdery mildew disease in several agricultural crops (Kiss, 2003).

10.4.2 ANTIBIOTICS PRODUCTION

Antibiotics are a low-molecular-weight containing group of organic compounds that are harmful to the growth and developmental activities of nearby microbial pathogens (Duffy, 2003). Growth inhibition of pathogen through the production of one or more antibiotics is the most common mechanism found in many antagonistic plant growth promoting rhizobacteria (PGPR) strains (Glick et al., 2007). Numerous antifungal volatile metabolites such as hydro cyanide (HCN), alcohols, sulfides, ketones and aldehydes, and nonvolatile nature such as 4-diacetylphloroglucinol (2 DAPG), phenazine-1-carboxylic acid, mupirocin, pyocyanin hydroxy phenazines, and pyrrolnitrin (Ahmad et al., 2008), are produced by biocontrol bacteria. lipopeptide containing antifungal antibiotics such as surfactin, bacillomycin, zwittermicin A, and iturins are mainly produced by fungicidal species of *Bacillus.*

In general, antibiotic compound inhibits the development of pathogens by inhibiting the cell wall synthesis, alteration in cell membrane, and blocking or preventing the process of protein synthesis. Many different species of bacterial genera such as *Bacillus*, *Pseudomonas*, *Streptomyces*, *Burkholderia*, *Pantoea*, *Lysobacter,* and *Enterobacter* are predominantly implicated in antibiotics production and antibiotics mediated fungal disease control. Diverse types of antibiotics, namely, DAPG, phenazines, pyoluteorin, pyrrolnitrin, pseudane, viscosinamide, amphisin produced by *Pseudomonas* species are recognized to have antagonisim toward fungal pathogens. *Bacillus* species are notorious for the production of numerous antibiotics including circulin, colistin, and polymyxin. Several pathogens belonging to plant pathogenic fungi, Gram-negative, and Gram-positive bacteria are susceptible to antibiotics of *Bacillus* (Maksimov et al., 2011). Similarly, antibiotics producing strains of *Pantoea* spp, *Lysobacter* spp, and *Enterobacter* spp have also shown the biocontrol potential against crop diseases. The partial list of some bioagents deployed biological suppression of crop disease has been summarized in Table 10.1.

10.4.3   PRODUCTION OF CELL WALL DEGRADING ENZYMES

The extracellular enzymes that can degrade wide range of compounds such as chitin, hemicelluloses, cellulose, protein, and DNA are produced by many BCAs. The production of lytic enzymes can help BCAs to parasitize and kill fungi. Biocontrol microbes having abilities to produce mucolytic enzymes have biocontrol potential (El-Tarabily, 2006). Chitin degrading enzymes are commonly distributed in the 1 bacterial genera of *Aeromonas, Arthrobacter, Bacillus, Clostridium, Chromobacterium, Enterobacter, Erwinia, Flavobacterium Klebsiella, Pseudomonas, Streptomyces, Serratia,* and *Vibrio.* PGPR that can produce hydrolytic enzymes is efficient BCAs against a broad range of phytopathogenic fungi including *F. oxysporum, R. solani, S. rolfsii, Phytophthora* spp., *P. ultimum,* and *Botrytis cinerea,* (Singh et al., 1999). The fungal cell wall, primarily composed of chitin and glucan, is degraded by cell wall lytic enzymes such as chitinases, β-1,3 glucanases, proteases, cellulases, and lipases, produced by BCAs. These enzymes also adversely affect conidial growth and germ-tube extension of fungal pathogens and destroy their oospores (El-Tarabily, 2006). Some strains of bacterial genera *Lysobacter* and *Myxobacteria* produce large quantity of lytic enzymes that were found effective in inhibiting growth of pathogens. The report of some studies has shown in vitro disintegration of pathogens cell walls by the action of chitinase or β-1,3-glucanase alone or in combination. Likewise, the production of β-1,3-glucanase enzyme by *Lysobacter enzymogenes* strain C3 significantly contributed to its biocontrol potential (Palumbo et al., 2005). It is clear that microbial production of these lytic enzymes is often needed to utilize the polymeric compound for obtaining carbon nutrition for their cellular growth.

10.4.4   PRODUCTION OF ANTIFUNGAL VOLATILE COMPOUNDS AND SECONDARY METABOLITES

Different volatile compounds such as HCN, ammonia, and other secondary metabolites produced by BCAs also curb and suppress pathogenic fungi growth. Ahmad et al. (2008) reported that bacterial strain capable of producing volatile antibiotic (HCN) may protect plants from fungal attack. Respiratory cytochrome oxidase pathway is effectively blocked by HCN and, therefore, it is extremely lethal to all aerobic organisms. HCN production by a few fluorescent pseudomonads is supposed to be primarily responsible for controlling root pathogens of agricultural crops. Under gnotobiotic conditions, the control of tobacco black root rot caused by *Thielaviopsis basicola,* was achieved through application of HCN producing *P. fluorescens* strain CHAO. This strain is widely recognized for the production of several antifungal metabolites such as HCN, DAPG, pyoluteorin, and pyoverdine (Maurhofer et al., 1995)

At present, actinobacteria produce 42% of total known secondary microbial metabolite compounds. As per report, actinomycete produces about 70%–80% of the known bioactive compounds. The largest numbers of secondary metabolites are secreted by a certain species of the genus *Streptomyces* (Berdy, 2005). Species of *Streptomyces* produce copious amounts of secondary metabolites having antibiotic and antifungal activities against *Colletotrichum gloeosporioides, Penicillium digitatum, Fusarium oxysporum, Sclerotium rolfsii, Aspergillus brassicicola, Fusarium lycopersici, Fusarium asparag,* and *Fusarium vasinfectum* (Khamna et al., 2009). Among them, antifungal metabolites produced such as kasugamycine from *Streptomyces kasugaensis,* Blastcidin-S from *Streptomyces griseochromogenes*, cycloheximide from *Streptomyces griseus*, and Rhizovit from *Streptomyces rimosus* have been identified and well familiar.

10.4.5   COMPETITION FOR LIMITING MINERAL NUTRIENT AND SPACE

In this process, pathogens are competitively excluded by utilization of limiting nutrients or by physical occupation of niche (Lorito et al., 1993). It is an indirect method of pathogen growth inhibition. In nature, the nutrient competition process occurs when BCAs reduce the availability of a specific mineral or other growth factors, thereby, limiting its access for pathogen growth. Competition for iron-mediated biocontrol mechanism has been reported and well-studied. Primarily, pathogens have a less competent iron uptake system than do for BCAs mainly due to the production of siderophores (Harman and Nelson, 1994). Siderophores are small-sized high-affinity iron chelaters produced by potential rhizospheric species of certain bacterial and fungal genera. In general, siderophores of BCAs have a high affinity for iron than do soil-borne fungal pathogen. The siderophore production is also feature of actinomycetes group (Khamna et al., 2009). Different kinds of siderophores produced by BCAs includes catechols produced by *Agrobacterium tumefaciens, Erwinia chrysanthemi*, pyoveridins of *Pseudomonas*, hydroxamates by *E. carotovora, Enterobacter cloacae*, and various fungi, and rhizobactin of *Rhizobium meliloti*.

Iron competition by siderophore producing *Pseudomonas* bacteria and their consequences on hindering soil-borne fungi growth are one of the mechanisms for pathogen biocontrol (Duijff et al., 1994). Siderophore-mediated nutrient competition caused by strain *P. putida* WCS358 was explained as major mode of action implicated in controlling the fusarium wilt disease of carnation. Similarly, chlamydospore germination of pathogenic *F. oxysporum* was reduced following the soil application of siderophore pyoverdine producing *Pseudomonas* strain. They correlated that the production of siderophore pyoverdine by *Pseudomonas* with pathogens chlamydospore inhibition and suggested a possible role of pyoverdines in disease management. Competition for occupying physical niche (site) and utilization of nutrients for growth may also occur between plant pathogens and microbial BCAs. Such a process competitively excludes low potential candidates from niche and thus, biocontrol occurs between bioagents and pathogenic fungi (Maloy, 1993). Therefore, the competition for infection site between two or more fungi is also crucial in the biological control process.

10.4.6 ELICITATION OF HOST DEFENSE TO PATHOGEN

In a response to pathogen attack, systemic acquired resistance (SAR) is often stimulated in the tissue of host plant. Several biochemical defenses are associated with induced systemic resistance including lignifications of host cell wall, callose deposition, activation of defense (e.g., chitinases, glucanases, peroxidases), and pathogenesis-related proteins and accumulation of phyto-alexins (Sticher et al., 1997). Induced systemic resistance (ISR) in plant tissue is triggered when plants are inoculated with PGPR strains. Rhizobacteria-triggered systemic resistance renders makes different uninfected parts of plant more resistant to invading fungal, bacterial, viral, nematodes phytopathogens as well as to insect pests of crop (Van Loon et al., 2007). Particularly, species of bacteria *Pseudomonas* and *Bacillus* are the most studied rhizobacteria for triggering ISR in host (Kloepper et al., 2004; Van Wees et al., 1997).

In ISR system, direct interaction between the PGPR and the pathogen is not needed (Bakker et al., 2007). Besides jasmonate and ethylene molecule, ISR in host plants is also activated by a number of rhizobacterial signaling molecules such as flagellar proteins, β-glucans, pyoverdine, lipopeptide, surfactants, lipopolysaccharide, *O*-antigenic side chain of the bacterial outer membrane protein, and salicylic acid. In addition, 2,3-butanediol, and other volatile substances of PGPR also function as chemical elicitors for induction of SAR and ISR (Ryu et al., 2005). Elicitor molecule for triggering ISR in plants varies from bacteria to fungi. For example, diverse signaling molecules elicitor for defense induction in host consists of cold shock proteins of diverse bacteria, lipopolysaccharides, and flagellin in Gram-negative bacteria; and in fungus, it involves β-glucans, transglutaminase and elicitins in Oomycetes, chitin, and ergosterol in most of fungi, xylanase in *Trichoderma* and invertase in yeast (Numberger et al., 2004). Various rhizo-sphere microbes have been notorious for the induction of plant host defense against invading plant pathogens. Many biocontrol strains of *Trichoderma* and *Pseudomonas* strongly induce host defenses against approaching fungal pathogens (Harman, 2006, Hase et al., 2001).

10.5 PLANT PARASITIC NEMATODES AND THEIR BIOLOGICAL MANAGEMENT

10.5.1 NEMATODE AS AN ETIOLOGICAL AGENT OF CROP DISEASE

In animal kingdom, phylum Nematoda is the largest phylum in terms of individuals and species. Most of them are free-living species; however, plant parasitic nematodes (PPNs) have received much interest owing to their nature and potential economic damage caused to important agricultural crops. Among the PPNs, root-knot (*Meloidogyne* spp.) and the cyst nematodes (CN) (*Heterodera* and *Globodera* spp.) are the most studied species due to their associated economic impact to several important crops. Root-knot nematodes (RKN) and CN are endoparasites and spend their entire life inside host roots (Gheysen and Mitchum, 2011; Jones et al., 2013). The life cycle of RKN and CN varies from two to three weeks to several months depending on the species of nematode, suitability of host, soil, and environmental factors. After hatching from eggs, second stage motile juvenile (J2) can locate the host, pierce and enter the roots and inject oesophageal gland enzymes into a selected host cell to form a sophisticated nematode feeding site (NFS). NFS is a metabolically active site. It acts as a "nutrient sink" and provides a constant supply of nutrients to growing nematode for its development and reproduction (Mhatre et al., 2015). Inside the roots, J2 undergoes three molts to become adult males (vermiform and motile) and females (sedentary, swollen, and nonmotile). Mature male comes out of the root, fertilize the females, and dies in the soil whereas, growing female's extracts constant nourishment from NFS to support production of several hundred eggs inside their own body (CN) or in a gelatinous matrix (RKN) secreted outside the body. After the death of *Heterodera* and *Globodera* females, their body hardens to form a protective cover called "cyst" (Kaushal et al., 2007). The cyst and gelatinous matrix protect eggs from unfavorable environmental conditions and helps in nematode survival for several months/years until their hatch under favorable host and environmental conditions.

Approximately, nematodes are responsible for 12.3% reduction in agricultural productivity and in monetary terms it is about $157 billion worldwide per annum (Mhatre et al., 2019). To avoid this huge loss, nematode management becomes immensely important. Several strategies that are being adopted for PPNs damage control including the use of crop rotation, intercropping, deep plowing, nematicides, resistant cultivars, and BCAs. However, abilities of PPNs to form protective cysts/gelatinous matrix and hydrobiotic/dauer stages allow them to survive in the soil for many years without host and thus making management approaches through crop rotation, intercropping and deep plowing as an unattractive. Use of resistant variety is one of the alternatives to manage PPNs but prevalence of different nematode pathotypes and incessant growth of resistant cultivars on the same land loses its resistance in due course of time. Chemical method is effective to control PPNs; however, its disadvantages are high cost, environmental and human health hazards. Hence, we should focus on the most efficient, economical, and eco-friendly management strategy aiming at safer crop protection and easy acceptance by the growers, biocontrol strategy meets all these requirements (Mhatre et al., 2019).

10.5.2 BCAS AND THEIR MECHANISMS OF FUNGICIDAL ACTION

In biocontrol strategy, several types of organism, viz. bacteria, fungi, viruses, nematodes, protists, and other invertebrates parasitizing parasitic nematodes have been used successfully for their management (Table 10.2). Most of the biocontrol agents have the potential to target and parasitize the nematodes and upon killing releases large amount of propagules in the soil which leads to the concept of "nematode suppressive soils" where nematodes could not establish themselves. But a single strategy could not be a solution for any problem, therefore, most of the time biocontrol agents are combined with other management strategies in integrated nematode management. Biocontrol agents exhibit different mechanisms of actions to reduce the population of PPNs and can be subdivided manly in two basic groups, namely. direct antagonism and indirect effect. Direct antagonism includes predation, parasitism, and production of toxins, cellular enzymes, and other antinematicidal metabolic products, whereas indirect effects involve regulation of nematode behavior that adversely affects host recognition and penetration, abnormalities in the nematode feeding sites, alteration of sex ratio, competition for nutrient/space, ISR and plant growth promotion via enhanced nutrient uptake, etc. (Mhatre et al., 2019).

10.5.3 BACTERIAL BIOAGENTS

Bacteria, being one of the most abundant soil organisms had been successfully utilized in management of PPNs due to their potential biocontrol ability (Tian et al., 2007a). Diverse kinds of nematophagus bacterial agents, isolated from soil, different stages of nematodes, and host-plant tissues have been identified. Among them, *Pasteuria, Pseudomonas,* and *Bacillus* are the most explored genera due to their potential role in management of PPNs. In general, bacteria affect PPNs by different ways, namely, direct parasitization, producing toxins or enzymes which are lethal to nematodes, competing for space and nutrients, inducing systemic resistance, and promoting plant growth (Siddiqui and Mahmood, 1999). Based on the mechanism of nematode suppression, nematophagous bacteria can be subdivided into different groups as follows.

10.5.4 OBLIGATE AND OPPORTUNISTIC PARASITIC BACTERIA

The *Pasteuria* is Gram-positive, dichotomously branched, endospore forming, obligate parasite of wide range of invertebrates originally observed parasitizing water flea, *Daphnia* spp. Till date, six species of *Pasteuria* parasitizing PPNs have been identified. The *Pasteuria* endospore attachment to the nematode cuticle is required for initiating the process of infection then it penetrates nematode cuticle by forming a germ tube. The life cycle of *Pasteuria* is synchronized with the nematode life cycle and after completion of life cycle, it releases endospores in environment upon rupturing the body of killed nematode. *Pasteuria* became the most promising biocontrol agent for many nematode species due to their potential role in suppression of nematode reproduction (Mankau et al., 1976; Mohan et al., 2012). Opportunistic parasitic bacteria are nonobligate parasitic bacteria associated with nematode cuticle, body cavity, gut, gonads, etc., with a potential to target the nematodes as one of the possible hosts by infecting and killing them, for example, *Brevibacillus laterosporus* and *Bacillus* sp. B16 (Oliveira et al., 2004; Niu et al., 2005).

10.5.5 RHIZOBACTERIA

The crop rhizosphere harbors many species of rhizobacteria which helps the host plant to fight against several soil-borne pests, diseases, and nematodes. The rhizobacteria exhibit different mechanism of nematode suppression that includes direct antagonism by parasitization, producing toxins, enzymes, and other antinematicidal metabolic compounds and indirect effect by manipulating nematode behavior, altering root exudates, producing repellents that affect the host recognition, altering the nematode sex ratio inside the root, competing for essential nutrients, promoting growth of plants by enhancing nutrient uptake, producing phytohormones, nitrogen fixation, phosphate and potassium solubilization, and ISR as a front line of defense against parasitic nematodes. Among them, *Bacillus* spp. and *Pseudomonas* spp. are the most exploited bacterial genus against several species of plant-parasitic nematodes (Gopalakrishnan et al., 2017; Mhatre et al., 2019).

10.5.6   ENDOPHYTIC AND SYMBIOTIC BACTERIA

Endophytic bacteria are associated with plant system and found inside plant parts, viz. root, shoot, fruits, etc. These bacteria do no harm host plant instead display a broad spectrum of antagonistic activity against several parasitic organisms including nematodes. This action is mediated by direct antagonism, repellent action, and by stimulating expression of genes involved in the plant defense. The examples are *Pseudomonas putida* against *Radopholussimilis*, *B. cereus* against *M. incognita*, *Streptomyces* spp. Against *M. javanica* (Arvind et al., 2009; Su et al., 2017; Hu et al., 2017). Certain species of bacterial genera *Photorhabdus* and *Xenorhabdus* are the symbionts of entomopathogenic nematodes *Heterorhabditis* spp., and *Steinernema* spp., respectively. Potential antagonistic effect of the symbiotic bacteria on various PPNs has been reported by several researchers (Lewis et al., 2001).

10.5.7   CRY PROTEIN-FORMING BACTERIA

B. thuringiensis (Bt) is the crystal inclusions (Cry or d-endotoxins) producing bacteria and are toxic to a wide array of insect species including nematodes. The mechanism of Bt crystal toxicity in nematodes is same as that of insect toxicity, that is, damage to the intestine by forming a pore. Six Cry proteins (Cry5, Cry6, Cry12, Cry13, Cry14, and Cry21) are potentially toxic to the juveniles of several free-living and PPNs (Wei et al., 2003; Ravari and Moghaddam, 2015).

10.5.8   FUNGAL ANTAGONIST

Several fungi had been studied to have nematophagous potential. These fungi exhibit different modes of action such as predator, parasite, antagonist, plant growth promotion, and induction of defense. Different types of fungal group used in plant-parasitic nematode control are discussed here:

Endoparasitic fungi are the obligate class of plant-associated nematodes parasites and spend their complete life inside the host. The infection process begins when fungal spores adhere and germinate while attached to nematode cuticle followed completion of their entire life cycle within the host tissue. The examples of endoparasitic fungi are *Nematophthora gynophila*, *Catenaria auxiliaries*, *Hirsutella rhossiliensis*, etc. (Timper and Brodie, 1994). The predacious fungi are the group of fungi that capture/trap and kill nematodes in soil. To capture nematodes, these fungi produce different kinds of structures/traps, such as adhesive hyphae, for example, *Stylopagehydrea,* adhesive branches, for example, *Dactylella clonopaga, D. lobate,* adhesive knobs, for example, *Dactylaria candida, D. hyptotyla, a*dhesive nets, for example, *Arthrobotritisoligospora, A. musiformis,* nonconstructing rings, for example, *D. lysipaga* and constructing rings, for example, *Dactylella bembicoides, A. dactyloides* (Stirling and Mani, 1995). Another important group is opportunistic fungi, which can colonize the reproductive organs of nematodes and seriously affect their fecundity. The nematode females exposed out of roots, cysts, and eggs released in the soil are the most vulnerable and important targets of opportunistic fungi. Among the different species of opportunistic fungi, *Paecilomyces lilacinus,* and *Verticillium chlamydosporium* are the most studied nematode biocontrol agents by many workers (Khan et al., 2006).

The endophytic fungi can colonize the plants without causing apparent disease symptoms and plays a multifactorial role in nematode suppression. The mechanism of action involved against nematodes is direct predatory behavior by attacking and trapping, competition for resources, production of nematostatic or nematicidal metabolites, producing plant hormones which help to augment the plant growth, tolerance, and hot defenses against nematodes. Based on ecology the endophytic fungi can be characterized as balanciaceous/grass endophyte and nonbalanciaceous endophytes. The grass endophytes comprise the genera *Balansia* and *Epichloe* within the ascomycetes, whereas nonbalanciaceous endophytes genera of ascomycetes include *Acremonium, Alternaria, Fusarium, Colletotrichum,* and *Trichoderma* (Schouten, 2016). In addition, vascular arbuscular mycorrhiza (VAM) is the obligate root symbionts and capable of protecting their host plant from the attack of several biotic stresses including PPNs. VAM plays a multifaceted role against plant-parasitic nematodes which include increased plant tolerance by higher nutrient acquisition and altered root morphology, competition for space/nutrients, stimulation of ISR, and alteration of rhizosphere interactions such as alteration of root exudates which affects host finding and rhizosphere microbiome. The examples of VAM are *Glomus intraradices, G. versiforme, G. mosseae,* etc. (Schouteden et al., 2015).

10.5.9   *PREDATORY NEMATODES AS BCAS*

Predatory nematodes can feed on rhizosphere associated soil microorganisms as well as parasitic nematodes. They are constantly associated with

PPNs in the rhizosphere and release nutrients available for plans, which provide tolerance to the plants under nematode burden on roots. Majority of the predatory nematodes belong to the orders Mononchida, Aphelenchida, Diplogasterida, and Dorylaimida. However, Diplogasterids are the most suitable nematode predators due to their short life span, prey-specificity, easy cultivation, chemotaxis sense, and hardy nature (Khan and Kim, 2007).

10.5.10   MONONCHID PREDATORS

They have a strong sclerotized buccal cavity, with one tooth, puncturing teeth, and number of grasping teeth to cut/puncture their prey. If the prey is smaller in size then they may completely swallow their preys whereas they may feed on larger preys by cutting into small pieces. Several studies demonstrated the significant reductions in PPNs with increasing population densities of Mononchid species such as *Mononchus aquaticus, Iotonchus monhystera, Mylonchulus sigmaturus, Parahadron chusshakily, P. punctatus, I. Tenuicaudatus,* etc. This shows the efficient predatory capabilities of Mononchids against PPNs. However, some flaws from practical point of view such as intolerance to environmental fluctuations, low fecundity, cannibalism, and relatively longer life cycle, limit their use as a successful biocontrol option.

10.5.11   DORYLAIMID PREDATORS

Dorylaimids predators are armed with long hollow stylet/odontostyle that is used for puncturing the prey and sucking up the food via oral aperture of odontostyle. They are omnivorous in nature. While feeding on nematodes, the odontostyle dislocates the internal organs of the worms and makes them immobile in a quick time. The examples of Dorylaimid predators are *Labronema vulvapapillatum, Aquatidesthornei, Eudorylaimus obtusicaudatus,* etc.

10.5.12   DIPLOGASTERID PREDATORS

The Diplogasterids predators of nematodes have a small buccal cavity equipped with teeth. Majority of them feed on nematodes and other soil organisms. Their life cycle is short (8–15 days) and hence, can be easily cultured in simple nutrient media. Compared to other groups of nematode predators, they are more prey-selective. Several studies demonstrated the biocontrol potential of Diplogasterid predators against different PPNs, viz. *Odontopharynx longicaudata, Mononchoides fortidens, M. longicaudatus,* and *M. gaugleri.* Diplogasterid predators can be used successfully in bio management of PPNs because of their chemotaxis sense, short life cycle, high predation and reproduction rate, prey specificity, easy of culture, and due to their survival adaptations under adverse environmental conditions (Bilgrami et al., 2005; Khan and Kim, 2005).

10.5.13   APHELENCHID PREDATORS

Aphelenchids are armed with a fine needle-like stylet which is used in penetrating the cuticle of prey and for injecting the digestive enzymes into the prey body. These enzymes paralyzes the prey instantly followed by ingestion of body content. Among Aphelenchids, few species of genus *Seinura* are predatory in nature. With a short life cycle of 3–6 days, capability of feeding on larger preys, high reproductive potential, and easy and rapid cultivability, this species can be a potential biocontrol agent with further studies at field and commercial level (Khan and Kim, 2007).

10.6   DEVELOPMENT OF BIOCONTROL FORMULATION AND ITS COMMERCIALIZATION

Usage and application of BCAs at commercial scale have limitations largely due to the variable performances under diverse field and environmental circumstances (Fravel, 2005). Many BCAs have success in the laboratory and greenhouse conditions but unable to perform in the field. However, the performance of many BCAs under these situations can be improved through an in-depth understanding of the various environmental factors influencing microbial antagonist's growth and metabolism (Wang et al., 2003). Additionally, the cost of developing, registering, and marketing of biocontrol products may also negatively affect investment in the development and manufacturing of commercial microbe based biocontrol formulation (Ardakani et al., 2009).

Formulations of BCAs can be prepared as dust, wet powders, granules, and liquid products using different carrier materials. Presently, many bio-based products are available in the market for the control of different fungal phytopathogens. Over the last few decades, many strains of biocontrol microbes have been identified for control of crop disease; however, only some of these have been formulated and commercialized into biocontrol products. Such product formulations mandatory requires the regulatory approval based on its disease control efficacy and third-party evaluation for the safety of human health and ecosystem. In India, large numbers of microbial BCAs are available for the control of pests and diseases of crop plants. Such products are either marketed as stand-alone products or formulated as mixtures with other microbial antagonistic agents. Moreover, a few products with biocontrol properties may be sold instead as plant growth promoters without any specific claims of crop disease control. To increase the application of BCAs at commercial level, more emphasis need to be given on several aspects including training of growers, formulation development process, and assessing the role of environmental factors on BCAs

10.7   CONCLUSION AND FUTURE PROSPECTS

Environmental pollution and rise of fungicide resistance in pathogens are some prime negative aspects related to the synthetic fungicide use in the crop disease control. Due to this, high focus is paid to eco-friendly technologies such as biological control that have minimum dependence on synthetic chemicals. Microbial mediated biological disease suppression is an emerging and alternative strategy for the control of many phytopathogens. This strategy is also concurrent with the goal and practices of agricultural sustainability. For a successful biocontrol approach, deeper know-how of cropping pattern, pathogens epidemiology, etiology, biology, ecology, and population dynamics of biocontrol strain and the overall interactions between these is necessary. This allows the rational basis for choice and construction of more efficient antipathogenic biocontrol organisms. Globally, large numbers of bioagents are available for usage, but further progress and their successful adoption will often necessitate a better indulgence of the complex interactions among crops, peoples, and the ecosystem. Further, advances in the new molecular techniques give insights into interactions between the antagonist and pathogen, ecological traits of antagonists and thus aid in improving the efficacy of microbial BCAs. Emergence of new pathogen strains and mechanisms their pathogenicity is different on host plants, for that reason, it is very immense to find out new and novel biocontrol microorganisms having different mechanisms of pathogen inhibition.

Though, in near future, biological control will be an alternative strategy for the crop disease control, however, other methods of pathogen control in IPM strategy are still needed in various prevailing environmental conditions because an agro-ecosystem is a variable and functioning system influencing crop and pathogens development. Therefore, integrative disease management strategies need to be considered and applied for an efficient reduction of crop disease severity and thereby improvement in the crop yield in the different cropping pattern.

TABLE 10.1 The Partial Example Biological Control Agents and Their Reported Mode of Action Against Targeted Crop Pathogens

Example of BCAs	Target Phytopathogens	Mechanisms of Antagonism	References
Pseudomonas chlororaphis O6	*Rhizoctonia solani* and *Fusarium graminearum*	Antibiotic (pyrrolnitirin) production	Park et al. (2011)
Pseudomonas fluorescence	*Sclerotium rolfsii*	Antibiotic (2,4 DAPG) production	Asadhi et al. (2013)
Bacillus subtilis RP24	Growth inhibition of many fungal pathogens	Antibiotic (Iturin) production	Grover et al. (2010)
Bacillus sp. A3F	*Sclerotium sclerotiorum*	Antibiotic (Bacillomycin) production	Kumar et al. (2012)
Lysobacter sp. strain SB-K88 A	*Aphanomyces cochlioides*	Xanthobaccin production	Islam et al. (2005)
Pantoea Agglomerans C9-1	*Erwinia amylovora*	Production of herbicolin	Sandra et al. (2001)
Trichoderma virens	*Rhizoctonia solani*	Antibiotic (Gliotoxin) production	Wilhite et al. (2001)
Pseudomonas PGC2	*R. solani* and *Phytophthora capsici*	Cell wall lytic enzyme production	Arora et al. (2008)
Bacillus alvei NRC 14	*Fusarium oxysporum*	Lytic enzyme production	Abdel-Aziz (2013)
P. fluorescens LPK2, *Sinorhizobium fredii* KCC5	*F. oxysporum* and *Fusarium udum*	Production of beta-glucanases and chitinase	Ramadan et al. (2016)
Pseudomonas spp.	*Colletotrichum falcatum*	Siderophores production	Viswanathan and Samiyappan (2007)
Brevibacter antiquum Acinetobacter tandoii Pseudomonas monteilii	*Macrophomina phaseolina*	Siderophore production	Gopalakrishnan et al. (2011)

TABLE 10.1   *(Continued)*

Example of BCAs	Target Phytopathogens	Mechanisms of Antagonism	References
P. fluorescens CHAO	*Thielaviopsis basicola*	HCN production	Ahmed et al. (2008)
P. fluorescens, P. corrugate, P. hlororaphis, P. aurantiaca	Inhibition of various pathogenic mycelium growth and spore germination	Volatile metabolites	Fernando et al. (2005)
Burkholderia cepacia, Ralstonia solanacearum	Detoxification of fusarium pathotoxin	Detoxification, inactivation, and degradation of pathogenicity factors	Scuderi et al. (2009)

TABLE 10.2   List of Biological Control Agents and Their Reported Mode of Action Against Targeted Nematodes of Crops

BCAs	Target Nematode	Mechanism of Action	Reference
Obligate parasitic bacteria			
Pasteuria penetrans, P. thornei, P. nishizawae; Pasteuria usage	323 nematode species of 116 genera	Parasitism	Tian et al. (2007a)
Opportunistic parasitic bacteria			
Bacillus sp. B16	*Panagrellus redivius*	Parasitism, production of enzymes and toxin	Tian et al. (2007b), Niu et al. (2005)
Brevibacillus Laterosporus	*Bursaphelenchus xylophilus*		
Rhizobacteria/Endophytic bacteria			
B. subtilis	*Rotylenchulus reniformis*	Interfering with recognition, production of toxin, nutrient competition, plant-growth promotion; induction of systemic resistance	Schmidt et al. (2010), Tabatabaei and Saeedizadeh (2017), Tiwari et al. (2017), Mhatre et al. (2019)
P. fluorescens	*M. javanica*		
P. putida	*R. reniformis*		
Rhizobium etli	*M. incognita; G. pallida*		
Bacillus tequilensis and *Bacillus flexus*	*M. incognita*		

TABLE 10.2 *(Continued)*

BCAs	Target Nematode	Mechanism of Action	Reference
Cry protein-forming bacteria			
B. thuringiensis	*Trichostrongylus colubriformis, Caenorhabditis elegans,* and *Nippostrongylus brasiliensis*	Cry proteins caused damage to the intestines of nematodes	Kotze et al. (2005), Wei et al. (2003)
Symbiotic bacteria			
Xenorhabdus and *Photorhabdus*	*B. xylophilus, M. incognita* and *their eggs*	Toxin production	Lewis et al. (2001)
Endoparasitic fungi			
Hirsutella rhossiliensi	*Heterodera schachtii and Meloidogyne javanica*	Parasitism	Tedford et al. (1993)
Predacious fungi			
Arthrobotrys irregularis	*Meloidogyne arenaria*	Predation	Vouyoukalou (1993)
Opportunistic fungi			
Paecilomyces lilacinus	*M. javanica, H. Avenae, Radopholus similis, H. Cajani, M. incognita*	Reduced nematode number significantly	Khan et al. (2006)
Verticillium chlamydosporium	*H. cajani, M. javanica, Heterodera trifolii*	Significantly reduced nematode populations	Siddiqui and Mahmood (1996)
Endophytic fungi			
Trichoderma harzianum	*M. javanica*	Suppressed nematode reproduction with increased host growth	Javeed and Alhazmi (2015)
VAM			
Funneliformismosseae	*M. incognita* and *P. penetrans*	Reduced infection due to induced systemic resistance	Vos et al. (2012)
R. irregularis	*R. similis* and *P. coffeae*	Systemic suppression of nematodes by ISR	Elsen et al. (2008)

TABLE 10.2   *(Continued)*

BCAs	Target Nematode	Mechanism of Action	Reference
F. mosseae	*R. similis* and *P. coffeae*	Suppression of nematodes by altering root morphology	Elsen et al. (2003)
Predatory nematodes			
Iotonchus spp.	*Rotylenchus robustus, Trichodorus obtusus, X. Americanum, H. multicinctus, H. oryzae, M. incognita, R. reniformis, X. Elongatum, T. semipenetrans, H. dihystera*	Predation by direct feeding on nematodes by swallowing, cutting in to small pieces, puncturing the body and sucking the body matter or by injecting enzymes to immobilize the nematodes	Khan and Kim (2007)
Dorylaimus spp.	*H. schachtii (*eggs*), A. tritici, H. indicus, H. mothi, H. oryzae, Longidorus, M. incognita, P. citri, Paratrichodorus, T. mashhoodi, X. americanum*		
Mononchoides gaugleri	*A. tritici, H. indicus, H. mangiferae, H. oryzae, H. indicus, L. attenuatus, M. incognita, P. christiei, T. mashhoodi, X. americanum*		

KEYWORDS

- **pathogenic fungi**
- **nematode**
- **biological control agents**
- **bodes of action**
- **commercialization**

REFERENCES

Abdel-Aziz, S.M. Extracellular metabolites produced by a novel strain, *Bacillus alvei* NRC-14 Multiple plant-growth promoting properties. *J. Basic Appl. Sci. Res.* 2013, *3*, 670–682.

Ahmad, F., Ahmad, I.; Khan, M.S. Screening of free-living rhizospheric bacteria for their multiple plant growth promoting activities. *Microbiol. Res.* 2008, *163*, 173–181.

Amkraz, N.; Boudyach, E.H.; Boubaker, H.; Bouizgarne, B.; Ait Ben Aoumar, A. Screening for fluorescent pseudomonades, isolated from the rhizosphere of tomato, for antagonistic activity toward *Clavibacter michiganensis subsp. michiganensis*. *World J. Microbiol. Biotech.* 2010, *26*, 1059–1065.

Ardakani, S.; Heydari, A.; Khorasani, N.; Arjmandi, R.; Ehteshami, M. Preparation of new biofungicides using antagonistic bacteria and mineral compounds for controlling cotton seedling damping-off disease. *J. Plant Prot. Res.* 2009, *49*, 49–55.

Arora, N.K.; Khare, E.; Oh, J.H.; Kang, S.C.; Maheshwar, D.K. Diverse mechanisms adopted by fluorescent Pseudomonas PGC2 during the inhibition of *Rhizoctonia solani* and *Phytophthora capsici. World J. Microbiol. Biotechnol.* 2008, *24*, 581–585.

Asadhi, S.; Reddy, B.V.B.; Sivaprasad, Y.; Prathyusha, M.; Krishna, T.M.; Kumar, K.V.K. Characterization, genetic diversity and antagonistic potential of 2,4-diacetylphloroglucinol producing *Pseudomonas fluorescens* isolates in groundnut-based cropping systems of Andhra Pradesh, India. *Arch. Phytopathol. Plant Prot.* 2013, *1*, 12.

Berdy, J. Bioactive microbial metabolites. *J. Antibiot.* 2005, *58*, 1–26.

Bilgrami, A.L.; Gaugler, R.; Brey, C. Prey preference and feeding behaviour of diplogastrid predator *Mononchoides gaugleri* (Nematoda: Diplogasterida). *Nematology.* 2005, *7*, 333–342.

De Vasconcellos, R.L.F.; Cardoso, E.J.B.N. Rhizospheric Streptomycetes as potential biocontrol agents of *Fusarium* and *Armillaria* pine rot and as PGPR for *Pinus taeda*. *Biocontrol.* 2009, *54*, 807–816.

Duffy, B. Pathogen self-defense: mechanisms to counteract microbial antagonism. *Ann. Rev. Phytopathol.* 2003, *41*, 501–538.

Duijff, B.J.; Bakker, P.A.H.M.; Schippers, B. Suppression of *Fusarium* wilt of carnation by *Pseudomonas putida* WCS358 at different levels of disease incidence and iron availability. *Biocontrol Sci. Tech.* 1994, *4*, 279–288.

Dukare, A,; Paul, S. Effect of chitinolytic biocontrol bacterial inoculation on the soil microbiological activities and fusarium population in the rhizosphere of pigeonpea (*Cajanascajan*). *Ann. Plant Protects. Sci.* 2018, *26*(1), 98–103.

Dukare, A.S.; Paul, S.; Nambi, V.E.; Gupta, R.K.; Singh, R.; Sharma, K.; Vishwakarma, R.K. Exploitation of microbial antagonists for the control of postharvest diseases of fruits: a review. *Crit. Rev. Food Sci. Nutr.* 2018, *58*(2), 1–16.

Dukare, A.S.; Prasanna, R.; Nain, L.; Saxena, A.K. Optimization and evaluation of microbe fortified composts as biocontrol agents against phytopathogenic fungi. *J. Microbiol. Biotechnol. Food. Sci.* 2013, *2*, 2272–2276.

El-Mougy, N.S.; Abdel-Kader, M.M. Salts Application for suppressing potato early blight disease. *J. Plant Prot. Res.* 2009, *49* (4), 353–361.

Elsen, A.; Beeterens, R.; Swennen, R.; DeWaele, D. Effects of an arbuscular mycorrhizal fungus and two plant-parasitic nematodes on Musa genotypes differing in root morphology. *Biol. Fertil. Soils.* 2003, *38*, 367–376.

Elsen, A.; Gervacio, D.; Swennen, R.; DeWaele, D. AMF-induced biocontrol against plant-parasitic nematodes in *Musa* sp.: a systemic effect. *Mycorrhiza.* 2008, *18*, 251–256.

El-Tarabily, K.A.; Sivasithamparam, K. Non-streptomycete actinomycetes as biocontrol agents of soil-borne fungal plant pathogens and as plant growth promoters. *Soil Biol. Biochem.* 2006, *38*, 1505–1520.

Ezziyyani, M.; Requena, M.E.; Egea-Gilabert, C.; Candel, M.E. Biological control of *Phytophthora* root rot of pepper using *Trichoderma harzianum* and *Streptomyces rochei* in combination. *J. Phytopathol.* 2007, *155*, 342–349.

Fernando, W.G.D.; Ramarathnam, R.; Krishnamoorthy, A.S.; Savchuk, S.C. Identification and use of potential bacterial organic antifungal volatiles in biocontrol. *Soil Biol. Biochem.* 2005, *37*, 955–964.

Fravel, R.D. Commercialization and implementation of biocontrol. *Annu. Rev. Phytopathol.* 2005, *43*, 337–359.

Gheysen, G; Mitchum, M.G. How nematodes manipulate plant development pathways for infection. *Curr. Opin. Plant Biol.* 2011, *14*, 415–421.

Glick, B.R.; Todorovic, B.; Czarny, J.; Cheng, Z.; Duan, J.; McConkey, B. Promotion of plant growth by bacterial ACC deaminase. *Crit. Rev. Plant. Sci.* 2007, *26*, 227–242.

Gopalakrishnan, S.; Humayun, P.; Keerthi, K.B.; Kannan, I.G.K.; Vidya, M.S.; Deepthi, K.; Rupela, O. Evaluation of bacteria isolated from rice rhizosphere for biological control of charcoal rot of sorghum caused by *Macrophomonia phaseolina* (Tassi) Goid. *World J. Microbiol. Biotechnol.* 2011, *27*, 1313–1321.

Gopalakrishnan, S.; Srinivas, V.; Samineni, S. Nitrogen fixation, plant growth and yield enhancements by diazotrophic growth-promoting bacteria in two cultivars of chickpea (*Cicer arietinum* L.). *Biocatal. Agric. Biotechnol.* 2017, *11*, 116–123.

Grover, M.; Nain, L.; Singh, S.B.; Saxena, A.K. Molecular and biochemical approaches for characterization of antifungal trait of a potent biocontrol agent *Bacillus subtilis* RP24. *Curr. Microbiol.* 2010, *60*, 99–106.

Harman, G.E. Overview of mechanisms and uses of *Trichoderma* spp. *Phytopathol.* 2006, *96*, 190–194.

Hase, C.; Moenne-Loccoz, Y.; Defago, G. Survival and cell culturability of biocontrol *Pseudomonas fluorescens CHAO* in lysimeter effluent water and utilization of a deleterious genetic modification to study the impact of the strain on numbers of resident culturable bacteria. *FEMS Microbiol. Ecol.* 2001, *37*, 239–249.

Heydari, A.; Pessarakli, M. A review on biological control of fungal plant pathogens using microbial antagonists. *J. Biol. Sci.* 2010, *10*, 273–290.

Hu, H.J.; Chen, Y.L.; Wang, Y.; Tang, Y.Y.; Chen, S.L.; Yan, S.Z. Endophytic *Bacillus cereus* effectively controls *Meloidogyne incognita* on tomato plants through rapid rhizosphere occupation and repellent action. *Plant Dis.* 2017, *101*, 448–455.

Javeed, M.T.; Alhazmi, A. Effect of *Trichoderma harzianum*on *Meloidogyne javanica* in tomatoes as influenced by time of the fungus introduction inti soil. *J. Pure Appl. Microbiol.* 2015, *9*(1), 535–539.

Jones, J.T.; Haegeman, A.; Danchin, E.G.J.; Gaur, H.S.; Helder, J.; Jones, M.G.K.; et al. Top 10 plant-parasitic nematodes in molecular plant pathology. *Mol. Plant Pathol.* 2013, *14*, 946–961.

Kaushal, K. K.; Srivastava, A.N.; Pankaj, C.G.; Singh, K. Cyst forming nematodes in India: a review. *Indian J. Nematol.* 2007, *3*, 1–7.

Khamna, S.; Yokota, A.; Lumyong, S. Actinomycetes isolated from medicinal plant rhizosphere soils: diversity and screening of antifungal compounds, indole-3-acetic acid and siderophore production. *World J. Microbiol. Biotechnol.* 2009, *25*, 649–655.

Khan, A.; Williams, K.L.; Nevalainen, H.K.M. Control of plant-parasitic nematodes by *Paecilomyces lilacinus* and *Monacrosporium lysipagum* in pot trials. *Biol. Control.* 2006, *51*, 643–658.

Khan, Z.; Kim, Y.H. The predatory nematode; *Mononchoides fortidens* (Nematoda, Diplogasterida); suppresses the root-knot nematode; *Meloidogynea renaria*; in potted field soil. *Biol. Control.* 2005, *35*, 78–82.

Khan, Z.; Kim, Y.H. A review on the role of predatory soil nematodes in the biological control of plant parasitic nematodes. *Appl. Soil Ecol.* 2007, *35*, 370–379.

Kiss, L. A review of fungal antagonists of powdery mildews and their potential as biocontrol agents. *Pest Manage. Sci.* 2003, *59*, 475–483.

Kloepper, J.W.; Ryu, C.M.; Zhang, S. Induce systemic resistance and promotion of plant growth by *Bacillus* spp. *Phytopathol.* 2004, *94*, 1259–1266.

Kobayashi, D.Y.; Reedy, R.M.; Bick, J.; Oudemans, P.V. Characterization of a chitinase gene from *Stenotrophomonas maltophilia* strain 34S1 and its involvement in biological control. *Appl. Environ. Microbiol.* 2002, *68*, 1047–1054.

Kotze, A.C.; O'Grady, J.; Gough, J.M.; Pearson, R.; Bagnall, N.H.; Kemp, D.H.; Akhurst, R.J. Toxicity of *Bacillus thuringiensis* to parasitic and free-living life stages of nematodes parasites of livestock. *Int. J. Parasitol.* 2005, *35*, 1013–1022.

Kumar, A.; Saini, S.; Wray, V.; Nimtz, M.; Prakash, A.; Johri, B.N. Characterization of an antifungal compound produced by *Bacillus* sp strain A3F that inhibits *S. sclerotiorum*. *J. Basic Microbiol.* 2012, *52*, 670–678.

Lewis, E.E.; Grewal, P.S.; Sardanelli, S. Interactions between *Steinernema feltiae–Xenorhabdus bovienii* insect pathogen complex and root-knot nematode *Meloidogyne incognita*. *Biol Control.* 2001, *21*, 55–62.

Lo, C.T.; Nelson, E.B.; Hayes, C.K.; Harman, G.E. Ecological studies of transformed *Trichoderma harzianum* strain 1295-22 in the rhizosphere and on the phylloplane of creeping bentgrass. *Phytopathol.* 1998, *88*, 129–136.

Lorito, M.; Harman, G.E.; Hayes, C.K.; Broadway, R.M.; Tronsmo, A.; Woo, S.L.; Di Pietro, A. Chitinolytic enzymes produced by *Trichoderma harzianum:* antifungal activity of purified endochitinase and chitobiosidase. *Phytopathol.* 1993, *83*, 302–307.

Maksimov, I.V.; Abizgil'dina, R.R.; Pusenkova, L.I. Plant growth promoting rhizobacteria as alternative to chemical crop protectors from pathogens *Appl. Biochem. Microbiol.* 2011, *47*, 333–345.

Maloy, O.C. Plant Disease Control: Principles and Practice. John Wiley & Sons, Inc., New York, USA. 1993; pp. 346.

Mankau, R.; Imbriani, J.L.; Bell. A.H. SEM observations on nematode cuticle penetration by *Bacillus penetrans*. *J. Nematol.* 1976, *8*, 179–181.

Maurhofer, M.; Keel, C.; Haas, D.; De'fago, G. Influence of plant species on disease suppression by *Pseudomonas fluorescens* strain CHA0 with enhanced antibiotic production. *Plant Pathol.* 1995, *44*, 40–50.

Mhatre, P.H.; Karthik, C.; Kadirvelu, K.; Divya, K.L.; Venkatasalam, E.P.; Srinivasan, S.; Ramkumar, G.; Saranya, C.; Shanmuganathan, R.. Plant growth promoting rhizobacteria (PGPR), a potential alternative tool for nematodes bio-control. *Biocatal. Agric. Biotechnol.* 2019, *17*, 119–128.

Mhatre, P.H.; Malik, S.K.; Kaur, S.; Singh, A.K.; Mohan, S.; Sirohi, A. Histopathological changes and evaluation of resistance in Asian rice (*Oryzasativa*) against rice root knot nematode; *Meloidogyne graminicola*. *Indian J. Genet. Plant Breed.* 2015, *75*, 41–48.

Mohan, S.; Mauchline, T.H.; Rowe, J.; Hirsch, P.R.; Davies, K.G. Pasteuria endospores from *Heterodera cajani* (Nematoda, Heteroderidae) exhibit inverted attachment and altered germination in cross-infection studies with *Globodera pallida* (Nematoda, Heteroderidae). *FEMS Microbiol. Ecol.* 2012, *79*, 675–684.

Niu, Q.H.; Huang, X.W.; Tian, B.Y.; Yang, J.K.; Liu, J.; Zhang, L.; Zhang, K.Q. *Bacillus* sp. B16 kills nematodes with a serine protease identified as a pathogenic factor. *Appl. Microbiol. Biotechnol.* 2005, *69*, 722–730.

Numberger, T.; Brunner, F.; Kemmerling, B.; Piater, L. Innate immunity in plants and animals: striking similarities and obvious differences. *Immunol. Rev.* 2004, *198*, 249–266.

Oliveira, E.J.; Rabinovitch, L.; Monnerat, R.G.; Passos, L.K.J.; Zahner, V. Molecular characterization of *Brevibacillus laterosporus* and its potential use in biological control. *Appl. Environ. Microbiol.* 2004, *70*, 6657–6664.

Palumbo, J.D.; Yuen, G.Y.; Jochum, C.C.; Tatum, K.; Kobayashi, D.Y. Mutagenesis of beta-1,3-glucanase genes in *Lysobacter enzymogenes* strain C3 results in reduced biological control activity toward Bipolaris leaf spot of tall fescue and Pythium damping-off of sugar beet. *Phytopathol.* 2005, *95*, 701–707.

Park, J.Y.; Oh, S.A.; Anderson, A.J.; Neiswender, J.; Kim, J.C.; Kim, Y.C. Production of the antifungal compounds phenazine and pyrrolnitrin from *Pseudomonas chlororaphis* O6 is differentially regulated by glucose. *Lett. Appl. Microbiol.* 2011, *52*, 532–537.

Paul, S.; Dukare, A.S.; Bandeppa, G.; Manjunatha, B.S.; Annapurna, K. Plant growth promoting rhizobacteria for abiotic stress alleviation in crops. In: *Advances in Soil Microbiology: Recent Trends and Future Prospects*, Volume 2: *Soil-Microbe-Plant-Interaction* (Eds. Adhya TK, Mishra BB, Annapurna K and Kumar U), Springer Nature, Singapore, 2017; pp. 57–79.

Ramadan, E.M.; AbdelHafez, A.A.; Hassan, E.A.; Saber, F.M. Plant growth promoting rhizobacteria and their potential for biocontrol of phytopathogens. *Afr. J. Microbiol. Res.* 2016, *10*, 486–504.

Ravari, S.B.; Moghaddam, E.M. Efficacy of *Bacillus thuringiensis* Cry14 toxin against root knot nematode; *Meloidogyne javanica*. *Plant Protect. Sci.* 2015, *51*(1), 46–51.

Ryu, C.M.; Hu, C.H.; Locy, R.D.; Kloepper, J.W. Study of mechanisms for plant growth promotion elicited by rhizobacteria in *Arabidopsis thaliana*. *Plant Soil.* 2005, *268*, 285–292.

Schmidt, L.M.; Hewlett, T.E.; Green, A.; Simmons, L.J.; Kelley, K.; Doroh, M.; Stetina, S.R. Molecular and morphological characterization and biological control capabilities of a *Pasteuria* spp. parasitizing *Rotylenchulus reniformis*; the reniform nematode. *J. Nematol.* 2010, *42*, 207–217.

Schouteden, N.; De Waele, D.; Panis, B.; Vos, C.M. Arbuscular mycorrhizal fungi for the biocontrol of plant-parasitic nematodes, a review of the mechanisms involved. *Front. Microbiol.* 2015, *6*, 1280.

Schouten, A. Mechanisms involved in nematode control by endophytic fungi. *Annu. Rev. Phytopathol.* 2016, *54*(1), 121–142.

Scuderi, G.; Bonaccorsi, A.; Panebianco, S.; Vitale, A.; Polizzi, G.; Cirvilleri, G. Some strains of *Burkholderia gladioli* are potential candida for post harvest biocontrol of fungal rots in citrus and apple fruits. *J. Plant Pathol.* 2009, *91*(1), 207–213.

Siddiqui, Z.A.; Mahmood, I. Biological control of *Heterodera cajani* and *Fusarium udum* on pigeonpea by *Glomusmosseae*; *Trichodermaharzianum* and *Verticillium chlamydosporium*. *Israel. J. Plant Sci.* 1996, *44*, 49–56.

Siddiqui, Z.A.; Mahmood, I. Role of bacteria in the management of plant parasitic nematodes, a review. *Bioresour. Technol.* 1999, *69*, 167–179.

Singh, P.P.; Shin, Y.C.; Park, C.S.; Chung, Y.R. Biological control of Fusarium wilt of cucumber by chitinolytic bacteria. *Phytopathol.* 1999, *89*, 92–99.

Soytong, K.; Srinon, W.; Rattanacherdchai, K.; Kanokmedhakul, S.; Kanokmedhakul, K. Application of antagonistic fungi to control anthracnose disease of grape. *Int. J. Agric. Tech.* 2005, *1*, 33–41.

Sterling, G.R. Biological Control of Plant Parasitic Nematodes: Progress, Problems and Prospects. C. A. B. International, Wallingford, U. K., 1991; pp. 282.

Stirling, G.R.; Mani, A. The activity of nematode trapping fungi following their encapsulation in alginate. *Nematologica.* 1995, *41*, 240–250.

Su, L.; Shen, Z.; Ruan, Y.; Tao, C.; Chao, Y.; Li, R. Shen, Q. Isolation of antagonistic endophytes from banana roots against *Meloidogyne javanica* and their effects on soil nematode community. *Front. Microbiol.* 2017, *8*, 2070.

Tabatabaei, F.S.; Saeedizadeh, A. Rhizobacteria cooperative effect against *Meloidogyne javanica* in rhizosphere of legume seedlings. *Hell. Plant Prot. J.* 2017, *10*, 25–34.

Tedford, E.C.; Jaffee, B.A.; Muldoon, A.E.; Anderson, C.E.; Westerdah, B.B. Parasitism of *Heterodera schachtii* and *Meloidogyne javanica* by *Hirsutella rhossiliensis* in microplots over two growing seasons. *J. Nematol.* 1993, *25*, 427–433.

Tian, B.; Yang, J.; Zhang, K.Q. Bacteria used in the biological control of plant-parasitic nematodes, populations; mechanisms of action; and future prospects. *FEMS Microbiol. Ecol.* 2007a, *61*, 197–213.

Tian, B.; Yang, J.; Lian, L.H.; Wang, C.Y,; Zhang, K.Q. Role of neutral protease from *Brevibacillus laterosporus* in pathogenesis of nematode. *Appl. Microbiol. Biotechnol.* 2007b. *74*, 372–380.

Timper, P.; Brodie, B.B.; Effect of *Hirsutella rhossiliensis* on infection of potato by *Pratylenchus penetrans*. *J. Nematol.* 1994, *26*, 304–307.

Tiwari, S.; Pandey, S.; Chauhan, P.S.; Pandey, R. Biocontrol agents in co-inoculation manages root knot nematode [*Meloidogyne incognita* (Kofoid and white) Chitwood] and enhances essential oil content in *Ocimum basilicum* L. *Ind. Crops Prod.* 2017, *97*, 292–301.

Van Loon, L.C. Plant responses to plant growth-promoting rhizobacteria. *Eur. J. Plant Pathol.* 2007, *119*, 243–254.

Van Wees, S.C.M.; Pieters, C.M.J.; Trisjssenaar, A.; Van't Westende, Y.A.M.; Hartog, F.; Van Loon, L.C. Differential induction of systemic resistance in Arabidopsis by biocontrol bacteria. *Mol. Plant. Microbe. Interact.* 1997, *10*, 716–724.

Viswanathan, R.; Samiyappan, R. Siderophores and iron nutrition on the pseudomonas mediated antagonism against *Colletotrichum falcatum* in sugarcane. *Sugar Technol.* 2007, *9*, 57–60.

Viterbo, A.; Inbar, J.; Hadar, Y.; Chet, I. Plant disease biocontrol and induced resistance via fungal mycoparasites. In: *Environmental and Microbial Relationships*, 2nd edn. *The Mycota IV*. (Eds. C.P. Kubicek and I.S. Druzhinina). Springer-Verlag, Berlin, Heidelberg, 2007; pp. 127–146.

Von Der Weid, I.; Duarte, G.; Van Elsas, J.D.; Seldin, L. *Paenibacillus brasilensis* sp. nov., a novel nitrogen-fixing species isolated from the maize rhizosphere in Brazil. *Int. J. Syst. Evol. Microbiol.* 2002, *52*, 2147–2153.

Vos, C.; Claerhout, S.; Mkandawire, R.; Panis, B.; deWaele, D., Elsen, A., Arbuscular mycorrhizal fungi reduce root-knot nematode penetration through altered root exudation of their host. *Plant Soil.* 2012, *354*, 335–345.

Vouyoukalou, E. Effect of *Arthrobotrys irregularis* on *Meloidogynea renaria* on tomato plants. *Fundam. Appl. Nematol.* 1993, *16*, 321–324.

Wei, J.Z.; Hale, K.; Carta, L.; Platzer, E.; Wong, C.; Fang, S.C.; Aroian, R.V. *Bacillus thuringiensis* crystal proteins that target nematodes. *Proc. Natl. Acad. Sci.* 2003, *100*, 2760–2765.

Weindling, R.; Fawcett, H.S. Experiments in the control of Rhizoctonia damping off of citrus seedling. *Hilgardia.* 1936, *10*, 1–16.

Weller, D.M. Biological control of soil borne plant pathogens in the rhizosphere with bacteria. *Ann. Rev. Phytopathol.* 1988, *26*, 379–407.

Yang, C.Y.; Ho, Y.C.; Pang, J.C.; Huang, S.S.; Tschen, J.S. Cloning and expression of an antifungal chitinase gene of a novel *Bacillus subtilis* isolate from Taiwan potato field. *Bioresour. Technol.* 2009, *100*, 1454–1458.

CHAPTER 11

Novel Forecasting, Bioindicator, and Bioremediation for Pesticides Contamination in Soil

MOHAMED S. KHALIL[1] and WAEL M. KHAMIS[2*]

[1]*Central Agricultural Pesticides Laboratory, Agricultural Research Center, Al-Sabhia, Alexandria, Egypt*

[2]*Plant Protection Research Institute, Agricultural Research Center, Al-Sabhia, Alexandria, Egypt*

Corresponding author. E-mail: melonema@gmail.com

ABSTRACT

The global concern has been increased drastically due to the environmental pollution over the public health. Pesticides are one of the main environmental contaminants of the 20th and 21st centuries. Only a small amount of the pesticides used actually reaches their targets. The contaminated pesticides in soil are moving into the deep layers of soil, accumulating, and transforming there. Pesticides may be found in unexpected places a long time after they were used. Despite pesticides are harmful to human health and environment, but there are certain benefits. The pesticides' physico-chemical traits are responsible for persistence, movement, and fate within the environment. The contamination of pesticides in soil lessens the quality, especially microbial diversity and enzymes. Also, soil fauna is promising bioindicator for contamination. Bioremediation is effective process to remove the accumulated hazardous chemicals from environment by microbes (microbial-remediation) or plants (phytoremediation). Phytoremediation is considered a novel forecasting method referring to plants usages for removing contaminated toxicants from different environments.

Certain mechanisms for phytoremediation were included: phytotransformation, phytodegradation, phytovolatilization, and rhizoremediation.

11.1 INTRODUCTION

Pesticides are the substance or mixture of substances, which are used to manage different genera of pests. Scientifically, classification of pesticides is depending on their chemical group, toxicity, environmental stability, targeted pests, or other features. The excessive use of pesticides caused many environmental and human health problems. Soil contaminated with pesticides is one of the most important problems that is being faced by human beings because this kind of contamination reaches to all live organisms (e.g., humans, plants, domestic animals, soil insects, wild animals, groundwater, soil animals, and microorganisms).

The toxicity of pesticides is affected by several factors including pesticide dose, exposure time, dose, organism characteristics, environmental traits, and pesticide properties. The persistence of pesticides within environment is related to either their physico-chemical properties or organisms' deficiency, which degraded them. Also, pesticides are affected by environmental conditions, for example, light, temperature, or humidity, which degrade or volatile it (Sibanda et al., 2011). Hence, certain scientists have suggested some solutions, which included bioremediation to eliminate contaminated pesticides from the soil.

Bioremediation concept is referring to any tactic that can be used to remove unwanted consequences of pollutants from environment. Bioremediation is the most effective solution for decontamination of environmental pollutants (Mervat, 2009). There are certain alternatives to remediate environmental pollutants by using microbes (the most common) or plants (Conesa, et al., 2012). Phytoremediation process means using green plants to remove or reduce the amount of any environmental pollutants (Thakur et al., 2016).

The word "phytoremediation" is coined from Greek and Latin words. "Phyto" from Greek "phutón" meaning plants, while "remediation" from Latin "remedium" meaning a medicine, remedy, cure, and restoring balance (Galadima et al., 2018). Phytoremediation has been classified based on the process into phytoextraction, phytotransformation, phytostimulation, phytostabilization, phytovolatilization, and phytofiltration (Galadima et al., 2018).

11.2   DEFINITION OF PESTICIDE

Firstly, we must know what pesticides are. In fact, pesticides mean any formulated substance(s) used singly or combined from natural or synthetic resources to kill, eliminate, prevent, repel, and decrease the pests that compete with human on food (FAO, 1989). These pests could be insects, weeds, fungi, bacteria, algae, and nematodes, as well as undesirable plants, animals, or mammals causing harm during production process. Also, these protection agents (pesticides) are used mainly to inhibit spoilage of crops in fields or after harvest to protect the commodity from deterioration during storage and transport (FAO, 2002). In addition to control vector pests on domestic animals and human beings, the definition of pesticides included attractant substances for the purpose of destroying pests (EPA, 2007). Pesticides can be applied at public places, homes, schools, parks, gardens, and agricultural fields. Therefore, pesticides are discovered in food, soil, water and air, and even in mothers' breast milk.

11.3   THE CLASSIFICATION OF PESTICIDES

Globally, pesticides are divided into different categories depending upon various bases. The most important and common bases included pest types or uses, mode of action, chemical structure (Drum, 1980). Also, pesticides can be divided depending on site of action, mode of formulation, toxicity levels, and others.

11.3.1   CLASSIFICATION OF PESTICIDES ACCORDING TO PEST TYPES

This classification includes insecticides, acaricides, fungicides, nematicides, herbicides, algicide, antifeedants, avicides, bactericides, bird repellents, chemosterillant, herbicide softeners, insect attractants, insect repellents, mammal repellents, mating disrupters, molluscicides, plant activators, virucides, plant growth regulators, rodenticides, synergists, and miscellaneous.

11.3.2   CLASSIFICATION OF PESTICIDES ACCORDING TO MODE OF ACTION

In this route, pesticides are classified as systemic and nonsystemic (contact) pesticides. Systemic pesticides mean that applied compounds can penetrate the tissues of treated plant and move through the vascular system. On contrary, the nonsystemic or contact pesticides mean those that do not penetrate the plant tissues and are not transported within the plant vascular system.

11.3.3   CLASSIFICATION OF PESTICIDES ACCORDING TO CHEMICAL STRUCTURE

In this section, classification covers organic and inorganic pesticides. The organic pesticides include different families such as organochlorine, organophosphorus, pyrethroids, carbamates, neonicotinoids, avermectins, spinosyns, milbemycins, fiproles, and diamides.

11.3.4   CLASSIFICATION OF PESTICIDES ACCORDING TO SITE OF ACTION

Under this type of classification, pesticides are classified as acetylcholin-esterase inhibitors, GABA-gate chloride channel blockers, sodium channel modulators, nicotinic acetylcholine receptors, mitochondrial ATP synthase inhibitors, octopamine receptor agonists, and ryanodine receptor modulators.

11.4   SOME MINOR CLASSIFICATIONS OF PESTICIDES

11.4.1   CLASSIFICATION OF PESTICIDES ACCORDING TO APPLICATION METHODS

This classification includes application pesticides as spray, dust, and fumigation.

11.4.2   CLASSIFICATION OF PESTICIDES ACCORDING TO THE MODE OF FORMULATION

The type of formulation is considered a base of classification, which includes the liquid formulations (emulsifiable concentrates and suspension concen-trate), solid formulations (granules, dust, and baits), and gaseous formula-tions, which are usually packaged under pressure and stored as liquids and/ or packaged as tablets or pellets that release gas when exposure to humidity.

11.4.3 CLASSIFICATION OF PESTICIDES ACCORDING TO THE TOXICITY LEVEL

This type of classification was developed according to the World Health Organization vision. It is based on the following five classes: (1) Class Ia means that pesticide is extremely hazardous, (2) Class Ib means that pesticide is highly hazardous, (3) Class II means that pesticide is moderately hazardous, (4) Class III means that pesticide is slightly hazardous, and (5) Class IV means that pesticide is unlikely to cause risks in normal use.

11.4.4 CLASSIFICATION OF PESTICIDES ACCORDING TO SPECTRUM ACTIVITY

In this type, the pesticides are divided into two groups including selective pesticides, which kill only specific pests and not selective pesticides, which kill different species and/or genus of pests.

11.4.5 CLASSIFICATION OF PESTICIDES ACCORDING TO MODE OF PENETRATION

This kind of classification includes stomach poisons, contact poisons, and gaseous poisons.

11.4.6 CLASSIFICATION OF PESTICIDES ACCORDING TO SOIL PERSISTENCE

This classification includes the following categories of pesticides: pesticides with half-life < 30 days are considered nonpersistent, pesticides with half-life from 30 to 100 days are considered moderately persistent and pesticides with half-life > 100 days are considered persistent.

11.5 BENEFITS AND RISKS OF PESTICIDES

From the mid-1940s until now and with pesticide development, farmers are becoming dependent on pesticides as a weapon against pests to achieve progress in foodstuff production (Maksymiv, 2015). The annual pesticide sales during 2009 were less than $2 billion in Africa and Middle East region, whilst in Europe and Asia sales were $10 billion. However, during 2012 the sales of pesticides in Africa and Middle East region were increased and reached to about $2 billion, while in Europe and Asia sales were more than $12 billion (FAO, 2015). The sales numbers reflect the importance of pesticides in modern agribusiness and imply that the demand for pesticides will increase in the nearest future (Maksymiv, 2015). Although there are benefits of pesticides as plant protection agents, but the expanded pesticide applications and the massive uses attributed to certain problems. These problems were incarcerated in humans' health and environmental risk problems. Soil contaminated with pesticides is one of the real dangerous problems because that means pesticides will reach humans' food, water, and animals.

11.5.1   BENEFITS OF PESTICIDES

The initial expectation for benefits of using pesticides in our environments is to protect human beings from pests that propagate infectious diseases. In addition, pesticides are protecting domestic animals from ticks and other pathogenic insects. There are tremendous benefits of pesticide usage, but the most important after protecting humans' life is to protect our crops from pests (namely, insects, fungi, bacteria, acari, and nematodes). In agricultural sector, pesticides are decreasing crops spoilage, improving productivity, and increasing the food quality. In public health sector, pesticides especially insecticides are playing basic role in killing vector-borne diseases that are transfused by insects. In the same context, in industry, pesticides are widely used to inhibit damages caused by insect pests, rodents, yeasts, fungi, algae, and bacteria in the manufacture of electrical equipment, fridges, carpets, paper, paints, cardboard, and food packaging materials (Garcia et al., 2012).

11.5.2   RISKS OF PESTICIDES

Although chemical pesticides are at the top of the most dangerous pollutant list, but we still depend on it to control or prevent pests and diseases. During the application or preparation processes of pesticides, the farmers or workers are exposed to high doses of pesticides, Therefore, yearly there are millions of people in the Third World countries suffering from poisoning (Alavanja, 2009). Many of applied pesticides are reaching to destinations far from their targeted including nontarget species or organisms, air, water, and soil (Miller,

2004). The excessive usage of pesticides is attributed to the contamination of these pesticides in food and soil as well as sediment, surface and groundwater, which create multiimpacts on aquatic fauna and flora, wildlife, and the human health (Cerejeira et al., 2003; Khalil, 2014; Ghorab and Khalil, 2015). Many risks are caused by pesticides exposure. We can summarize some as follows:

1. Acute toxicity.
2. Chronic toxicity.
3. Immunotoxicity.
4. Reproductive effects.
5. Cytogenetic impacts.
6. Teratogenic effects.
7. Mutagenic toxicity.
8. Carcinogenic effects.
9. Effects on tissue carboxyesterase.
10. Porphyrin effects.
11. Effects on lipid metabolism.

11.6 THE INTERACTION BETWEEN PESTICIDE PROPERTIES AND ITS CONTAMINATION IN AGRO-SOIL

The properties of pesticides play a great role in their persistence, movement, and fate within the environment. Pesticides' properties are different from compound to another depending on their chemical nature and formulation. Moreover, physical properties of a pesticide, in particular, determine the pesticide mode of action, dosage, mode of application, and the subsequent environmental chemodynamics. We have to understand the relationship between pesticide physical properties and environmental characteristics, especially the agricultural soil because it helping us to minimize the soil contamination. The most important characteristics of pesticides are mentioned as follow:

11.6.1 SOLUBILITY

Solubility means the maximum amount of any substance, which dissolves in water (Barrigossi et al., 2005; Martins et al., 2013). Solubility of pesticides is measured in part-per-million (ppm), which is the same as milligrams per liter (mg/L). If pesticide is highly soluble in waster, this means that the transpose of this agrochemical is very bearable to greater depth in soil than the less soluble (Oliveira and Brighenti, 2011; Martins et al., 2013). The less soluble agrochemical means more contamination in soil and water as well. However, there are certain factors that affect solubility such as polarity, pH, temperature, chemical structure, functional group, and molecular mass (Capel, 1993; Martins et al., 2013). Certain categories of agrochemicals (pesticides) solubility in water was suggested by Dias (2010) as follow: insoluble (< 1 ppm), very low (1–10 ppm), low (11–50 ppm), moderate (51–150 ppm), high (151–500 ppm), very high (500–5000 ppm), and extremely high (> 5000 ppm).

11.6.2 VAPOR PRESSURE

Vapor pressure simply means the air solubility or volatile ability of pesticides and turns into vapor (gas state) (Barrigossi et al., 2005). The volatility of pesticides from soil in some cases may be considered as an advantage because soil has disposed it. However, pesticides with high vapor pressure may drift the applied pesticide(s) causing environmental pollution. Pesticides with low vapor pressure do not move into air, so they could get accumulated in soil if they are water soluble. In contrast, if pesticide is not water soluble it may accumulates in soil or biota. The usually used unit (s) to measure vapor pressure is pascal (Pa) or expressed as mol m^{-3} or atm (Mackay et al., 2006). Pesticides with high vapor pressure are sometimes not preferred because they may get lost before reaching their targets.

Furthermore, volatility was divided into categories depending on vapor pressure logarithmic (log VP) according to Silva and Fay (2004) as follow: log VP = −3 (very high volatility), log VP = −4 to −3 (high), log VP = −5 to −4 (moderate), log VP = −6 to −5 (low), log VP = −7 to −6 (very low) and log VP = −7 (extremely low). Volatilization of pesticides from plant surfaces depends on the temperature of atmosphere. However, pesticides in soil are affected by many factors such as temperature, adsorption of pesticide molecules, moisture in soil, soil PH, soil textures, soil organic matter (OM), applied pesticide concentrations, vapor pressure, absorption, and product solubility (Gevao et al., 2000; Silva and Fay, 2004; Minguela and Cunha, 2011).

11.6.3 HENRY'S LAW CONSTANT (K_H)

In moist soils, the volatility of pesticides is described by Henry's law constant (K_h), which is usually expressed as Pa m^3 mol^{-1}. Henry's law calculates the volatility in solution depending on vapor pressure, solubility, and molecular weight of pesticides (Cabrera et al., 2008; Oliveira and Brighenti, 2011). When Henry's law constant is less than 10^{-1}, this means that pesticides are low volatile from water to the soil. In contrary, constant higher than 10^2 means that the probability of pesticides volatilization is higher (Mackay et al., 2006).

11.6.4 SOIL ADSORPTION COEFFICIENT (K_{OC}/K_D)

Adsorption of pesticides to soil, sediments, and groundwater is the main factor to determine pesticides' destination in the environment. The adsorption of pesticides in soil is estimated by using sorption coefficient (K_d) and coefficient of organic carbon-water partition (K_{oc}) (Barrigossi et al., 2005; Mackay et al., 2006; Minguela and Cunha, 2011). Coefficient of soil colloid adsorption (K_d) is the ratio of adsorbed pesticide concentration in the soil (the OM content was neglected) to the concentration of pesticide in water. Another way to estimate adsorbed pesticides is by using coefficient of organic carbon-water partition (K_{oc}), which is calculated by measuring the ratio of pesticide concentration that is adsorbed by soil organic carbon to the concentration of pesticide in water and this is the preferred criterion to estimate the soil capability of adsorption pesticides. The value of K_{oc} depends on content of OM in soil, soil pH, and the polarity of pesticides. According to Gebler and Spadotto (2004), pesticides are classified depending on the strength of sorption by OM (K_{oc}) as follows: very strong sorption (> 5000), strong (4900–600), moderate (599–100), and weak (0.5–99).

11.7 PESTICIDES CONTAMINATION IN AGRICULTURAL SOIL

About 317 samples were collected from agricultural soils to estimate the residues of 76 pesticides and their metabolites, which are currently approved in the EU markets. These samples were taken from 11 EU Member States and 6 main cropping systems. Results showed that about 83% of soil samples contained pesticide residues as well as multiple residues were estimated in 58% of the samples. The highest total pesticide content in soil reached up to 2.87 mg/kg. About 166 different mixtures of pesticides represent the great variation among individual soil samples. The major pesticide mixture in soils was commonly detected in the form of glyphosate and its main metabolite aminomethylphosphonic acid that is almost shared in the total pesticide content in soil. Moreover, the mixture of glyphosate and its main metabolites of aminomethylphosphonic acid and phthalimide were commonly detected (Silva et al., 2019).

Eventually, an established method was carried out by Hwang et al. (2018) to evaluate the allowed permissible concentrations of contaminated pesticides in soil (Cs, permissible) and uptake traits of endosulfan, including α-, β-isomers, and sulfate-metabolites at time of planting carrot and potato to curb the maximum residue level below standard limits. Endosulfan residues were analyzed in treated soils at concentrations of 2 and 10 mg (kg soil)$^{-1}$ in carrots and potatoes plants respectively, at age of 60–90 days. Both plants indicated that there is a dissipation of endosulfan in soil. However, the concentrations of endosulfan in different plant parts exceeded the maximum residue limits. The bioconcentration factor (the ratio of pesticide concentration in soil to concentration in edible plant parts at harvest time) of endosulfan was gradually decreased in soil over time, while increased in plant roots (edible part of each plant) at harvest day. This study was performed as a prediction tool to secure crops from pesticide residues in soil.

11.8   AMENDMENTS ARE NEEDED FOR SOIL QUALITY

More validations are needed for the predicted environmental levels (PECs) established by European Food Safety Authority, which came as consequence of the high content of individual pesticide residues in soil and exceeded the PEC levels. Up till now, soil contamination by total or individual pesticide residues has not been adequately expressed in quality standards of soil. Moreover, EU legislation has no distinct threshold limits or soil protection policies that reasonably could protect the soil fauna form this hidden threat (Silva et al., 2019).

The modifications process in activity of soil enzymes may be linked with the soil microbial diversity and its sensitivity to pesticides (Micuți et al., 2018). For example, using fungicides such as Ridomil Gold WG (metalaxyl-M 4% and mancozeb 68%) and Bravo 500 SC (chlorothalonil) were found to be minimizing soil enzymes like amylase and urease, while Ridomil Gold increased the level of cellulase. With respect to insecticides, Mospilan (acetamiprid 20% SG) and Vertimec (abamectin 1.8% EC) recorded enhancement in cellulase and xylanase activities. Therefore, the excessive usage of pesticides for long term, short term and temporary could result in drastic changes on agricultural soil in terms of nutrient content, predominant soil species, structural and functional diversity of microbial populations as well as soil enzyme's activities (Micuți et al., 2018).

Otherwise, soil amendments like organic fertilizers (composts and manures) and biocontrol agents have been established as eco-friendly and preferred additives to restore the quality and fertility, in addition to further forward with continuously approach. The fertility and quality of soil are tightly linked with the level of microbial diversity in soil, which may influence certain properties and functions of soil that are attributed to a threat to global food security (Prashar and Shah, 2016). Although soil biota is involved in energy and nutrient cycle, as well as an important and unstable fraction of OM, dynamic characteristics of soil such as microbial biomass, enzyme activity, and respiration are more responsible than changes in crop management practices (Doran et al., 1996; Malik et al., 2017b).

In Slovakia, Jozef and Koco (2018) developed the indices for agricultural soil quality based on database of immobilization indexes based and maps of soil criteria related to production, environmental characteristics, and risk factors (Barančíková et al., 2010). The agricultural soil quality index was divided into five different categories as in table 11.1 and the quality values were ranged from 1.87 to 18.78.

TABLE 11.1 The Quality Index of Agricultural Soil Ability to Immobilize Organic Pollutants

Values	Immobilize Ability	Index
≤14.36	Very high	1
10.12–14.35	High	2
6.75–10.11	Medium	3
4.17–6.74	Low	4
≤4.16	Very low	5

Nevertheless, the mentioned maps and databases elicited the soil potential to transform organic pollutants. Therefore, Barančíková et al. (2010) suggested the transformation index for this parameter (the sum of biotransformation and abiotic transformation) was categorized into five groups ranging from 18.66 to 79.87 (Table 11.2).

TABLE 11.2 The Quality Index of Agricultural Soil Ability to Transform Organic Pollutants

Values	Transformation Ability	Index
≤46.36	Very high	1
38.40–46.35	High	2
32.62–38.39	Medium	3
27.48–32.61	Low	4
≤27.47	Very low	5

Multistage process has been demonstrated to prioritize and analyze the most suitable physical soil quality indicators (SQIs) for monitoring soil quality and function. Referring to soil functions elicited from current soil and environmental policy in the United Kingdom, the prioritized SQIs can be related to soil processes, functions, and consequently directed to improve ecosystem. These efforts structure the future of soil and environmental policy in the United Kingdom, which helps in improving programs that aim to monitor the physical quality of soil (Corstanje et al., 2017).

11.9 DIVERSITY OF SOIL FAUNA IN DIFFERENT SOIL TYPE

Different studies were carried out on the diversity and variability of the soil fauna as a biological indicator of healthy soil via different soil types. Soil fauna population may be associated with physical and chemical soil parameters. These thoughts were sided with the findings that the biological indicators of soil quality have been related with OM content, terrestrial arthropods, fauna, lichen, microbial biomass (carbon or nitrogen), metabolic products (ergosterol and glomalin), and microbial respiration of soil and soil enzyme production (Malik et al., 2017b; Gorain and Paul, 2017). In this regard, recent studies have investigated the relationship between dominant soil microflora (e.g., fungi: *Yeast* spp., *Aspergillus* spp., *Mucor* spp., *Rhodotorula* spp., *Penicillium* spp., *Rhizopus* spp., and bacteria: *Bacillus* spp., *Pseudomonas* spp., *Klebsiella* spp., *Staphylococci aureus*, *E. coli*, *Streptococci* spp.) and soil quality parameters (pH, temperature, and OM content). There was positive correlation between soil carbon or nitrogen and the corresponding microbial load even though it was not statistically significant ($P > 0.05$). These relationships can be used as an index of soil health and fertility (Mohammed and Zigau, 2016).

Micro-arthropods populations hosted selected locations that represented loamy sand, silt loam, and clay agriculture soils in Egypt, were extracted and identified during the two growing seasons of 2016 and 2017. The taxonomic groups of these micro-arthropods comprised seven types identified as *Symphyla, Pauropoda, Gamasida, Collembola, Oribatida, Actinedida,* and *Psocoptera.* Throughout the two successive seasons, the distribution of these taxonomic groups was varied according to the types and depths of these soils. The topsoil layers (0–10 cm) were featured by some taxonomic groups. In season of 2016, the overall populations per 942.48 cm^3 of soil samples prevailed in *Oribatida* (4.14) and *Actinedida* (3.95) in loamy sand soil. The clay fauna was featured by *Psocoptera* (4.05) and *Collembola* (3.95). In season of 2017, the most prevalent micro-arthropods in loamy sand soil were 3.86 for each of *Oribatida* and *Actinedida,* besides, *Psocopetra* (3.71) and *Collembola* (3.95) in clay soil. Meantime, silt loam soil was featured by *Collembola* (3.52), *Poscoptera* (3.43),* and *Oribatida* (3.71). Eventually, *Collembola* and *Psocoptera* in clay soil and *Oribatida* and *Actinedida* in loamy sand soil are almost maintained at their highest significant levels during the two seasons. The lower soil layer (10–20 cm), during the season of 2016, was featured by *Oribatida* at levels of 1.62 and 2.62 in silt loam and loamy sand soils, respectively. Clay soil samples were featured by *Collembola* (1.76) and *Pscoptera* (2.95). In season of 2017, high significant populations of *Oribatida* (1.43) and *Actinedida* (1.67) mediated silt loam soil. At the same time, *Oribatida* reached high levels of 2.14 in loamy sand soil. In addition, abundant populations of *Collembola* (2.14) and *Pscoptera* (2.62) settled clay soil (Khamis, 2018).

11.10   SOIL FAUNA AS BIOINDICATOR FOR PESTICIDES CONTAMINATION

The forecasting of survival soil micro-arthropods population could be used as bioindicator for pesticide residues in soil. This connotation was confirmed by the field study carried out on some neonicotinoids application via loamy sand, silt loam, and clay soils in Egypt at two growing seasons (2016 and 2017) of tomato seedlings. The treated soil depths of 0–10 cm and 10–20 cm adjacent to root system of the plant in loamy sand and silt loam soils with thiamethoxam were safer to soil micro-arthropods incidence. In addition, thiacloprid was safe only in loamy sand soils by the end of the 7th week post-treatments. Treated micro-arthropods with the thiamethoxam and thiacloprid could restore their initial recovery levels, especially in the lower depths. However, these applications were not safe for micro-arthropods in the high OM-clay soil (Khamis, 2018). Recovery levels of soil micro-arthropods populations may be declined in high OM soil content. These findings were sided with mobility retardation in high OM content-soil that controlled sorption–desorption characteristics, which may decrease the neonicotinoids degradation (Zhang et al., 2018).

Rapid dissipation of thiamethoxam was concurrent with the rapid recovery of micro-arthropods at the end of the 7th week in aerated soils of loamy sand and silt loam than terrible clay soil where micro-arthropods failed to restore their recovery levels. Mobility and leaching properties in laboratory studies of clothianidin, thiamethoxam, and imidacloprid elution profiles were high in sandy soil but moderate in loam soil and slow in silty and cancel clay soil. Nevertheless, recoveries of effluent solution of thiamethoxam had high rapidity than clothianidin in sandy soil correlated with their water solubility (Toscano and Byrne, 2005; Mortl et al., 2016). Water retention could be enhanced in high electric conductivity (EC) values of soil that increase osmotic potential of the soil, and hence contact time for adsorption subsequently increases. Therefore, adsorption in sandy loam soil may be declined due to the scare of vacant sites in its sand particles compared to silt or clay (Shar et al., 2016).

On the other hand, in Pakistan, a study was carried out in open field to evaluate the effects of four conventional insecticides; bifenthrin, chlorpyrifos, endosulfan, and imidacloprid at their recommended doses on soil arthropods. All these insecticides drastically reduced soil-arthropod populations of spiders, black spiders, ants, field cricket, snow bug, insect larvae, and silverfish at 1 day post-treatment. So far, a slow recovery for these soil arthropods populations up to 14 days post-treatment had occurred gradually at levels below the untreated plots. Maximum residues were 3.21, 7.94, 5.89, and 3.01 ppm for bifenthrin, chlorpyrifos, endosulfan, and imidacloprid, respectively, at 1 day post-treatment. The minimum residues were obtained at 14 days post-treatment for all insecticides treatments. However, bifenthrin (0.57 ppm) had the least value and endosulfan (3.59 ppm) had the highest value among the tested insecticides at 14 days post-treatment, the reduction percentages in soil arthropods populations had the order of bifenthrin > chlorpyrifos > imidacloprid > endosulfan (Shar et al., 2016).

11.11   RAPID FORECASTING OF PESTICIDES CONTAMINATION IN SOIL

Fortunately, stochastic model is the first analytical solution to a transitory probability distribution that can be applied now as a rapid forecast to the concentrations and withholding periods of pesticide (like herbicides) and its residue in soil and managing water quality as well. Withholding period for particular crops may now be managed more actively in accordance with their established phytotoxic thresholds. Moreover, the Bayesian model demonstrated the uncertain reactions between two parameters: climate (storm depth and rainfall vent frequency) and chemical (sorption and degradation). The stochastic model can be calculated easily allowing a rapid completion of Bayesian model calibration and forecasting to adjust the linear degradation or dissipation rate of pesticides in soil. With the ability to quantify concentrations in the field the time between measurement and a forecast can be reduced to minutes. In addition, geo-location services allow determining the variable climate parameters along the season from handheld devices in the field (Gavan et al., 2019).

11.12   SOIL BIOREMEDIATION

Bioremediation as a definition means the process that involves the utilization of living organisms to minimize or remove hazardous chemicals accumulated in an environment (Fingerman and Nagabhushanam, 2016). The naturally occurring organisms used in bioremediation included fungi, bacteria, algae, planktons, plants, and protozoans. Also, genetically modified organisms can potentially be used (Watanabe, 2001; Sudip et al., 2002). The organisms can extirpate organic chemicals, whereas the contaminated metals can either be directly removed or converted to a stable form (Watanabe, 2001). Bioremediation contains three main bases including biosorption (Park et al., 2010), bioaccumulation (Peng et al., 2008), and biocrystalization (Mathew, 2005).

11.12.1   SOIL PHYTOREMEDIATION

Phytoremediation is a subcategory of bioremediation that is the usage of plants for removing contaminated toxicants in atmosphere. Phytoremediation is considered as a rapid and novel forecasting method. Certain studies

(Malik et al., 2017a; Kadu et al., 2017; Nadaf et al., 2017) clarified that phytoremediation mechanisms include uptake, phytotransformation (conversion of more toxic metals to relatively lower toxic metals), phytodegradation (metabolizing organic contaminant via plant enzymes), phytovolatilization (volatilization of organic contaminants via plant leaves), and rhizoremediation (at rhizospheric zone, organic compounds are released through root exudation and activate microbiome leading to improved rhizospheric pollutant detoxification).

11.13   THE CLEANUP OF CONTAMINATED SOIL WITH PESTICIDES

11.13.1   PHYTODEGRADATION

Phytodegradation is also expressed as "phytotransformation". It is defined as a degradation (breakdown) of complex organic molecules (contaminants) after they have been taken up by the plant to simple molecules or conjunction of these molecules into plant tissues. In general, plant uptake occurs through the process of phytoextraction and phytovolatilization only when the solubility and hydrophobicity of contaminants reach a certain acceptable range. Phytodegradation has been entitled to organic contaminants remediation like chlorinated solvents, herbicides, and munitions in soil, sediment, and groundwater (Galadima et al., 2018; Malik et al., 2017a; Conesa et al., 2012).

11.13.2   RHIZODEGRADATION

Rhizodegradation is also expressed as "Phytostimulation" and "Rhizodegradation". It is defined as a breakdown of contaminants that may be executed by bacteria or other microorganisms within the plant rhizosphere. The plant rhizosphere exudes enzymes, sugars, amino acids, and other compounds to stimulate bacterial growth. In addition, the roots also provide additional space area for microbes growing and a pathway for oxygen transfer. Rhizodegradation is primarily successful in treating organic chemicals including petroleum hydrocarbons, polycyclic aromatic hydrocarbons, chlorinated solvents, pesticides, polychlorinated biphenyls, benzene, toluene, ethylbenzene, and xylenes (Galadima et al.,2018; Malik et al., 2017a).

11.14   ADVANTAGES AND DISADVANTAGES OF PHYTOREMEDIATION

11.14.1   ADVANTAGES OF PHYTOREMEDIATION

- Phytoremediation is significantly lower than traditional processes in costs.
- Facilitate monitoring plants during soil cleanup process.
- Phyto-mining companies could recover and reuse valuable metals.

11.14.2   DISADVANTAGES OF PHYTOREMEDIATION

- Phytoremediation is limited to rhizosphere area.
- Long term needed for slow growth and low biomass.
- Could not resolve the problem of contaminants leaching into groundwater.
- The plants used in remediation process are affected by toxicity of contaminants, which then pass into food chain to consumers.

11.15   GENETICALLY MODIFIED SOIL MICROORGANISMS

The application of molecular-based approaches in bioremediation using isolated resistant microbes is a novel technology for the biodegradation of toxic pesticides in soil. Some microorganisms may benefit from pesticides that contaminate soil as a source of nutrients to obtain their needful energy. In the field or under laboratory conditions, various species of bacteria and fungi have been distinguished by their performances in the biodegradation of carbamate pesticides. Environmental meta-genomic information from soil and sea could be an important source of genes. Genetically modified microorganisms could be considered an alternative pathway for biodegradation of pesticides. Moreover, biomolecular engineering system should be developed more to enhance degradative enzyme activities in soil. Great efforts are needed to establish comprehensive knowledge of the molecular basis for catabolic sequences, to protect enzyme activity at high threshold concentration of toxic organic contaminants (Mustapha et al., 2019).

11.16   CONCLUSION

Excessive use of high persistence pesticides in the agricultural soils over the last few decades has led to soil contamination with pesticide residue and toxicity to nontarget soil biota species (Silva et al., 2019; Khamis, 2018). Therefore, this chapter aimed to show the novel solutions adequate for monitoring and remediating programs of pesticide residues and/or contamination in a wider range of soil species. In this regard, we focused on the role of bioindicators (micro-arthropods), phytoremediation that should be activated more appropriately to curb the dilemma of long period-contaminated soil with pesticide residues. Likewise, the soil enzyme activities should be improved for pesticide degradations by developing technology of molecular basis for catabolic sequences using genetically modified microorganisms. New approaches for applied models should be updated in particular crops for forecasting pesticide residues and withholding periods in soil regarding phytotoxic thresholds. Thus, calls for evolution of thresholds and quality standards for pesticide residues in soil in the EU legislation became a must to protect soil biota from definite threats, especially in the absence of soil protection policies.

KEYWORDS

- phytoremediation
- pesticide residues
- phytodegradation
- rhizodegradation
- phytostimulation

REFERENCES

Alavanja, M. C. Introduction: Pesticides Use and Exposure Extensive Worldwide. *Rev. Env. Health.* **2009**, *24*(4), 303–309.

Barančíková, G.; Koco, Š.; Makovníková, J.; Torma, S. Filter and Transport Functions of Soil. *Bratislava: Slovak, Soil Sci. Conserv. Res. Inst.* **2010**, *14*(02), 68–76.

Barrigossi, J. A. F.; Lanna, A. C.; Ferreira, E. Inseticidas Registrados Para a Cultura Do ArrozEanálise de ParâmetrosIndicadores de SeuComportamento No Ambiente. *Santo Antônio de Goiás: EmbrapaArroz e Feijão, Circular Técnica.* **2005**, *74*, pp 4.

Cabrera, L.; Costa, F. P.; Primel, E. G. Estimativa de Risco de Contaminação das Águaspor Pesticidas Na Região Sul do Estado do RS. *Química Nova*. **2008**, *31*(8), 1982–1986.

Capel, P. D. Organic Chemical Concepts. In: Alley, W. M. (Ed.). *Regional Ground-Water Quality*. New York: Van Nostrand Reinhold. **1993**, pp. 155–180.

Cerejeira, M. J.; Viana, P.; Batista, S.; Pereira, T.; Silva, E.; Valerio, M. J. Pesticides in Portuguese Surface and Ground Waters. *Water Res*. **2003**, *37*(5), 1055–1063.

Conesa, H. M.; Evangelou M. W. H.; Robinson, B. H.; Schulin, R. A. Critical View of Current State of Phytotechnologies to Remediate Soils: Still a Promising Tool? *Sci. World J.* **2012**, doi:10.1100/2012/ 173829.

Corstanje, R.; Mercer, T. G.; Rickson, J. R.; Deeks, L. K.; Newell-Price, P. L.; Holman, I.; Kechavarsi, C.; Waine, T. W. Physical Soil Quality Indicators for Monitoring British Soils. *Solid Earth*. **2017**, *8*, 1003–1016.

Dias, A. C. R. Comportamento No Ambiente e Propriedades Físico-Químicas de Pesticidas. *30 set. Palestra*. **2010**.

Doran, J. W.; Sarrantonio, M.; Liebig, M. A. Soil Health and Sustainability. *Adv. Agron*. **1996**, *56*, 1–54.

Drum, C. *Soil Chemistry of Pesticides*, PPG Industries, Inc. USA. **1980**.

EPA. *What are Pesticides and How Do They Work?* **2007**, http://www.environment.nsw.gov. au/pesticides/pestwhatrhow.htm.

FAO. *Food and Agriculture Organization of the United Nations*, Rome, Italy. **2015**, pp. 65.

FAO. *International Code of Conduct on the Distribution and Use of Pesticides*. Rome, Italy. **1989**.

FAO. *International Code of Conduct on the Distribution and Use of Pesticides*. Revised Version, Adopted by the Hundred and Twenty-Third Session of the FAO Council in November 2002, Rome, Italy. **2002**, http://www.fao.org/fileadmin/templates/agphome/ documents/PestsPesticides/Code/code.pdf.

Fingerman, M.; Nagabhushanam, R. *Bioremediation of Aquatic and Terrestrial Ecosystems*. CRC Press. **2016**, p. 622.

Galadima, A. I.; Mohammed, S.; Abubakar, A.; Deba, A. A. Phytoremediation: A Preeminent Alternative Method for Bioremoval of Heavy Metals from Environment. *J. Adv. Res. Appl. Sci. Eng. Technol*. **2018**, *10*(1) 59–71.

Garcia, F. P.; Ascencio, S. Y. C.; Oyarzun, J. C. G.; Hernandez, A. C.; Alavarado, P. V. Pesticides: Classification, Uses And Toxicity Measures of Exposure and Genotoxic Risks. *J. Res. Env. Sci. Toxicol*. **2012**, *1*(11), 279–293.

Gavan, M. P.; Rao, P. S. C.; Mellander, P.; Kennedy, I.; Rose, M.; Zwieten, L. V. Real-Time Forecasting of Pesticide Concentrations in Soil. *Sci. Total Env*. **2019**, *663*, 709–717.

Gebler, L.; Spadotto, C. A. ComportamentoAmbiental dos Herbicidas. In: Vargas, L.; Roman, E. S. (Eds.). *Manual de Manejo e Controle de PlantasDaninhas*. *Bento Gonçalves: Embrapa Uva Vinho*. **2004**, 57–87.

Gevao, B.; Semple, K. T.; Jones, K. C. Bound Pesticide Residues in Soils: A Review. *Env. Pollution*. **2000**, *108*(1), 3–14.

Ghorab, M. A.; Khalil, M. S. Toxicological Effects of Organophosphates Pesticides. *Inter. J. Env. Monit. Analysis*. **2015**, *3*(4), 218–220.

Hwang, J.; Zimmerman, A. R.; Kim, J. Bioconcentration Factor-Based Management of Soil Pesticide Residues: Endosulfan Uptake by Carrot and Potato Plants. *Sci. Total Env*. **2018**, *627*, 514–522.

Jozef, V.; Koco, S. Integrated Index of Agricultural Soil Quality in Slovakia. *J. Maps*. **2018**, *14*(2), 68–76.

Khalil, M. S. Fate of Pesticides in the Agricultural Environment. *Biol. Med.* **2014**, *7* (3), 1000E117.

Khamis, W. M. Residuals Efficacy of Thiamethoxam and Thiaclopride Soil Applications against Whitefly, *Bemisiatabaci* (Homoptera; Aleyrodidae) and Their Impact on Soil Microarthropods. *Egypt. Acad. J. Biolog. Sci.; F. Toxicol. Pest Control*. **2018**, *10*(1), 69–84.

Mackay, D.; Shiu, W. Y.; Ma, K. C.; Lee, S. C. *Handbook of Physical-Chemical Properties and Environmental Fate for Organic Chemicals*. 2nd ed. Vol. 1. Boca Raton, FL, USA: CRC Press. **2006**, pp. 925.

Maksymiv, I. Pesticides: Benefits and Hazards. *J. Precarpathian Natl. Univ.* **2015**, *2*(1), 70–76

Malik, C. P.; Wani, S. H.; Bhati-Kushwaha, H.; Kaur, R. Advanced Technologies for Crop Improvement and Agricultural Productivity. In: Gorain B.; Paul, S. (Eds.), Important Biological Indicators for Monitoring Soil Quality. *Magazine Agric. Biolog. Sci.* **2017a**, *15*(12), pp. 148.

Malik, C. P.; Wani, S. H.; Bhati-Kushwaha, H.; Kaur, R. Advanced Technologies for Crop Improvement and Agricultural Productivity. In: Kadu, J. B.; Nadaf, J. M.; Kondvilkar, N. B.; Annapurna, M. V. V. I. (Eds.), Photoremediation and its Mechanisms. *Magazine Agric. Biolog. Sci.* **2017b**, *15*(12), pp. 148.

Martins, C. R.; Lopes, W. A.; Andrade, J. B. Solubilidade das SubstânciasOrgânicas. *Química Nova*. **2013**, *36*(8), 1248–1255.

Mathew, A. M. Phytoremediation of Heavy Metal Contaminated Soil. PhD Dis*sertation,* Oklahoma State University. **2005**.

Mervat, S. M. Degradation of Methomyl by the Novel Bacterial Strain *Stenotrophomonas maltophilia* M1. *Electron J. Biotechnol.* **2009**, *12*(4):1–6.

Micuți, M.; Bădulescu, L.; Israel-Roming, F. Effect of Pesticides on Enzymatic Activity in Soil. *Bulletin UASVM Animal Sci. Biotechnol.* **2018**, *75*(2), 80–84.

Miller, G.T. Sustaining the Earth. 6th ed., Thompson Learning, Inc. Pacific Grove. **2004**, pp. 211–216.

Minguela, J. V.; Cunha, J. P. A. R. Manual de Aplicação de Produtos Fitossanitários. *Viçosa: Aprenda Fácil Editora*. **2011**, 588 pp.

Mohammed, U. A.; Zigau, Z. A. Influence of Soil pH and Temperature on Soil Microflora. *Gashua J. Sci. Humanities*. **2016**, *2*(1), 39–47.

Mortl, M.; Kereki, O.; Darvas, B.; Klatyik, S.; Vehovszky, A.; Gyori, J.; Szekacs, A. Study on Soil Mobility of Two Neonicotinoid Insecticides. *J. Chem.* **2016**, 4546584, pp. 9.

Mustapha, M. U.; Halimoon, N.; Lutfi, W.; Johar, W.; Abd-Shukor, M. Y. An Overview on Biodegradation of Carbamate Pesticides by Soil Bacteria. *Pertanika J. Sci. Technol.* **2019**, *27*(2), 547–563.

Oliveira, M. F.; Brighenti, A. M. Comportamento dos Herbicidas No Ambiente. In: Oliveira J. R.; Constantin, J.; Inoue, M. H. (Eds.). *Biologia e Manejo de PlantasDaninhas*. Curitiba-PR: Omnipax. **2011**, pp. 263–304.

Park, D.; Yun, Y. S.; Park, J. M. The Past, Present, and Future Trends of Biosorption. *Biotechnol. Bioprocess Eng.* **2010**, *15*(1), 86–102.

Peng, K.; Luo, C.; Lou, L.; Li, X.; Shen Z. Bioaccumulation of Heavy Metals by The Aquatic Plants *Potamogeton Pectinatus* L. and *Potamogeton Malaianus* Miq. and Their Potential

Use for Contamination Indicators and in Wastewater Treatment. *Sci. Total Env.* **2008**, *392*(1), 22–29.

Prashar, P.; Shah, S. Chapter 8: Impact of Fertilizers and Pesticides on Soil Microflora in Agriculture. In: Eric Lichtfouse (Ed.), *Sustainable Agric. Rev.* **2016**, *19*, 331–361.

Shar, Z. U.; Rustamani, M. A.; Nizamani, S. M.; Shar, M. U.; Shar, T. Effect of Pesticide Residues on Soil Arthropods in Okra Crop. *Eur. Acad. Res.* **2016**, *3*(10).

Sibanda, M. M.; Focke, W. W.; Labuschagne, F. J.; Moyo, L.; Nhlapo, N. S.; Maity A.; Muiambo, H.; Massinga Jr., P.; Crowther, N. A. S.; Coetzee, M.; Brindley, G. W. A. Degradation of Insecticides Used for Indoor Spraying in Malaria Control and Possible Solutions. *Malaria J.* **2011**, *10*, pp 307.

Silva, C. M. M. S.; Fay, E. F. Agrotóxicos: Aspectos Gerais. In: Silva, C. M. M. S.; Fay, E. F. (Ed.) *Agrotóxicos e Ambiente. Brasília: Embrapa Informação Tecnol.* **2004**, 17–73.

Silva, V.; Mol, H. G. J.; Zomer, P.; Tienstra, M.; Ritsema, C. J.; Geissen, V. Pesticide Residues in European Agricultural Soils—A Hidden Reality Unfolded. *Sci. Total Env.* **2019**, *653*, 1532–1545.

Sudip, S. K.; Singh, O.V.; Jain, R. K. Polycyclic Aromatic Hydrocarbons: Environmental Pollution and Bioremediation. *Trends Biotechnol.* **2002**, *20*(6), 243–248.

Thakur, S.; Singh, L.; Wahid, Z. A.; Siddiqui, M. F.; Atnaw, S. M.; Din, M. F. Plant-Driven Removal of Heavy Metals from Soil: Uptake, Translocation, Tolerance Mechanism, Challenges, and Future Perspectives. *Env. Monit. Assess.* **2016**, *188*, (4), pp. 206.

Toscano, N. C.; Byrne, F. J. Laboratory and Field Evaluations of Neonicotinoid Insecticides against the Glassy-Winged Sharpshooter, in Proceedings of the Pierces Disease. *Res. Symp.* **2005**, 380–383.

Watanabe, K. Microorganisms Relevant to Bioremediation. *Curr. Opin. Biotechnol.* **2001**, *12*, (3), 237–241.

Zhang, P.; Ren, C.; Sun, H.; Min, L. Sorption, Desorption and Degradation of Neonicotinoids in Four Agriculture Soil and Their Effects on Soil Microorganisms. *J. Sci. Total. Env.* **2018**, *615*, 59–69.

CHAPTER 12

Pesticide Pollution: Risk Assessment and Vulnerability

SAIMA HAMID*, ALI MOHD YATOO, MOHAMMAD YASEEN MIR, and AZRA N KAMILI

Centre of Research for Development/P.G. Department of Environmental Sciences, University of Kashmir 190006, Jammu and Kashmir, India

Corresponding author. E-mail: cord.babasaima4632@gmail.com

ABSTRACT

In crop production, pesticides are commonly used to avoid or combat pests, illnesses, bacteria, and other crop pathogens to decrease or eliminate loss of output and preserve good product value. While pesticides are established with very rigorous regulatory procedures to work with sensible assurance, with minimal environmental and human health effects, severe health hazards arising from occupational exposure and residues in food and drinking water have been identified. In the event of industrial employees in the open field and greenhouses, pesticide employees often experience occupational exposure to pesticides. Work-related exposure to pesticides is often the situation for open-field farm employees, greenhouse and pesticide employees, and home pest exterminators. The overall population is mainly exposed to pesticides via the consumption of meat and drinking water with pesticide residues contaminated, whereas significant damage can happen within or around the household. Regarding negative environmental impacts (e.g., leaching, rivers and spray drifting, water, land, and air contamination, as well as negative impacts on fauna and fish, crops, and other nontarget species), the strength, steps made for its use, dosage used, adsorption to ground colloids, the climate circumstances after implementation, and how long the pesticide will persist in the setting are all factors that influence most of these impacts. As a consequence of different phases and concentrations of exposure, the kinds used (concerning toxicity and persistence), and the features of the setting, threat assessments of the effect of pesticides on human health or the atmosphere are not a simple and specifically precise method. New instruments or methods that are more reliable than current ones are thus required to assess pesticide prospective dangers, and thus help to reduce the negative impacts on human health and the atmosphere.

12.1   INTRODUCTION

A pesticide is a poisonous chemical product or a combination of deliberately released drugs or biological agents into the setting to prevent, regulate, and/ or kill and destroy populations of animals, plants, or other damaging pests. Pesticides are of enormous benefit to people when correctly used but their indiscriminate use can, however, pose significant health and climate risks. There are a vast number of health issues identified by constant exposure of pesticides such as births, neonatal fatalities, congenital birth defects, blindness, depressed breathing, cardiac disease, cancer, and tumor. Problems with pesticide use are generally worse in emerging nations where many class I WHO medicines are still in use. The use of pesticides in modern cultivation has increased the effect of these chemicals on the atmosphere from an economic and health perspective (Debenest et al., 2009, 2010; Kouser and Qaim, 2011). An acute poisoning-induced pesticide is a health hazard to farmers in an elevated danger of toxicity following exposure to the pesticide (Lekei et al., 2014). Nearly three million individuals have been intoxicated and there are reports of more than 200,000 deaths every year from pesticide toxicity around the world (Shaikh, 2011).

Global use of pesticides is growing, and in freshwater and marine habitats, there is a growing prevalence of pesticides, including herbicides, fungicides, and insecticides. The worldwide use of pesticides has risen from 2.27 billion to 2.36 billion kg from 2000 to 2007, respectively (US EPA, 2012). More than 5 billion kg of chemicals have been used in the United States alone in 2001 (Kiely et al., 2004) and pesticides have been identified in over 90% of watersheds established (as any watershed predominating in agriculture, rural, or blended property use). It is also essential to acknowledge that intermediate pesticide products can have a significant effect on the santé of marine habitats, in relation to the knowledge of the impact of parent pesticide compounds. The intermediate pesticide is the decomposition product generated from the parent pesticide compound via photolytic, hydrolytic, or micro-induced decomposition. Many of those intermediate agents are not inert biologically and can have widened or totally different (positive or négative) effects as compounds of the parent (Kralj et al., 2007; Zeinat et al., 2008). The use of pesticides and their consequent intermediates can also be additionally beneficial for nontarget animals, as most pesticides are intended for the purpose of a specific pesticide or pest unit (Rohr et al., 2006). Pesticide exposure, for example, has resulted in a decrease in the biodiversity, toxic to certain algases, diatoms, and alterations of the food webs of the ecosystem, including heterotrophic activity (Debenest et al., 2010; Beketov et al., 2013; Malaj et al., 2014). A WHO study has revealed that over 200,000 deaths occur each year due to the poisoning of these harmful chemicals. Indeed, the number of casualties does not verify the actual image of pesticide toxicity, but about three million instances of intoxication have been recorded every year. Therefore, the exposure to these pesticides or harmful chemicals results in several health conditions such as asthma, neck rashes, and chronic conditions such as emphysema and carcinogenic diseases. Therefore, it is necessary to reduce the adverse health effects of pesticides as well as to impose restrictions on their use and human health when they are found to be dangerous above the maximum level (James et al., 2008).

12.2   PESTICIDE USE AND THEIR CLASSIFICATION

Pesticides are a heterogeneous category of natural or synthetic chemical substances designed for the prevention of pesticides, weeds, or plant diseases. There are 50,000 plant pathogens, 9000 mites and insect species, and 8000 weed species that can damage plants. Annual crop production may be affected by pesticides up to 30% (Saleem and Ashfaq, 2004). According to the United Nations Food and Agriculture Organisation, the worldwide potential loss of food is around 55%, including 35% pre-harvest and 20% post-harvest loss. About 14% of loss is reported as a result of insect pests, 13% as a result of plant pathogens, and 13% as a result of weeds (Pimentel, 2009). Without the use of pesticides, losses of vegetables, fruit, and grains would be 54%, 78%, and 32%, respectively (Cai, 2008). Approximately one-third of agricultural products are produced using pesticides (Liu et al., 2002) in agricultural production, which is so indispensable. Tens of thousands of different types of chemicals are used worldwide to control crop losses by pests, insects, and diseases. Some of the pesticides contain a number of active substances to fight two or more parasites, making their rating truly challenging. More than 1100 pesticide substances are currently registered in the status list of active substances commercially available in the European Union (EU) (Húšková et al., 2008). These active substances are classified on the basis of (1) the active molecular structure, (2) the formulation, and (3) the biological target.

12.3 PESTICIDE AND THE ENVIRONMENT

Apart from its potential adverse effects on human health, pesticides also have adverse environmental effects (water, soil, and air pollution, toxic effects on nontarget organisms; Bürger et al., 2008; Mariyono, 2008). In particular, improper use of pesticides was related to (1) adverse effects on nontarget organisms (e.g., reducing beneficial populations), (2) mobile pesticide or pesticide drift water contamination, (3) volatile pesticide air pollution, and (4) herbicide drift injuries to nontarget plants, (5) fro-rotational crops injuries, (6) crop injury due to high application rates, and improper timing of use or adverse environmental conditions at and after pesticide application. Most adverse environmental effects of pesticides depend on the physicochemical interactions in pesticide (vapor pressure, stability, and solubility pKa), soil adsorption, and other physical properties. Furthermore, the toxicity, the dosage used, the weather conditions following application, and the persistent environmental duration of the pesticide can take account of their adverse environmental effects.

Soil factors and weather conditions have long been recognized as major factors affecting the fate and activity, selectivity, and adverse environmental effects of the pesticide in the environment (Reus et al., 2001). However, as these factors vary year after year from site to site, the results from any field study on the pesticide's fate and behavior are specific for a given place and season. The behaviors and the destiny of a pesticide are thus initially assessed by calculating the predicted ambient concentration (PEC) for environmental risk assessment; Initially, computations of the environmental concentration predicted, known in the United States as the estimate of the environment concentration, are used to evaluate the behavior and the fate of the pesticide (Matthews, 2006). The soil, water, sediments, and air concentrations are calculated, and validation will be conducted by comparison with data from the three tests (available for approval-registration purposes) to assess the toxicity of pesticides in key nontarget organisms. In addition, the TER is also calculated to determine whether or not the risk is acceptable to the organism (Eleftherohorinos, 2008). The TER of a PEC divided organism of susceptibility, relevant to that in which the organism lives, shall be calculated by an equivalent measures LC50 or (LD50) (no observing effect concentration). A detailed higher level risk assessment (Boxall, 2001) is generally necessary when TER is lower than 100, whereas in TER < 10 a chronic risk assessment is required (Matthews, 2006).

Although the farm soil is the main beneficiary of pesticides, aquatic bodies adjoining agricultural areas are usually the final beneficiary of pesticide residues (Pereira et al., 2009). This problem is why European authorities need data on the risk of nontarget terrestrial and aquatic organisms (before plaguicide marketing in Europe) for potential adverse environmental effects. Given the negative effects of pesticides in agriculture, it now appears that the use of criteria in the selection of pesticides is effective, cost-efficient, and safe for operator and environment (Reus et al., 2002; Bockstaller et al., 2009). Furthermore, the use of some environmental risk indicators as alternatives to the measurement of a direct pesticide impact is linked to methodical problems or for reasons of practice (i.e., time, costs; Bockstaller et al., 2009) was also a reality. Reus et al. (2002) and Bockstaller et al. (2010) have already applied the indicators to assess the potential risks of aquatic pesticides, soil organisms, bees, air emissions, bioaccumulation, and human health. The environmental indicators used in the studies were calculated based on the pesticide persistence in the soil (half-life, DT_{50}), soil mobility (coefficient for the adsorption of organic carbon), water toxicity (lethal concentration of aquatic organisms, LC_{50}), and organisms from the soil (no observing effect concentration). With respect to the contributions made by Reus et al. (2002) in assessing 15 individual pesticide applications using eight indicators, environmental indicators for pesticide selection showed the following:

(1) Some of the 15 pesticide applications had a high level of environmental impact (higher environmental impacts) with all indicators used, but they were significantly different in their ranking when the score for the environment was concerned as a whole,

(2) The score was not related to most other product rankings of the pesticide risk indicator based on the indicator kilograms of an active ingredient,

(3) Same rankings of 15 pesticide applications for the individual areas of surface water, groundwater, and soil pollution were established by the pesticide risk indicators used.

The latter was largely determined through the pesticide toxicity to aquatic organisms by the values for surface water pollution, while the values for groundwater contamination were determined largely by DT_{50} and KoC. However, two pesticides found toxic or mobile but used at extremely low rates were the exception. These results show that new and more reliable indicators to anticipate the potential risk of pesticides are needed to help reduce the environmental impacts of pesticides (Reus et al., 2002).

12.4   VULNERABILITY ASSOCIATED WITH PESTICIDE USE

Their benefit has been exceeded by the risks associated with pesticide usage. Pesticides have drastic effects on nontarget species and have an impact on the biodiversity, aquatic- and land-based food webs, and ecosystems of animals and plants. Majewski and Capel (1995) report that approximately 80%–90% of the pesticides applied can volatilize within a few days from their application. It is very common and can happen with sprayers. Volatilized pesticides evaporate into the air and can damage a nontarget organism subsequently. An excellent example here is the use of herbicides that volatilize the treated plants and their vapors cause serious damage to other plants (Straathoff, 1986). Uncontrolled pesticide use has led to several species of animals and plants being reduced in land and aquatic environments. Some of the rare species like the bald eagle, the peregrine falcon, and osprey have been also threatened for survival (Helfrich et al., 2009). Furthermore, air, water, and soil systems are contaminated to toxic levels by these chemicals. Insecticides are considered to be the most toxic among all pesticide categories, while fungicides and herbicides are the second and third categories on the toxicity list. Depending on their solubility, pesticides enter natural ecosystems with two different means. Water-soluble pesticides disintegrate into water and enter soil water, streams, rivers, and lakes, and therefore harm nontarget species. On the other hand, by a process called "bio-amplification," fat-soluble pesticides enter into animals, as shown in Figure 12.1. They are absorbed into animal fatty tissue, which leads to pesticide persistence in food chains for longer periods (Warsi s.d.).

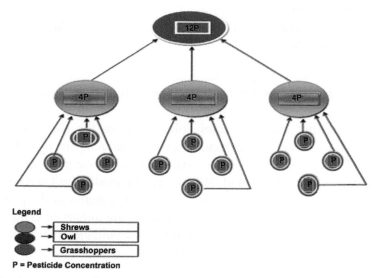

Legend

Shrews
Owl
Grasshoppers

P = Pesticide Concentration

FIGURE 12.1 Biomagnifications of pesticide.

The biomagnification process can be described as follows:

1. The animal bodies in the nutrient chain such as Grasshopper (primary user) receive small concentrations of pesticide.
2. Shrews (secondary consumers) eat a lot of hoppers and therefore the pesticide concentration in their bodies will increase.
3. When the high-level predator, like an owl, eats shrews and other prey, the concentration of the pesticide ultimately increases many folds in his body.

The higher the level of the trophies, the higher the concentration of the pesticide, called bioamplification, is. This process disrupts the entire ecosystem, since more species in high trophic concentrations die because their bodies are more toxic. In the end, this will increase the secondary consumers population (shrews) and decrease the primary consumers population (grasshoppers) (Warsi n.d.).

12.5 PESTICIDE RESIDUES

Residues are very low quantities of pesticides that can remain for years in the blood, tissue, or on a plant, which are bioaccumulated throughout the food chain. All foods do not contain pesticide residues and are usually found at low levels whenever they occur. Residues of pesticides also include pesticide breakdown products.

12.5.1 PESTICIDE RESIDUES IN FRUITS AND VEGETABLES

In samples of fruit and vegetables collected from the Karachi markets, pesticide residues were also detected. For residual analysis, only 45 goods were tested. It has been found that most organochlorine pesticide residues such as heptachlorine, BHC, and DDT were contained in fruit and vegetable products that is quite harmful and alarming for the health of the people, as they are basic human needs. These food products are important to the human being. The fruit and vegetables used by humans should therefore be carefully washed for any dangerous incidents (Azmi et al., 2011).

12.5.2 PESTICIDE RESIDUES IN HUMAN BLOOD

Studies on the presence of pesticides in human biological material have been extensively done around the world. A few reports have already been published in that regard by Krauthacker et al. (1980), Saxena et al. (1980), Mercedes and Thiel (1986), Sabbah et al. (1987), and Krawingel et al. (1989). A study was performed in Pakistan. Far from all the samples reported here, organo-chlorine (OC) contamination was found. However, more pesticides were found in the samples obtained from two centers. Workers who are engaged in chlorination (third center) have more tenakil compounds and other OC compounds. Only a few samples have been identified as having deltamethrin (pyrethroid) and malathy (organophosphate) due to fresh exposure.

Blood samples were taken from people between 6 and 36 years of age. Only two of these pesticides were discovered in the highest concentration, malathion (30.0 µg/mL) and Aldrin (18.0 µg/mL). One sample was from a girl who was just 6 years of age, her serum sample had 0.035 µg/mL DDE. She may have received DDE, through her mother who subsequently degraded to DDE. The reports by Chikuni (1991) and Kanja et al. (1992) are as follows. Azmi and Naqvi (2005) carried out a same type of study in which blood samples of people (especially farmworkers and sprayers) engaged in different fruits were collected and of the individuals involved in various farm stations in Gadap (rural area) Karachi, Pakistan were collected (specially farmworkers and man-made spraying). A total of 287 blood samples were taken from individuals exposed and controlled for the residual analyses by HPLC and 29.6% organophosphate, 29.6% organochlorine, and 48.1% pyrethriode were reported as residual samples. It therefore clearly shows that more pesticides such as cypermethrin, deltamethrin profenophos, diazinon, monocrotphos, DDT, DDE, and autorin are available to people who are active in field work as a result of increased exposure to these pesticides during spraying time in these areas. In this connection, the highest concentration of pesticides (12.22 mg/mL) was reported by Jahan (1995) and Bissacot and Vassilieff (1997b) in case of deltamethrin and slightly high concentration of 0.025–5.0 µg/mL was reported by Musshoff et al. (2002) in case of diazinon. In case of monocrotophos, slightly high level of concentration 0.025–50 mg/g was reported by Musshoff et al. (2002). 0.27 mg/mL$^{\wedge\wedge}$ by Kocan et al. (1994), 0.9 mg/L by Guardino et al. (1996), 4.71 and 38.13 mg/L by Dua et al. (1996), 1.9 ppb by Luo et al. (1997), and 0.78 mg/kg by Waliszewski et al. (2000) was reported in the blood samples. Low levels of DDT 0.22, 0.25, and 0.30 mg/L was also reported by Heudorf et al. (2003). In the case of DDE, high concentration of 8.0 mg/L was reported by Guardina et al. (1996), 9.10 mg/mL by Rubin et al. (2001), and 380 mg/kg by Ntow (2001) in the exposed persons. Residues of DDE were also detected; 5.2, 6.2, and 2.5 ng/g by Ahmed et al. (2002) and 3.99 and 1.42 mg/mL by Butler et al. (2003). High level of DDE was also reported by Van Ooastdam et al. (2004).

12.5.3 PESTICIDE RESIDUES IN GLANDULAR TISSUE

Pesticides (i.e., DDT, DDE, aldrin, dieldrin, malathion, and deltamethrin) were tested in thyroid gland tissue. This study was based on glandular tissue samples collected during operations from Jinnah Medical Center. DDT is much less than maximum residual limit (MRL) in both of the samples. The amount of aldrin however exceeds MRL that appears to be health-hazardous as organochlorine pesticides are lipophilic. In addition, pesticides have an impact on hormones and endocrine glands (adrenal cortex) as reported by Vilar and Tullner (1959); Kuservitsky et al. (1970) on thyroid gland; and by Desola et al. (1998); Jarrer et al. (1998); You et al. (1998) and Padungtod et al. (1998) on hormones. The effect of dieldrin (Cyclodiene) has been reported by Wakeling et al. (1972) on 5-dihydrotestosterone binding with specific protein receptors, as 33% inhibition. This indicates that pesticides affect the hormones and endocrine glands.

12.6   ASSESSMENT OF HUMAN EXPOSURE WITH PESTICIDES

The residual level of a pesticide in humans is an exposure index. The presence of different residual pesticide levels in human blood clearly indicates the extent to which the person has been exposed by the use of the chromatograms to quantify the level of residues in the blood using the fluid and gas (GC) technique of high-performance chromatography. The impact of pesticide exposure can be acute, work related, or accessory. The residual level in general population measures accidental exposure and the levels of persistent pesticides that remain together in the soil and tissues for years and their bioaccumulation takes place via food chain. Unfortunately, in urine, fecal matter, bile, and air, the body is able to bio-transform and excretes parts of these compounds. But when the absorption rate and fat deposition exceeds the elimination rate, their increased concentrations lead to toxic and pathological effects. Given this, EPA continually monitors the pollution of pesticides in developed countries through regular monitoring and does not hesitate to destroy large food stocks. Long-standing publication is available on this subject in those countries (Heath, 1961; Krauthacker et al., 1980; Mercedes and Theil, 1986; Anna et al., 1988; Alawi et al., 1992; Cantor et al., 1992; Ferrer et al., 1992; Matuo et al., 1992; Saady et al., 1993; Swaen et al., 1994). In developing countries, however, Mughal and Rehman did little work, as in Pakistan: The Pakistan Agriculture Research Center (PARC), in 1973 however, has worked on crops in particular (Naqvi and Jahan, 1996). The investigation of residues of pesticides in the human blood and their pathological effects is therefore urgently needed.

12.7   IMPACTS OF PESTICIDE USE ON HUMAN HEALTH

Studies show that pesticides can be linked to different diseases, including cancer, leukemia, and asthma. The risk of health risks from exposure to pesticide products not only depends on the toxicity of the ingredients, but also on the exposure level. Moreover, certain people, like children, pregnant women, or aging populations, may be more vulnerable than others to the effects of pesticides. The general health impact types caused by exposure to pesticides are depicted in Figure 12.2.

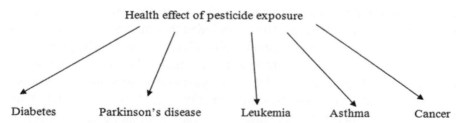

FIGURE 12.2 HUMAN HEALTH IMPACT OF PESTICIDE EXPOSURE

12.7.1   CANCER

Many studies have reported the link between pesticides and cancer. The prospective cohort study results with 57,310 pesticide applicators indicated in the United States that two imidazolinone (Imazethapyr and Imazaquin) herbicides were associated with bladder cancer (Koutros et al., 2015). The increase in the risk of bladder cancer was associated with pesticide exhibition (odds ratio (OR) = 1.68, 95% confidence interval (CI): 1.23–2.19), a dose-defined basis in another case–control study (953 cases and 881 controls) of male agricultural workers in Egypt (Amr et al., 2015). Samanic et al. (2008) reported that in the hospital case–control study with 462 gliomas and 195 meningioma patients in the United States, women with occupational herbicides exposure had a significant increase in the risk for meningioma in comparison to those who were never exposed (OR = 2.4, 95% CI: 1.4–4.3). A population-based case–control study showed significant association with brain tumors, with 221 incidents and 442 individually matched controls in France. The redox imbalance that altered the antioxidant defense system in breast cancer cells was found to be present in chlorpyrifos (CPF) in pesticides (Ventura et al., 2015). A case study in Brazil interviewed 110 women (aged 20–35) diagnosed with breast cancer, showing that their risk of breast cancer was increased by residential use of adult pesticides (Ortega-Jacome et al., 2010).

12.7.2   ASTHMA

A combination of pesticide exposure and bronchial hyper-reactiveness symptoms and asthma was reported in several clinical and epidemiological studies. Exposure to a pesticide may contribute to asthma aggravation by irritation, inflammation, immunodepression, or endocrine disorder (Hernández et al., 2011; Amaral, 2014). Raanan et al. (2015) also looked at the links between early-life exposure to surgeries and breathing outcomes for 359 moms and children in the United States. They concluded that the breathing symptoms of this exposure could be consistent with children's asthma. The prevalence of ocular nasal disease in a cross-sectional study of women farmworkers (n = 211) in Africa. The prevalence of eye–nasal symptoms was positive when a pesticide-spray area was entered (OR = 2.97, 95% CI: 0.93–9.50; Ndlovu et al., 2014). Most pesticides are weakly immunogenic, however, to a limited extent to the sensitivity of airways in exposed populations, while only some pesticides are sufficiently powerful to cause bronchial mucosal damage (Hernández et al., 2011). Any use of farm pesticide was found to be associated with atopic asthma in a survey involving 25,814 farmer women in the United States (OR = 1.46; 95% CI: 1.14–1.87; Hoppin et al., 2008).

12.7.3   DIABETES

Emerging scientific evidence shows that exposure to environmental pollutants should be affected by diabetes. An increased risk for developing types 2 diabetes and its comorbidities, including organic chlorine and metabolite, is suspected of exposure to pesticides (Azandjeme et al., 2013). A systematic literature review indicated that several pollutant concentrations (e.g., polychlorine dibenzodioxins and dibenzofurans (PCDD/Fs), PCBs, several organochlorine pesticides (Jaacks and Staimez, 2015) were positively linked to diabetes (DDT) and serum concentrations (DDE) oxychlordane, trans-nonachlor, hexachlorobenzene, and hexachlorocyclohexane). However, the actual data sets have been significantly limited as most studies have been cross-sectional. Only a few studies dealt with selection differences and the confusing effect, whereas most estimates were based on exceptionally large intervals. Positive associations of T2DM risk with organochlorine exposure were seen with various populations (Everett et al., 2007; Turyk et al., 2009) in some epidemiological studies. In a cross-sectional study, 116 pesticide sprayers and 92 unexposed controls were conducted in Bolivia and an abnormal glucose regulation was found (defined as HbA1c to be 5.6%) of the sprayers in 61.1% compared to 7.9% of the unexhibited controls (Hansen et al., 2014).

12.7.4   PARKINSON'S DISEASE

Studies of epidemiology suggest that the risk of Parkinson's disease (PD) may increase occupational exposure to pesticides. A case–control survey

(133 cases and 298 controls) on a French population examined quantitative aspects of pesticide-related occupational exposure in relation to PD (Moisan et al., 2015) and concluded that PD was associated with exposure to pesticides in vineyards (OR = 2.56, 95% CI: 1.31–4.98). The risk for PD in Colorado Medicare Beneficiary Datable Base in the United States was also found to be increased by 3% for every 1.0 μg of L-001 pesticide in underwater (OR = 1.03; 95%CI: 1.02–1.04; Beneficiary Database in Colorado Medicare, USA; James and Hall, 2015). The cohort study in the Holland showed a possible link between PD death and pesticide occupational exposure (Brouwer et al., 2015) in which 58,279 males and 62,573 females (aged 55–69 years) were enrolled. Due to dose-dependent growth observations in cell αS, Chorfa et al. (2016) reported that PD had a relationship with the use of pesticides (e.g., paraquate, rotenone, and maneuvers), insecticides (e.g., organophosphates and three pyrethroids), and fungicides (e.g., thiophanate-methyl, fenhexamide, and cyprodinill).

The development of PD at a small age in patients with no family history of the disease was associated with chronic exposure to metals and pesticides (Ratner et al., 2014). Moreover, the exposure time was also a key factor controlling the magnitude of this effect. Based on cohort and case–control data meta-analyzes, the risk to PD increased due to exposure to pesticides, herbicides, and solvents of all kinds, and more specifically, the risk of PD was approximately double-fold due to paraquat or maneuver/manozeb exposure (Pezzoli and Cereda, 2013).

12.7.5 LEUKEMIA

One of the major causes of acute leukemia is exposure to pesticides. The effect of pesticide exposure on children's leukemia was investigated in some earlier studies. ORs for acute lymphoblastic leukemia (ALL) for three pesticide-exposure types, shortly before conception, during pregnancy, and after birth were 1.39, 1.43, and 1.36, respectively, of 12 case–control leukemia studied by Bailey et al. (2015).

In case–control studies in Iran, an occupational farmer was significantly more likely than other jobs to develop acute leukemia, especially for children because of pesticide exposure (Maryam et al., 2015). Meta-analysis from 40 studies in France showed that in children with prenatal exposure to their mom, lymphoma and leukemia were substantially increasing (OR = 1.53; 95% CI: 1.22–1.91 and OR = 1.48; 95% CI = 1.26–1.75; Vinson et al., 2011). During pregnancy, exposure was positively associated with childhood leukemia with unspecified domestic pesticides, insecticides, and herbicides. Turner et al. (2011) found that in their systematic review of previous observational epidemiological trials meta-analysis, such exposures during pregnancy are positively linked with children's leukemia.

12.8   MITIGATING THE EFFECTS OF PESTICIDE

Despite continuously disagreeing with pesticide risk, it would appear that the use of impact on human health and environmental quality has been increasingly worried (Damalas, 2009). These increased concerns were largely due to a reduction in trust in agricultural and industrial production methods and the regulations of the Authority aimed at protecting both human health and the environment. Therefore, when assessing the use of pesticide safety, the presence of several uncertainties should be taken into account, scientific data, policy guidelines, and professional judgment should be considered.

It is very low for the producer to believe that the reduction of the risk involves either reduced production or increased input, resulting in substitution of the input of pesticides (Paul et al., 2002). Policies aimed at reducing the risks of pesticide use will therefore impose a cost on the farming community, which in turn has implications for agricultural commodity prices. This has been confirmed by Paul et al. (2002), a cost-function production model that showed that the environmental reduction requirements would impose substantial expenses on the agriculture sector.

These costs are directly linked to increased demand for effective pesticides at a given farm output level and imply incentive innovation for increased costs in relation to pesticide quality.

Concerns about the environmental and human health implications of pesticide usage have led the European Union to develop its "Sustainable Sustainable Pesticide Use Theme Strategy" (CEC 2006). In addition, agricultural scientists began developing alternative crops management systems, in order to minimize the harmful impacts on the environment and human health of agriculture (mainly based on pesticides for crop protection). The Integrated Cultural Management (ICM) in particular includes guidelines used by farmers' unions as a means of enforcing actions for safe and environmentally friendly production of agricultural products (Baker et al., 2003). Moreover, ICM covers good agricultural practices implementation (GAP) actions, safety and hygiene of workers, product safety, full traceability of measures, and specific environmental conservation actions (Chandler et al., 2008). ICM encourages the use of complementary methods of pesticide management in the control of plagues (such as insecticide and fungal crop resistance, biological control, and other cultural methods) to reduce the population of animal pests or weeds below their level of economic harm and minimize the impacts of pesticides on other agroecosystem components (Kogan, 1998; Way et al., 2000). As regards pesticide use, ICM permits pesticide use solely through an Integrated Pest Management Program (IPM) (Mariyono, 2008; Nwilene et al., 2008; Chandler et al., 2008), where some criteria for pesticides are used, specific instructions for application on crops have been followed and residual analysis has been used as one of its tools.

Pesticides selected for use in IPM are:

(1) Biologically effective (high selectivity, rapid impact, optimal residual effect, good plant tolerance, and low resistance risk).

(2) User friendly (low acute toxicity and low chronic toxicity, optimum formulation, safe packaging, easy application, long store stability)

(3) Environmentally friendly/compatible (low nontarget toxicity, optimum formulation, safe packaging, easy application method, long store stability)

(4) Economically viable/profitable, competitive, patentable (Palacios, 2009; good cost/profit ratio for the farmers, large spectrum of activities, applicable in IPM, innovative product features).

Specific instructions that are followed in the application on crops include (1) using the pesticide with the recommended dose when the pesticide is detected or considered necessary for precautionary treatment; (2) optimizing pesticide use in economic efficacy by adjusting dose levels according to population density of the pests.

Regarding the analysis of the level of ingredients or the amount expended on a pesticide, such variables should only be used as an initial estimation as they do not have a close relation with the environment, while the dosage of active ingredients is often more costly and environmentally friendly and innovative compound than the outdated dangerous. All the above indicates clearly that introducing IPM would significantly reduce the effect of pesticides on human health and the environment without affecting crop productivity or increasing the chance of losses of crops (Mariyono, 2008; Berger et al., 2008).

12.9 CONCLUSION

Pesticides have played an essential role in ensuring the reliable delivery of farm products at affordable prices for consumers, enhancing product quality, and ensuring farmers high profits. While pesticides are developed to function with reasonable certainty and minimal health and environmental risk, many studies raised concerns about health risks from exposure of farmers and nonprofessionals to the food and drinking water residue of people. Several indicators have been used to evaluate the potential health and environmental risks of pesticides. They have nevertheless shown a reduced certainty in their use, suggesting the need for developing alternative indicators that should make pesticide risk assessment accurate and reliable, thus reducing possible adverse effects on human health and the environment of pesticides.

The development and implementation of pesticide-related alternative cropping solutions that rely less on pesticides could minimize pesticide exposure and the undesirable human health effects of exposure. In addition, the use of appropriate and properly maintained pulverizing equipment, together with all necessary precautions, could also reduce pesticide exposure at all stages of pesticide handling. The general optimization of pesticide management strictly within the rules and consideration of public concerns about pesticide residues in food and drinking water may help to reduce the adverse environmental and health effects of pesticides. All these might sound difficult but appear to be a good way of providing safe food production in a viable system for agricultural production.

KEYWORDS

- **environmental impacts**
- **health effects**
- **pesticide safety**
- **pesticide toxicity**
- **risk assessment**

REFERENCES

Ahmed, M. T.; Loutfy, N.; El Shiekh, E. Residue levels of DDE and PCBs in the blood serum of women in the Port Said region of Egypt. *J. Hazard. Mater.* **2002**, *89*, 41–48.

Alawi, M. A.; Ammari, N.; Al-Shuraiki, Y. Organochlorine pesticide contaminations in human milk samples from women living in Amman, Jordan. *Arch. Environ. Contam. Toxicol.* **1992**, *23*, 235–239.

Amaral, A. F. Pesticides and asthma: challenges for epidemiology. *Front. Public Health* **2014**, *2*, 1–3.

Amr, S.; Dawson, R.; Saleh, D. A. A.; Magder, L. S.; St. George, D. M.;El-Daly, M.; Loffredo, C. A. Pesticides, gene polymorphisms, and bladder cancer among Egyptian agricultural workers. *Arch. Environ. Occup. Health* **2015**, *70*, 19–26.

Anna, P.; Nandoor, P; Ildiko, F. Pesticide use related to cancer incidence as studied in a rural district of Hungary. *Sci. Total. Environ.* **1988**, *73*, 229–244.

Arrebola, J. P.; Belhassen, H.; Artacho-Cordón, F.; Ghali, R.; Ghorbel, H.; Boussen, H.; Olea, N. Risk of female breast cancer and serum concentrations of organochlorine pesticides and polychlorinated biphenyls: a case–control study in Tunisia. *Sci. Total Environ.* **2015**, *520*, 106–113.

Azandjeme, C. S.; Bouchard, M.; Fayomi, B.; Djrolo, F.; Houinato, D.; Delisle, H. Growing burden of diabetes in sub-Saharan Africa: contribution of pesticides? *Curr. Diabetes Rev.* **2013**, *9*, 437–449.

Azmi, M. A.; Naqvi, S. N.; Akhtar, K.; Parveen, S.; Parveen, R.; Aslam, M. Effect of pesticide residues on health and blood parameters of farm workers from rural Gadap, Karachi, Pakistan. *J. Environ. Bio.* **2005**, *30*, 747–756.

Bailey, H. D.; Infante-Rivard, C.; Metayer, C.; Clavel, J.; Lightfoot, T.; Kaatsch, P.; Milne, E. Home pesticide exposures and risk of childhood leukemia: findings from the Childhood Leukemia International Consortium. *Int. J. Cancer* **2015**, *137*, 2644–2663.

Baker, B. P.; Benbrook, C. M.; III, E. G.; Benbrook, K. L. Pesticide residues in conventional, integrated pest management (IPM)-grown and organic foods: insights from three US data sets. *Food Addit. Contam.* **2003**. *19*, 427–446.

Beketov, M. A.: Kefford, B. J.: Schäfer, R. B.: Liess, M. Pesticides reduce regional biodiversity of stream invertebrates. *Proc. Nat. Acad. Sci. USA.* **2013**. *110*, 11039–11043.

Bissacot, D. Z.; Vassilieff, I. HPLC determination of flumethrin, deltamethrin, cypermethrin, and cyhalothrin residues in the milk and blood of lactating dairy cows. *J. Analy. Toxicol.* **1997**, *21*, 397–402.

Bockstaller. C.; Guichard, L.; Keichinger, O.; Girardin, P.; Galan, M.B.; Gaillard, G. Comparison of methods to assess the sustainability of agricultural systems. A review. *Agron. Sustain. Dev.* **2009**, *29*, 223–235.

Boxall, R. A. Post-harvest losses to insects—a world overview. *Int. Biodeter. Biodegr.* **2001**, *48*, 137–152.

Brouwer, M.; Koeman, T.; van den Brandt, P. A.; Kromhout, H.; Schouten, L. J.; Peters, S.; Vermeulen, R. Occupational exposures and Parkinson's disease mortality in a prospective Dutch cohort. *Occup. Environ. Med.* **2015**, *72*, 448–455.

Bürger, J.: de Mol, F.: Gerowitt, B. The "necessary extent" of pesticide use—thoughts about a key term in German pesticide policy. *Crop Prot.* **2008**, *27*, 343–351.

Butler, W.J.; Saddon, L.; McMullen, E.; Houseman, J.; Tofflemire, K.; Corriveau, A.; Weber,J.P.,; Mills, C.; Smith, S.; Van Oastdam, J. Organochlorine levels in maternal and umbilical cord blood plasma in Arctic Canada. *Sci. Total. Environ.* **2003**, *2*, 27–52.

Cai, D. W. Understand the role of chemical pesticides and prevent misuses of pesticides. *Bul. Agri. Sci. Technol.* **2008**, *1*, 36–38.

Cantor, K. P.; Blair, A.; Everett, G.; Gibson, R.; Burmeister, L. F.; Brown, L. M.; Dick, F. R. Pesticides and other agricultural risk factors for non-Hodgkin's lymphoma among men in Iowa and Minnesota. *Cancer Res.* **1992**, *52*, 2447–2455.

Chandler, D.; Davidson, G.; Grant, W. P.; Greaves, J.; Tatchell, G. M. Microbial biopesticides for integrated crop management: an assessment of environmental and regulatory sustainability. *Trends Food Sci. Technol.* **2008**, *19*, 275–283.

Chikuni, A. Residues of organochlorine pesticides in human milk from mothers living in the greater Harare area of Zimbabwe. *Cent. Afri. J. Med.* **1991**, *37*, 136–141.

Chorfa, A.; Lazizzera, C.; Bétemps, D.; Morignat, E.; Dussurgey, S.; Andrieu, T.; Baron, T. A variety of pesticides trigger in vitro α-synuclein accumulation, a key event in Parkinson's disease. *Arch. Toxicol.* **2016**, *5*, 1279.

Commission of the European Communities. Proposal for a regulation of the European Parliament and of the Council Concerning the Placing of Plant Protection Products on the Market. Commission of the European Communities; Brussels, Belgium: **2006**.

Damalas, C. A. Understanding benefits and risks of pesticide use. *Sci. Res. Essays.* **2009**, *4*, 945–949.

Debenest, T.L.; Silvestre, J.; Coste, M.; Pinelli, E. Effects of pesticides on freshwater diatoms. In *Reviews of Environmental Contamination and Toxicology,* David M. Whitacre (Ed.) **2010**, 87–103. Springer, New York.

Desola, S.R.; Bisshop, C.A.; Van der Kiraak, G.J.; Brook, R.J. Impact of organochlorine contamination on levels of Dex hormones and external morphology of common snapping turtle in Ontario, Canada. *Environ. Health. Perspect.* **1998**, *106*, 253–260.

Dua, V. K.; Pant, C. S.; Sharma, V. P. Determination of levels of HCH and DDT in soil, water and whole blood from bioenvironmental and insecticide-sprayed areas of malaria control. *Ind. J. Malario.* **1996**, *33*, 7–15.

Eleftherohorinos, I. G. *Weed Science: Weeds, Herbicides, Environment, and Methods for Weed Management.* AgroTypos, Athens, Greece, **2008** .

Everett, C. J.; Frithsen, I. L.; Diaz, V. A.; Koopman, R. J.; Simpson Jr, W. M.; Mainous III, A. G. Association of a polychlorinated dibenzo-p-dioxin, a polychlorinated biphenyl, and DDT with diabetes in the 1999–2002 National Health and Nutrition Examination Survey. *Environ. Res.* **2007**, *103*, 413–418.

Ferrer, A.; Bona, M. A.; Castellano, M.; ToFigueras, J.; Brunet, M. Organochlorine residues in human adipose tissue of the population of Zaragoza (Spain). *Bull. Environ. Centam.* **1992**, *48*, 561–566.

Gilliom, R. J. Pesticides in US streams and groundwater. *Environ. Sci. Technol.* **2007**, 3408–3414.

Guardino, X.; Serra, C.; Obiols, J.; Rosell, M. G.; Berenguer, M. J.; Lopez, F.; Brosa, J. Determination of DDT and related compounds in blood samples from agricultural workers. *J. Chromatogr. A.* **1996**, *719*, 141–147.

Hansen, M. R.; Jørs, E.; Lander, F.; Condarco, G.; Schlünssen, V. Is cumulated pyrethroid exposure associated with prediabetes? A cross-sectional study. *J. Agromed.* **2014**, *19*, 417–426.

Heath, D.F. *Organophosphorus Poisons.* Pergamon Press Oxford **1961**.

Helfrich, L. A.; Weigmann, D. L.; Hipkins, P. A.; Stinson, E. R. *Pesticides and Aquatic Animals: A Guide to Reducing Impacts on Aquatic Systems,* VCE Publications, **2009**.

Hernández, A. F.; Parrón, T., and Alarcón, R. Pesticides and asthma. *Curr. Opin. AllergyClin. Immunol.* **2011**, *11*, 90–96.

Heudorf, U.; Angerer, J.; Drexler, H. Current internal exposure to pesticides in children and adolescents in Germany: Blood plasma levels of pentachlorophenol (PCP), lindane (γ-HCH), and dichloro (diphenyl) ethylene (DDE), a biostable metabolite of dichloro (diphenyl) trichloroethane (DDT). *Int. J. Hyg. Environ. Health* **2003**, *206*, 485–491.

Hoppin, J. A.; Umbach, D. M.; London, S. J.; Henneberger, P. K.; Kullman, G. J.; Alavanja, M. C.; Sandler, D. P. Pesticides and atopic and nonatopic asthma among farm women in the Agricultural Health Study *Am. J. Respir. Crit. Care Med.* **2008**, *177*, 11–18.

Húšková, R.; Matisová, E.; Kirchner, M. Fast GC–MS pesticide multiresidue analysis of apples. *Chromatographia* **2008**, *68*, 49–55.

Jaacks, L. M.; Staimez, L. R. Association of persistent organic pollutants and non-persistent pesticides with diabetes and diabetes-related health outcomes in Asia: a systematic review. *Environ. Int.* **2015**, *76*, 57–70.

Jahan, M. Pesticide residues in random blood samples of human population in Karachi. *J. Coll. Physicians Surg. Pak.* **1995**, *6*, 151–153.

James P.J.; Cramp A.P.; Hook S.E. Resistance to insect growth regulator insecticides in populations of sheep lice as assessed by a moulting disruption assay. *Med. Vet. Entomol.* **2008**, *22*, 326–330.

James, K. A.; Hall, D. A. Groundwater pesticide levels and the association with Parkinson disease. *Int. J. Toxicol.* **2015**, *34*, 266–273.

Jarrer, J.; Goomen, A.; Foster, W.; Brant, R.; Cham, S.; Sevcik, M. Evaluation of reproductive outcomes in women inadvertently exposed to hexachlorobenzene in southern Turkey in 1980s. *Reprod. Toxicol.* **1998**, *12*, 469–476.

Kanja, L. W.; Skaare, J. U.; Ojwang, S. B. O.; Maitai, C. K. A comparison of organochlorine pesticide residues in maternal adipose tissue, maternal blood, cord blood, and human milk from mother/infant pairs. *Arch. Environ. Contam. Toxicol.* **1992**, *22*, 21–24.

Kiely, T.; Donaldson, D.; Grube, A. H. Pesticides industry sales and usage: 2000 and 2001 market estimates. Biological and Economic Analysis Division, US Environmental Protection Agency, **2004**.

Kocan, A.; Petrik, J.; Drobna, B. Chovancova, J. Levels of PCB's and some organochlorine pesticides in human population of selected areas of Slovak Republic. I. Blood. *Chemosphere* **1994**, 29, 2315–2325.

Kogan, M. (Integrated pest management: historical perspectives and contemporary developments. *Ann. Rev. Entomol.* **1998**, *43*, 243–270.

Kouser, S.; Qaim, M. Impact of Bt cotton on pesticide poisoning in smallholder agriculture: A panel data analysis. *Ecol. Econ.* **2011**, *70*, 2105–2113.

Koutros, S.; Silverman, D. T.; Alavanja, M. C.; Andreotti, G.; Lerro, C. C.; Heltshe, S.; Beane Freeman, L. E. Occupational exposure to pesticides and bladder cancer risk. *Int. J. Epidemiol.* **2015**, *45*, 792–805.

Kralj, M. B.; Černigoj, U.; Franko, M.; Trebše, P. Comparison of photocatalysis and photolysis of malathion, isomalathion, malaoxon, and commercial malathion—products and toxicity studies. *Water Res.* **2007**, *41*, 4504–4514.

Krauthacker, B.; Alebić-Kolbah, T.; Kralj, M.; Tkalčević, B.; Reiner, E. Organochlorine pesticides in blood serum of the general Yugoslav population and in occupationally exposed workers. *Inter. Occup. Environ. Health* **1980**, *45*, 217–220.

Krawinkel, M. B.; Plehn, G.; Kruse, H.; Kasi, A. M. Organochlorine residues in Baluchistan/ Pakistan: blood and fat concentrations in humans. *Bull. Environ. Contam. Toxicol.* **1989**, *43*, 821–826.

Kusevitsky, I.A.; Kirlich, A.Y.; Khovayera, L.A. Inhibitory effect of organochlorines on thyroid gland. *Veterinanariya* **1970**, *4*, 73–77.

Lekei, E.; Ngowi, A. V.; London, L. Hospital-based surveillance for acute pesticide poisoning caused by neurotoxic and other pesticides in Tanzania. *Neurotoxicology* **2014**, *45*, 318–326.

Liu, B.; McConnell, L. L.; Torrents, A. Herbicide and insecticide loadings from the Susquehanna River to the northern Chesapeake Bay. *J. Agri. Food Chem.* **2002**, *50*, 4385–4392.

Luo, X. W.; Foo, S. C.; Ong, H. Y. Serum DDT and DDE levels in Singapore general population. *Sci. Environ.* **1997**, *208*, 97–104.

Majewski, M. S.; Capel, P. D. *Pesticides in the Atmosphere: Distribution, Trends, and Governing Factors*. CRC Press, Cleveland, OH. **1995**.

Malaj, E.; Peter, C.; Grote, M.; Kühne, R.; Mondy, C. P.; Usseglio-Polatera, P.; Schäfer, R. B. Organic chemicals jeopardize the health of freshwater ecosystems on the continental scale. *Proce. Nat. Acad. Sci. USA.* **2014**, *111*, 9549–9554.

Mariyono, J. Direct and indirect impacts of integrated pest management on pesticide use: A case of rice agriculture in Java, Indonesia. *Pest Manag. Sci.* **2008**, *64*, 1069–1073.

Maryam, Z.; Sajad, A.; Maral, N.; Zahra, L.; Sima, P.; Zeinab, A.; Davood, M. Relationship between exposure to pesticides and occurrence of acute leukemia in Iran. *Asian Pac. J. Cancer Prev.* **2015**, *16*, 239–244.

Matthews, G. *Pesticides: Health, Safety and the Environment*. John Wiley & Sons, Hoboken, NJ. **2006**.

Matuo, Y. K.; Lopes, J. N. C.; Casanova, I. C.; Matuo, T. Organochlorine pesticide residues in human milk in the Ribeirão Preto region, state of São Paulo, Brazil. *Arch. Environ. Contam. Toxicol.* **1993**, *22*, 167–175.

Mercedes, B.; Thiel, R. DDT and polychlorinated biphenyl residues in human milk. *Rev. Costrarric Cienc. Med.* **1986**, *7*, 1333–1336.

Moisan, F.; Spinosi, J.; Delabre, L.; Gourlet, V.; Mazurie, J. L.; Bénatru, I.; Elbaz, A. Association of Parkinson's disease and its subtypes with agricultural pesticide exposures in men: a case–control study in France. *Environ. Health Perspect.* **2015**, *123*, 1123–1129.

Mughal, H. A.; Rahman, M. A. Organochlorine pesticide content of human adipose tissue in Karachi. *Arch. Environ. Health* **1973**, *27*, 396–398.

Musshoff, F.; Junker, H.; Madea, B. Simple determination of 22 organophosphorous pesticides in human blood using headspace solid-phase microextraction and gas chromatography with mass spectrometric detection. *J. Chromatogr. Sci.* **2002**, *40*, 29–39.

Naqvi, S.N.H.; Jahan, M. Detection of some pesticides in blood of some people of Karachi by HPLC. *JCPSP.* **1996**, *6*, 151–153.

Naqvi, S.N.H.; Jahan, M. Detection of some pesticides in blood of some people of Karachi by HPLC. *JCPSP.* **1995**, *6*, 151–153.

Ndlovu, V.; Dalvie, M. A.; Jeebhay, M. F. Asthma associated with pesticide exposure among women in rural Western Cape of South Africa. *Am. J. Ind. Med.* **2014**, *57*, 1331–1343

Ntow, W. J. Organochlorine pesticides in water, sediment, crops, and human fluids in a farming community in Ghana. *Arch. Environ. Contam. Toxicol.* **2001**, *40*, 557–563.

Nwilene, F. E.; Nwanze, K. F.; Youdeowei, A. Impact of integrated pest management on food and horticultural crops in Africa. *Entomol. Exp. Appl.* **2008**, *128*, 355–363.

Ortega Jacome, G. P.; Koifman, R. J.; Rego Monteiro, G. T.; Koifman, S. Environmental exposure and breast cancer among young women in Rio de Janeiro, Brazil. *J. Toxicol. Environ. Health A* **2010**, *73*, 858–865.

Padungtod, C.; Lasley, B.G.L.; Christiani, D.C.; Raayan, L.M.; Xu, X. Reproductive hormone profile among pesticide factory workers. *J. Occup. Environ. Med.* **1998**, *40*, 1038–1047.

Palacios Xutuc, C. N. Manual to Train Trainers on Safe and Correct Use of Plant Protection Products and Integrated Pest Management (IPM) Crop Life Latin America. Guatemala City, Guatemala. **2009**.

Paul, C. J.M.; Ball, V. E.; Felthoven, R. G.; Grube, A.; Nehring, R. F. Effective costs and chemical use in United States agricultural production: using the environment as a "free" input. *Am. J. Agr. Econ.* **2002**, *84*, 902–915.

Pereira, J. L.; Antunes, S. C.; Castro, B. B.; Marques, C. R.; Gonçalves, A. M.; Gonçalves, F.; Pereira, R. Toxicity evaluation of three pesticides on nontarget aquatic and soil organisms: commercial formulation versus active ingredient. *Ecotoxicology* **2009**, *18*, 455–463.

Pesticides Safety Directorate. Revised Assessment of the Impact on Crop Protection in the UK of the "Cut-off Criteria" and Substitution Provisions in the Proposed Regulation of the European Parliament and of the Council Concerning the Placing of Plant Protection Products in the Market; Pesticides Safety Directorate: York, UK, **2008**.

Pezzoli, G.; Cereda, E. Exposure to pesticides or solvents and risk of Parkinson disease. *Neurology* **2013**, *80*, 2035–2041.

Pimentel, D. Environmental and economic costs of the application of pesticides primarily in the United States. In: D.R. Peshin, A.K. Dhawan (eds) *Integrated Pest Management.* Springer, Berlin, **2009** 89–111.

Raanan, R.; Harley, K. G.; Balmes, J. R.; Bradman, A.; Lipsett, M.; Eskenazi, B. Early-life exposure to organophosphate pesticides and pediatric respiratory symptoms in the CHAMACOS cohort. *Environ. Health Perspect.* **2015**, 123, 179–185.

Ratner, M. H.; Farb, D. H.; Ozer, J.; Feldman, R. G.; Durso, R. Younger age at onset of sporadic Parkinson's disease among subjects occupationally exposed to metals and pesticides. *Interdiscip. Toxicol.* **2014**, *7*, 123–133.

Reus, J.; Leendertse, P.; Bockstaller, C.; Fomsgaard, I.; Gutsche, V.; Lewis, K.; Alfarroba, F. Comparison and evaluation of eight pesticide environmental risk indicators developed in Europe and recommendations for future use. *Agr. Ecosys. Environ.* **2002**, *90*, 177–187.

Rohr, J. R.; Kerby, J. L.; Sih, A. Community ecology as a framework for predicting contaminant effects. *Trends Ecol. Evol.* **2006**, *21*, 606–613.

Rubin, C. H.; Lanier, A.; Socha, M.; Brock, J. W.; Kieszak, S.; Zahm, S. Exposure to persistent organochlorines among Alaska Native women. *Int. J. Circumpolar. Health* **2001**, *60*, 157–169.

Saady, J. J.; Fitzgerald, R. L.; Poklis, A. Measurement of DDT and DDE from serum, fat, and stool specimens. *J. Environ. Sci. Health Part A, Environ. Sci. Eng.* **1993**, *27*, 967–981.

Sabbah, S.; Jemaa, Z.; Bouguerra, M.L. Gas chromatography study of organochlorine pesticide residues in human milk and umbilical cord and adult blood. *Analysis* **1987**, *15*, 399–403.

Saleem, M. A.; Ashfaq, M. *Environmental Pollution and Agriculture*. B.Z. University Press, Multan, 2004.

Samanic, C. M.; De Roos, A. J.; Stewart, P. A.; Rajaraman, P.; Waters, M. A.; Inskip, P. D. Occupational exposure to pesticides and risk of adult brain tumors. *Am. J. Epidemiol.* **2008**, *167*, 976–985.

Saxena, M. C.; Seth, T. D.; Mahajan, P. L. Organo chlorine pesticides in human placenta and accompanying fluid. *Int. J. Environ. Anal. Chem.* **1980**, *7*, 245–251.

Shaikh, M. A. Mortality in patients presenting with organophosphorus poisoning at Liaquat University of Medical and Health Sciences, **2011**.

Straathoff, H. Investigations on the phytotoxic relevance of volatilization of herbicides. Mededelingen **1986**, *51*, 43–438.

Swaen, G. M.; van Vliet, C.; Slangen, J. J.; Sturmans, F. Cancer mortality among licensed herbicide applicators. *Scand. J. Work Environ.***1994**, *18*, 201–204.

Turner, M. C.; Wigle, D. T.; Krewski, D. Residential pesticides and childhood leukemia: A systematic review and meta-analysis. *Environ. Health Peres.* **2011**, *16*, 1915–1931.

Turyk, M.; Anderson, H.; Knobeloch, L.; Imm, P.; Persky, V. Organochlorine exposure and incidence of diabetes in a cohort of Great Lakes sport fish consumers. *Environ. Health Perspect.* **2009**, *117*, 1076–1082.

U. S. Environmental Protection Agency. Human Health Benchmarks for Pesticides. EPA-822-F-12-001. **2012.**

Van Oostdam, J. C.; Dewailly, E.; Gilman, A.; Hansen, J. C.; Odland, J. O.; Chashchin, V.; Soininen, L. Circumpolar maternal blood contaminant survey, 1994–1997 organochlorine compounds. *Sci. Total. Environ.* **2004**, *330*, 55–70.

Ventura, C.; Venturino, A.; Miret, N.; Randi, A.; Rivera, E.; Núñez, M.; Cocca, C. Chlorpyrifos inhibits cell proliferation through ERK1/2 phosphorylation in breast cancer cell lines. *Chemosphere* **2015**, *120*, 343–350.

Vilar, O.; Tullner, W.W. Effect of DDT on sex hormones. *Endocrinology.* **1959**, *65*, 80–84.

Vinson, F.; Merhi, M.; Baldi, I.; Raynal, H.; Gamet-Payrastre, L. Exposure to pesticides and risk of childhood cancer: a meta-analysis of recent epidemiological studies. *Occup. Environ. Med.* **2011**, *68*, 694-702.

Wakeling, A.E.; Schmidt, T.J.; Visek, W.J. Inhibition of 5-dihydrotestosterone binding with specific protein receptors by dieldrin treatment. *Fed. Proc.* **1972**, *31*, 725.

Waliszewski, S. M.; Aguirre, A. A.; Infanzon, R. M.; Silva, C. S.; Siliceo, J. Organochlorine pesticide levels in maternal adipose tissue, maternal blood serum, umbilical blood serum, and milk from inhabitants of Veracruz, Mexico *Arch. Environ. Contam. Toxicol.* **2000**, *40*, 432–438.

Warsi, F. How do pesticides affect ecosystems. *Pesticides.* Available from http://farhanwarsi. tripod. com/id9.html. Accessed January 16, 2015.

Way, M. J.; Van Emden, H. F. Integrated pest management in practice—pathways towards successful application. *Crop Prot.* **2000**, *19*, 81-103.

You, I.; Casonova, M.; Archibeque-Engle, S.; Sar, M.; Fam, I.Q.; Heck, H.A. Impaired male sex development in prenatal spargue Dawley and long Evans hooded rats, exposed to utero and loctationally functional samples to P′ P′ DDE. *Toxicol. Sci.* **1998**, *45*, 162–173.

Zeinat Kamal, M.: Nashwa, A. H.: Mohamed, A. I.: Sherif, E. N. Biodegradation and detoxification of malathion by of *Bacillus thuringiensis* MOS-5. *Aust. J. Basic Appl. Sci.* **2008**, *2*, 724–732.

CHAPTER 13

Utilization of Biosurfactant Derived from Beneficial Microorganisms as Sustainable Bioremediation Technology for the Management of Contaminated Environment: Panacea to a Healthy Planet

CHARLES OLUWASEUN ADETUNJI,[1*] OSIKEMEKHA ANTHONY ANANI,[2] and CHUKWUEBUKA EGBUNA[3,4]

[1]*Applied Microbiology, Biotechnology and Nanotechnology Laboratory, Department of Microbiology, Edo State University Uzairue, PMB 04, Auchi, Edo State, Nigeria*

[2]*Laboratory of Ecotoxicology and Forensic Biology, Department of Biological Science, Faculty of Science, Edo State University Uzairue, Edo State, Nigeria*

[3]*Africa Centre of Excellence in Public Health and Toxicological Research (ACE-PUTOR), University of Port-Harcourt, Rivers State, Nigeria*

[4]*Department of Biochemistry, Faculty of Natural Sciences, Chukwuemeka Odumegwu Ojukwu University, Anambra State 431124, Nigeria*

Corresponding author. E-mail: adetunjicharles@gmail.com

ABSTRACT

Biosurfactants have been recognized as surface-active biological molecules fabricated by different microorganisms. They possess hydrophobic and hydrophilic moiety. The structure of biosurfactant enables them to decrease interfacial tension or generate micro-emulsion. They possessed several unique properties that single them out for several applications in various sectors most especially for the cleanup of the heavily contaminated environment. Some of these special features include low toxicity, eco-friendly, dispersing properties, and enhanced biodegradability, wetting, and emulsifying properties. Their significance has also been identified in the bioremediation of degradation of hydrocarbons from water and soil, breaking down of degrading polycyclic aromatic hydrocarbons, eco-restoration of the heavily polluted environment. This chapter highlights various reports on the application of biosurfactant as a biotechnological tool for the bioremediation of heavily polluted soil and water. Moreover, several in vitro, in situ, and ex situ application of biosurfactant were highlighted with details. Also, the modes of action, biochemical responses, genetic applications as well as their advances in agriculture through which biosurfactant execute their bioremediation activity was provided in details.

13.1   INTRODUCTION

In today's environmental watch globally, inorganic and organic contaminants such as heavy metals and pesticides are chief issues of discourse; because they have affected many ecosystems, causing serious health issues and environmental distortions. Water, soil, and sediment have been seriously contaminated and compromised, thereby displacing or endangering the biota that thrive therein.

In Nigeria, many agrochemical industries have produced varieties of chemicals (pesticides and heavy metals) in the quest of combating pests and weeds. Pesticides and heavy metals have been reported to endanger both the human and environmental health (Anani and Olumokoro, 2017; Adetunji et al., 2017a, 2017b; Anani and Olumokoro, 2018a, 2018b). Consequent to these impacts, they are being controlled by environmental acts that have also incited the necessity to generate a more eco-friendly, less poisonous, and biodecomposable surfactants (Nitschke et al., 2005; Soberón-Chavez et al., 2005).

"Surface active agents" (surfactants) are constituents (saturating agents, emulsifiers, effervescing agents, and dispersants) that reduce the superficial stiffness in the middle of two liquids or solids. Generally, any constituents that affect the interfacial surface stiffness can be termed a "surfactant" (Karanth et al., 1999; Ji et al., 2016). Surfactants can be used to decontaminate or remediate polluted soil and water (Adetunji et al., 2017a, 2017b). Chemically manufactured surfactants such as amphiphiles that produce micelles are the most used surface-active agents (Claudia et al., 2015). These surfactants have been gradually replaced by biotechnology-improved created products, obtained from both by enzymatic or microbial production; via natural means (Adetunji et al., 2017b). These resultant entities of biomolecules derived from these improved products are clades called "biosurfactants" (Arora et al., 2016; Adetunji et al., 2017a, 2017b).

The advancements and advantages of biosurfactants over surfactants are seen in their simple biodegradability, the viability of small-large scale manufacturing, low-cost substrates, sophisticated effervescing, low noxiousness, and environment friendly (Arino et al., 1996; Karanth et al., 1999; Abouseoud et al., 2008; Fontes et al., 2012; Claudia et al., 2015; Ji et al., 2016; Adetunji et al., 2017a, 2017b).

The utilization of biosurfactant derived from beneficial microorganisms as a sustainable bioremediation technology for the management of contaminated environment is the focus of this study. Bioremediation is one of the best solutions to degrade environmental pollutants. Biosurfactants manufactured from eco-friendly isolates have a likely role in the agrochemical markets (Adetunji et al., 2017a, 2017b). These comprise microbes have proven to have high decontaminate potentials, low molecular mass complexes and devoid of any detrimental end product(s) (Mulligan, 2005; Zhang et al., 2005; Pirôllo et al.2008; Adetunji et al., 2017a, 2017b).

Many microorganisms and rhizopores produce biosurfactants that contain biomolecules that play an important function in microbes and plant collaborations. Rhamnolipids are one of such biosurfactants. Its uses have been clearly proposed by Cameotra and Makkar, (2004), Ciancio and Mukerji (2007), and Adetunji et al. (2017b) as possible extracts for green decontamination and as a natural control constituents. This is the main reason it is used in agriculture to remove flora disease-carrying organisms and to improve the nutrient carrying capacity of nitrogen-fixing bacteria associated with the root nodules of leguminous plants. This also shows that biosurfactants have the potential to improve water and soil qualities via decontamination which is the purpose of this research. This will reduce the huge sum of money spent on artificial surfactant globally (Adetunji et al., 2017a, 2017b).

13.2   BIOSURFACTANTS VERSUS BIOREMEDIATION

The significant role of pesticides in agriculture has increased food production globally. However, most of the commonly used pesticides are considered to be recalcitrant and persistent noxious chemicals that can remain in the ecosystem for a long time. The concern for this environmental pollution episode in the ecosystem, land, and water, has resulted in accumulation rate via the food chain to humans.

Currently, consequent of the challenges faced in the use of pesticides in agricultural purposes, many green technologies used in the remediation or removal of pesticides from water and soil are being employed. Bioremediation of pollutants in water and soil by microbes has been shown to be very economical. The efficacy of microbiologically produced surfactants (biosurfactants) on the biodegradation of contaminated sites has been proven to be the best panacea for the removal or decontamination of water and soil contaminated with pesticides.

Adetunji et al. (2017b) tested a vigorous biosurfactant-fabricator in agricultural surroundings obtained mainly from rhizosphere of a wheat plant that was used for the manufacture of rhamnolipid. The isolated microbial isolates used for this study were acknowledged to be *Pseudomonas aeruginosa* from 16S rDNA analysis (C1501: KF976394). The findings of the tested results revealed that *P. aeruginosa* has the capacity to propagate and decrease superficial tightness below an extensive array of carbon point source and pH. The biologically tested strain gotten from C1501 was shown to be efficient in biodegrading the crude oil with the resulting proportion; 33%, 71.3%, and 96% on days 5, 10, and 20, correspondingly. In conclusion, the biosurfactant did not show an inhibitory impact on the tested crops, however, showed some wide-ranging antimicrobial actions against rot-inducing, Gram-negative, and Gram-positive fungus when related to the artificial surfactant from Tween-20. The study, however, recommends the utilization of strain C1501 biosurfactant as a potent green biosurfactants for the decontamination of pollutants in agricultural soil.

Rufino et al. (2014) evaluated the classification and the assets of the biosurfactant produced by *Candidalipolytica* (UCP 0988). The biotechnological procedures used in this study were duly enhanced. The results of their study showed that the yeast cultivated within 72 h had a superficial strain of the cell-free bouillabaisse, which was decreased from 55 to 25 mN/m. The yeast produced biosurfactant of 8.0 g/L with a CMC of 0.03%, which was branded as an anionic lipo-peptide made up of 8% carbohydrates, 50% protein, and 20% lipids. In summary, the tested isolates revealed no noxiousness compared to various products. The characteristics of the biosurfactant manufactured proposes its possible utilization in commercial purposes that will be cost-effective.

Bustamante et al. (2012) reviewed the usefulness of biosurfactants as a bioremediation tool for contaminated sites. They suggested the use of biologically enhanced surfactant; microbes in the remediation negatively impacted area. The reason was based on the natural organic compounds microbes produced which plummet superficial and interfacial tensions of solutions that cannot mix and aggregate the solubility potential and sorption of water-disliking inorganic and organic constituents. Also, it also aids in degrading pathogenic microorganisms such as the white-rot fungi. In conclusion, the general idea indicated the abundant possibility of the use of biosurfactants in the decontamination of polluted areas.

Cameotra and Makkar (2010) evaluated microbes as enhanced biosurfactant of hydrophobic pollutants, that they have molecules that have water-liking and disliking fields and have the ability in reducing the superficial strain and the inter-facial pressures of any isolated media. Consequent to this, they are non-noxious biomolecules that are decomposable and environmentally friendly. Biosurfactants show robust blend with other stable suspensions because of their glycolipids, fatty acids, lipo-peptides, and neutral lipids formation structures. They have been shown to degrade pesticides and highly carcinogenic substances like polycyclic aromatic hydrocarbons (PAHs), Pb, Cd, and Cr. Hence, microbial manufactured surfactants can improve the bioavailability of certain water-liking and disliking compounds for the decontamination of pollutants. This has proven their probable in biotech applications.

Dhara et al. (2013) reported the use of biosurfactant to boost agronomic efficiency in the need to meet the teeming human population worldwide. Their evaluation high points the massive use of artificial agrochemical fertilizers as a panacea for agricultural boosting and their attendant health and environmental impacts and proposed the use of microbes as biosurfactants because; they are green amalgams that are environmentally friendly. Green surfactants like fungi, bacteria, and yeast have been reported to be less noxious when used as a fertilizer for sustainable agriculture. In addition, when biosurfactants are been used in agriculture, they can eradicate plant-causing diseases and increase the bioavailability of vital nutrients to plants. Their summation proposed an enhanced molecular technique like metagenomics in exploring biosurfactants in uncultured microbes.

Hussain et al. (2009) reviewed the bioremediation of pesticides using phytoremediation. They proposed the utilization of genetically improved or natural microbes and flora to decontaminate contaminants in situ. However, previous studies have proven that microorganisms and genetic modified plants can produce biomolecules-enzymatic constituents that can decontaminate pesticides and mineralize them with their metabolites into a sustainable form. In other to enhance the ability of plant growth, they proposed that microorganisms and transgenic plants depict and assign pesticide-degrading genes, metabolites, and cell superficial enzymes when acting on pollutants such as pesticides. Hence, a panacea for a green agricultural economy is the utilization of biosurfactants as against the artificial forms which are not eco-friendly.

Sáenz-Marta et al. (2015) reviewed the advance methods in bioremediation of wastewater and contaminated soil. The report focuses on a sustainable means of remediation using microbes with Amphiphiles (biosurfactants). Amphiphiles are chemical composites having both hydrophilic (having a propensity to blend with or melt in a media) and lipophilic (inclining to combine with or melt in lipids or fats) properties. They also stated that chemically manufactured surfactants are gradually replaced by biotech surfactants gotten from genetic engineered microbial isolates blend. Biosurfactants have been tested to be eco-friendly and low cost-effective. In conclusion, they proposed yeast as a possible biosurfactant-manufacturing microbe for the reason that of their exceptional organizations can be subjugated commercially.

Usman et al. (2016) reported the application of biosurfactants in the decontaminating of crude oil and heavy metal. They noticed that the conservative way to moderate, decontaminate, and eliminate toxic using artificial methods is not economical and eco-friendly. However, the uses of bioemulsifiers/biosurfactants have become a credible alternative for a green ecosystem that make them vital in the enhancement and utilization of crude oil recovery and decontamination of pollutants. In conclusion, they proposed at full-scale assessment and more evidence on the behavior and prototype of biosurfactant role on bioremediation of pollutants.

Das and Chandran (2011) reviewed the microbial degradation of petroleum hydrocarbon contaminants. These contaminants have been known to belong to the group of cancer-causing agents and neuron-noxious contaminants. The use of a nonenvironmental sustainable approach such as incineration and landfilling of the contaminants has been invoked that can provoke certain environmental and health anomalies and are very expensive. They proposed a more sustainable and green technology to remediate contaminants-biosurfactants. This method is very economical and will result in thorough mineralization of contaminants into CO_2, H_2O, inorganic complexes, and complex organic impurities. In conclusion, they recommend microbes as a bioremediation tool for degrading petroleum hydrocarbon pollutants.

Rashedi et al. (2005) tested the isolation and productivity of biosurfactant from *Pseudomonas aeruginosa*. The biological control test involved 50 strains of microbes. Out of these 50 strains, about 12 showed hemolytic actions and were considered worthy biosurfactant manufacturers. Two (2) of the microbes displayed the maximum biosurfactant production when cultivated in glycerol and paraffin carbon sources. Consequent to biosurfactant production, the superficial pressure of the medium was decreased from 73 to 32 mN/m in values. Further tests confirmed the microbial strain to be *P. aeruginosa*-a rhamnolipid generating bacterium. The findings of the study revealed a higher yield of the production of rhamnolipid as relative to dry weight (Yp/x = 0.65 g/g) when glycerol (C/N ratio of 55/1) and sodium nitrate (4.2 g/L) were used as nitrogen base compounds. In addition, the physical and chemical characteristics of the expended bouillabaisse with and without microbes were evaluated. It was concluded that low life-threatening micelle levels were observed.

Ibukun et al. (2018) reported the role of biosurfactants in the sustained effort for ecosystem restoration. Base on their advantages; low noxiousness and high actions in extreme temperature, biosurfactants gotten from microbes are friendly with the ecosystem when used to biodegrade pollutants. This clean-up process is a green sustainable way for bioremediation of pollutants via renewable waste products as substrates, waste decline, and probably recycle of the preserved waste. In summary, they proposed the use of biosurfactants as an environmental sustainable tool for the clean-up of environmental contaminants, and, for that reason, it increases ecosystem services and protection.

Santos et al. (2018) tested the fabrication and categorization of a biosurfactant formed by *Streptomyces* sp. (DPUA 1559) which was isolated from group of fungus and algae (lichens) of the Amazon district. The authors stressed that surfactants have the capability to reduce the inter-facial strain of any substance because they have amphipathic mixtures comprising both hydrophobic and hydrophilic clades. The biological controlled experiment of the microbes was cultured in 1%inorganic and organic substrate of left over scorching soybean oil which is the main carbon basis. This was complemented by plummeting the surface tension from 60 to 27.14 mN/m thereby bioemulsifying the microbial-surfactant hydrophobic constituents. The results of their study indicated that the isolated biosurfactant was about 1.74 g/L, in sum to the tremendous ability of plummeting the superficial tension of 25.34 mN/m of a pH value 8.5 at 28 °C. The micelle level of the biosurfactant was estimated as 0.01 g/mL. Their findings revealed that the microbial-surfactant was able to tolerate a high-temperature series, showed high pH stability vis-à-vis the superficial strain and were also tolerant against high saline levels. More so, the strains of biosurfactants showed no noxious level against tested fauna (*Artemiasalina*) and floras (*Lactucasativa* L. and *Brassica oleracea* L). They further recommend biosurfactant as a panacea modern industrial biodecontaminant substrates because of the biochemistry of its chemical mass about 14.3 kDa, a solitary protein assemblage, and an acid oddity, which signified its glycoproteic properties.

Mouafo et al. (2018) tested the latency of three native microbial strains (*Lactobacillus delbrueckii* N2, *Lactobacillus cellobiosus* TM1, and *Lactobacillus plantarum* G88) for the fabrication of biosurfactants with sugar-cane syrup as substrates. The results of the biological controlled experiment revealed diverse biosurfactants gotten from the syrup substrate to demonstrate high superficial strain decrease (72 mN/m) with a range of value of 7.50 ± 1.78–41.90 ± 0.79 mN/m and high mixture range index (49.89±5.2–81.00 ± 1.14%). *Lactobacillus* strain often produces antimicrobial quality against *Candida albicans* anytime is been used. The findings of their study showed that the earliest classification of crude biosurfactants discloses that they are primarily glycol-lipids and glycol-proteins with syrup and glycerol as based substances, correspondingly. They thus recommend sugarcane syrup or glycerol as efficiently be used by *Lactobacillus plantarum* as cheap materials to intensification their biosurfactants large scale commercial production.

13.3   MODE OF ACTIONS, BIOCHEMICAL RESPONSES, AND GENETIC APPLICATIONS OF BIOSURFACTANTS

Claudia et al. (2015) reported the mode of action of biosurfactants in re-engineering a sustainable ecosystem. They opined the reason biosurfactants are used particularly in the oil, pharmacological, and food manufacturing industries are because of the hidden potentials; sulfate acid esters, carboxylates, which make them modern desirable appeal compared to the artificial surfactants. More so, their mode of actions is usually based on their chemical constituents; a mixture of amphiphilic (lipopeptides, glycolipids, and phospholipids) poly-saccharides, lipo-poly-saccharides, lipoproteins, or proteins. Consequent to these, many mixtures are enhanced at diminishing the superficial strain-biosurfactants as well, others are able to yield constant suspensions bioemulsifiers.

Some biosurfactants have lipid-containing particles, which are manufactured in intense oxygen (aerobic) settings (Kosaric, 2001). Biosurfactants with low molecular mass are very competent in dropping superficial and interfacial strains. On the other hand, biosurfactants with higher molecular mass are more competent at alleviating oil-in-water mixture (Pacwa-Plociniczak et al., 2012). Microbial producing surfactants in varied chemicals are normally known as polymeric-microbial-surfactants (Al-Araji et al. 2007). One of the functional metabolic roles of surfactants created by microbes comprises antimicrobial action and the capability to generate substrates willingly obtainable for absorption by the body tissue/cells in antithetical environmental situations.

Das et al. (2008) reported the mode of actions of microbes used as surfactants in various environmental, therapeutic, and commercial applications. The authors examined their molecular genome and biochemistry of their mode of action and synthesis of several enzymes. Their physiological metabolic pathways need extracellular production of surfactin; a potent cyclic lipopeptide biosurfactant with antimicrobial properties that produces a nonribosomal biosynthesis excited by catalyzing multi-enzyme peptide synthetase (surfactinsynthetase). The authors opined that biosynthesis exhibit a great gradation of morphological resemblance among one another of the same clades of microbial strains. In conclusion, the authors recommended that biosurfactants such as putisolvin, amphisin, and viscosin manufactured by some *Psuedomonas* strains have limited genetic characteristics. To understand them fully, their mode of actions, biochemistry, and genetic characteristics will help to re-engineer and manufacture them and develop competence of application in inexpensive agricultural and industrial effluent such as solid, semi-solid or liquid wastes as substrates base.

13.4 ADVANCES OF BIOSURFACTANTS IN AGRICULTURE

Agronomic efficiency is geared toward meeting the need of the teeming human population. The need to boost agricultural output globally is in a dire condition because artificial fertilizers have not only improved some aspect of plant development and growth but have also contributed immensely to the pollution of the ecosphere, atmosphere, and biosphere, endangering the living organisms therein. Hence, there is the need to advance beyond artificial surfactants such as fertilizers in boosting crop yields and substitute it for a green, eco-friendly, and sustainable microbial surfactants-biosurfactants.

Dhara et al. (2013) did an extensive review of the use of biosurfactants in farming. They highlighted that biosurfactants have been described to be produced by microbes (fungi, yeasts, and bacteria)– green surfactants that have been considered to be eco-friendly, cheap, and have hidden abilities to be applied in widely commercial purposes as a promising role in the agro-industry. The authors noted that some rhizosphere and flora associated microorganisms manufacture biomolecules (biosurfactants) that show vivacious function in the biofilm development, motility, and signaling flora and microbial collaboration. The authors concluded by proposing biosurfactants as a promising biotool to substitute the punitive artificial surfactant widely used in agriculture. More so, there is apt concern to explore new biosurfactant from uncultured microorganisms in the ecosphere, atmosphere, and biosphere by using progressive procedures like well-designed meta-genomics.

Darne et al. (2016) reviewed the promising use of biosurfactants molecules in degrading petroleum wastes as advanced biotechnology. In the decontamination of petroleum contaminants, several artificial surfactant complexes act as activating agents. Nevertheless, biosurfactants have numerous prospective utilization compared to the chains of artificial formulations; anti-corrosive, biocides for sulfate-plumme ting microbes, blending or demulsifying agents, and fuel design ground-breaking applications. The authors believe that biosurfactants will have an important part in some prospect applications in the oil industries and in this review, therefore, we highlight recent important relevant utilization in petroleum and associated industries because of their green nature, sustainable use, and environmentally friendly.

Hélvia et al. (2019) reported the successive utilization and sustainable use of biosurfactant gotten from *Serratiamar cescens* from UCP 1549 comprising *Manihote sculenta* (cassava) flour wastewater (CWW), by the use of full-factorial strategy as viable use in agrochemical formulations. The results of the study revealed that *Serratiamar cescens* displayed a higher decrease of superficial strain (25.92 mN/m) in the fresh medium comprising 0.2%, 5%, and 6% (lactose, corn waste oil, and cassava flour wastewater) after 72 h of fermentation at 28 °C and 150 rpm established high permanence below dissimilar pH, temperatures, and salinity. Findings from their study indicated that the biomolecules (biosurfactants) were able to determine the permanency and noxiousness of pollutants against cabbage seeds. In conclusion, they proposed that the formulations will be able to excite the ability of seed sprouting for agriculture utilization and oil-spill bioremediation management.

Singh et al. (2018) reviewed the economic prospects and strategies in employing biosurfactants as a 21st-century biotech mixture. They recounted the ineptness of bioprocessing in the large-scale production of the biomolecules. Though a scarcity of literature on the future bioprocess optimization (the exploit of the most operational resources), approaches and their achievements are still not widely accepted. The proposed fresh advances of germ and strictures for commercial biosurfactant manufacture that would expose new-fangled paths on the research of biosurfactant fabrication.

Akbari et al. (2018) did an extensive review of a novel leading edge for social and ecological security. They reported some general features of biosurfactants such as its compatibility with other substrates, high biodegradable ability, little venomousness, ecological competence, and eco-friendly. The authors also reported the human-health safety and efficacy of this green sustainable biotechnology in ecosystem remediation. They recommended biosurfactant as a first-class biomolecules in the following industries; agricultural, food, pharmaceutical, petroleum, cosmetic, wastewater treatment, and textile.

Santos da Silva et al. (2018) reviewed on biosurfactant fabrication by fungi strains as a viable alternative to other surfactants especially the conventional artificial ones. The authors recounted the use of extensive diversity of microbes such as fungi for both biosurfactants (BS) and bioemulsifiers (BE), utilization for renewable resources. The variety and amount of microbial BS or BE formed rest on chiefly on the producer species as well as many influences like temperature, trace elements, pH, nitrogen, and carbon bases and aeration that make them readily applicable in different industrial spheres- in eco-friendly contamination control and decontamination. The authors concluded that these assortment of the BS and BE make microbes an optimized and sustainable technology in treating oily environmental contamination.

Randhir et al. (2011) reviewed on the frontiers in application of renewable substrates for biosurfactant manufacture. They opined that biosurfactants have the ability to reduce the superficial strain of two surfaces because they possess hydro-phobic and hydro-philicmoieties that aid in the partitioning of the gas/liquid, solid/liquid or liquid/liquid boundaries base on their amphiphilic molecules. Such features enable them to be used as a foaming, emulsifying, dissolving, and detergency material goods. Their mode of actions makes them possess little noxiousness and eco-friendly and the varied range of probable commercial utilizations in all ecospheres. They

further recommend large-scale manufacturing and utilization of biosurfactants because of symbiotic connections with living cells and nonliving
surroundings.

13.5 CONCLUSION

This study has provided a detailed and practical application of biosurfactant
as a biotechnological tool for eco-restoration of a heavily polluted environment. There is a need to utilize several agro-industrial wastes for the mass
production of biosurfactant to reduce the cost involved in the mass production of biosurfactant. Also, there is a need to explore several strains that
have not been exploited before searching for a strain with a high yield of
biosurfactant. Moreover, the application of several biotechnological techniques will enhance the biosurfactant producing attributes of selected strain
using genetic engineering and strain improvement using random and genetic
modification.

KEYWORDS

- **biosurfactant**
- **microorganisms**
- **bioremediation**
- **ecorestoration**
- **genetic modification**

REFERENCES

Abouseoud, M.; Maachi, R.; Amrane, A.; Boudergua, S.; Nabi, A. Evaluation of different
    carbon and nitrogen sources in production of biosurfactant by *Pseudomonas fluorescens*.
    Desalination **2008**, *223* (1–3), 143–151.
Adetunji, C.O.; Kumar, J.; Swaranjit, A.; Akpor, B. Synergetic effect of rhamnolipid
    from *Pseudomonas aeruginosa* C1501 and phytotoxic metabolite from *Lasiodiplodia
    pseudotheobromae* C1136 on *Amaranthus hybridus* L. and *Echinochloa crus-galli*
    weeds. *Environ. Sci. Pollut. Res.* **2017a**, *24*(15), 13700–13709. http://dx.doi.org/10.1007/
    s11356-017-8983-8.

Adetunji, C.O.; Oloke, J.K.; Pradeep, M.; Jolly R.S.; Anil, K.S.; Swaranjit, S.C.; Bello, OM. Characterization and optimization of a rhamnolipid from *Pseudomonas aeruginosa* C1501 with novel biosurfactant activities. *Sus. Chem. Pharm.* **2017b**, *6*, 26–36.

Akbari, S.; NourAbdurahman, H.; Yunus, R.M.; Fayaz, F.; Alara, O.R. Biosurfactants-a new frontier for social and environmental safety: a mini review. *Biotech. Res. Inno.* **2018**, *20*, 10.

Al-Araji, L.; Rahman, R.N.Z.; Basri, M.; Salleh, AB. Minireview: microbial surfactant. Asia Pacific J/*Mol. Biol. Biotech.* **2007**, *15*(3), 99–105.

Anani, O.A.; Olomukoro, J.O. The Evaluation of heavy metal load in benthic sediment using some pollution indices in Ossiomo River, Benin City, Nigeria. *Funai J. Sci. Technol.* **2017**, *3*(2), 103–119.

Anani, O.A.; Olomukoro, J.O. *Trace metal residues in a tropical Watercourse sediment in Nigeria: Health risk implications.* IOP Conference Series: Earth Environ. Sci. **2018b**, *210*, 012005.

Anani, O.A.; Olomukoro, J.O. Health risk from the consumption of freshwater prawn and crab exposed to heavy metals in a tropical river, Southern Nigeria. *J. Heavy Metal Tox. Dis.* **2018a**, *3*, 2, 5. doi: 10.21767/2473-6457.10024.

Arino, S.; Marchai, R.; Van Decasteel, JP. Identification and production of a *rhamnolipidic*biosurfactant by a *Pseudomonas* Species. *Appl. Microbiol. Biotechnol.* **1996**, *45*, 162–168.

Arora, A.; Cameotra, S.S.; Kumar, R.; Balomajumder, C.; Singh, A.K.; Santhakumari, B.; Kumar, P.; Laik, S. Biosurfactant as a promoter of methane hydrate formation: thermodynamic and kinetic studies. *Sci. Rep.* **2016**, *6* ,1–10. http://dx.doi.org/10.1038/srep20893.

Bustamante, M.; Durán, N.; Diez, MC. Biosurfactants are useful tools for the bioremediation of contaminated soil: a review. *J. Soil Sci. Plt .Nutr.* **2012**, *12*(4), 667–687.

Cameotra, S.S.; Makkar, RS.Recent applications of biosurfactants as biological and immunological molecules. *Curr. Opin. Microbiol.* **2004**, *7*, 262–266.

Cameotra, S.S.; Makkar, R.S. Biosurfactant-enhanced bioremediation of hydrophobic pollutants. *Pure Appl. Chem.* **2010**, *82*(1), 97–116. doi:10.1351/PAC-CON-09-02-10.

Ciancio, A.; Mukerji, K.G. General concepts in integrated pest and disease management. *Springer.* **2007**, *295*–304.doi: 10.1007/978-1-4020-6061-8.

Claudia I.S.M.; de Lourdes Ballinas-Casarrubias, M.; Blanca E.; Rivera-Chavira, Nevárez-Moorillón, GV.*Biosurfactants as Useful Tools in Bioremediation. Adv. Bioremed. Wastewater Polluted Soil.* **2015**, *5*, 93–108. http://dx.doi.org/10.5772/60751.

Darne, G.; De Almeida, Rita de Cássia, F.; Da Silva, S,; Juliana, M.; Luna, Raquel, D.; Rufino, Valdemir, A.; Santos, Ibrahim, M.; Banat, Leonie, A. Sarubbo. Biosurfactants: promising molecules for petroleum biotechnology advances. *Front. Microbiol.* **2016**, *7*, 1718.

Das, N.; Chandran, P. Microbial Degradation of Petroleum Hydrocarbon Contaminants: An Overview. SAGE-Hindawi. *Biotechnol. Res. Intern.* **2011**,*13*. doi:10.4061/2011/941810.

Das, P., Mukherjee, S.; Sen, R. Genetic regulations of the biosynthesis of microbial surfactants: an overview. *Biotechnol. Genetic Engin Rev.* **2008**, *25*(1), 165–186.doi: 10.5661/bger-25-

Dhara, P.; Sachdev, Swaranjit, S. Biosurfactants in agriculture. *Cameotra Appl. Microbiol. Biotechnol.* **2013**, *97*, 1005–1016. doi 10.1007/s00253-012-4641-8.

Fontes, G.C.; Ramos, N.M.; Amaral, P.F.F.; Nele, M.; Coelho, M.A.Z. Renewable resources for biosurfactant production by Yarrowiali polytica. *Braz. J. Chem. Eng.* **2012**, *29*(3), 483–493.

Hélvia, W.C.; Araújo, Rosileide, F.S.; Andrade, Montero-Rodríguez, D.; Rubio-Ribeaux, D.; Carlos A.; da Silva, A.; Galba. M.; Campos-Takaki. Sustainable biosurfactant produced by Serratiamarcescens UCP 1549 and its suitability for agricultural and marine bioremediation applications. *Microb. Cell. Fact.* **2019**, *18*, 2. doi: 10.1186/s12934-018-1046-0.

Hussain, S.; Siddique, T.; Arshad, M., Saleem, M. Bioremediation and phytoremediation of pesticides: recent advances. *Crit. Rev. Environ. Sci. Tech.* **2009**, *39*, 843–907.doi: 10.1080/10643380801910090.

Ibukun, O.; Olasanmi, Ronald, W.; Thring. The role of biosurfactants in the continued drive for environmental sustainability. *Sustainability.* **2018**, *10*, 4817. doi:10.3390/su10124817 www.mdpi.com/journal/sustainability.

Ji, F.; Li, L.; Ma, S.; Wang, J.; Bao, Y. Production of rhamnolipids with a high specificity by *Pseudomonas aeruginosa* M408 isolated from petroleum-contaminated soil using olive oil as sole carbon source. *Ann. Microbiol.* **2016**, *66*, 1145. http://dx.doi.org/10.1007/s13213-016-1203-9.

Karanth, N.G.K.; Deo, P.G.; Veenanadig, NK. Microbial production of biosurfactants and their importance. *Curr. Sci.* **1999**, *77*(1), 116-126.

Kosaric, N. Biosurfactants and their application for soil bioremediation. *Food Techn. Biotech.* **2001**, *39*(4), 295-304.

Mouafo, T.H.; Mbawala, A.; Ndjouenkeu, R. Effect of different carbon sources on biosurfactants' production by three strains of *Lactobacillus* spp. *Hindawi BioMed. Res. Intern.* **2018**, *15*. https://doi.org/10.1155/2018/5034783.

Mulligan, CN. Environmental applications for biosurfactants. *Environ. Pollut.* **2005**, *33*(2), 183–198.

Nitschke, M.; Siddhartha, G.V.A.; Costa, O.; Contiero, J. Rhamnolipid surfactants: an update on the general aspects of these remarkable biomolecules. *Biotechnol.Prog.* **2005**, *21*, 1593-1600.

Pacwa-Plociniczak M., Płaza G.A., Piotrowska-Seget Z., Cameotra S.S. Environmental applications of biosurfactants: recent advances. *Int. J. Mol. Sci.* **2012**,*12*,633-654.

Pirôllo, M.P.S.; Mariano, A.P.; Lovaglio, R.B.; Costa, S.G.V.; Walter, A.O.; Hausmann, R.; Contiero, J. Biosurfactant synthesis by *Pseudomonas aeruginosa* LBI isolated from a hydrocarbon-contaminated site. *J. Appl. Microbiol.* **2008**, *105*(5), 1484.

Randhir, S.; Makkar, Swaranjit, S.; Cameotra, Ibrahim, M.; Banat. Advances in utilization of renewable substrates for biosurfactant production. *AMB Express.* **2011**, *1*, 5. doi: 10.1186/2191-0855-1-5.

Rashedi, H.; amshidi, E.; Mazaheri Assadi, M.; Bonakdarpour, B. Isolation and production of biosurfactant from *Pseudomonas aeruginosa* isolated from Iranian southern wells oil. *Int. J. Environ. Sci. Tech.* **2005**, *2*(2), 121-127.

Rufino, R.D.; de Luna J.M.; de Campos Takaki G.M.; Sarubbo, LA. Characterization and properties of the biosurfactant produced by *Candidali polytica* UCP 0988. *Electron. J. Biotech.* **2014**,*17*, 34-38.

Sáenz-Marta, C.I.; de Lourdes Ballinas-Casarrubias, M.; Blanca, E.; Rivera-Chavira, Nevárez-Moorillón, GV. Biosurfactants as Useful Tools in Bioremediation. *Advances in Bioremediation of Wastewater Polluted Soil.* **2015**, 1–19. http://dx.doi.org/10.5772/6075.

Santos da Silva, A.C.; dos Santos PPN.; Silva, T.A.; Andrade, R.F.S.; Campos-Takaki, GM.Biosurfactant production by fungi as a sustainable alternative.*rq. Inst. Biol.* **2018**, 85. http://dx.doi.org/10.1590/1808-1657000502017.

Santos, A.P.P.; Silva, M.D.S.;   Costa, E.V.L.; Rufino, R.D.; Santos, V.A.; Ramos, C.S.; Sarubbo, L.A.; Porto, ALF. Production and characterization of a biosurfactant produced by Streptomyces sp. DPUA 1559 isolated from lichens of the Amazon region. *Brazil. J. Med. Biol. Res.* **2018**, *51*(2), 6657. doi: 10.1590/1414-431X20176657.

Singh, P.; Patil, Y.; Rale, V. Biosurfactant production: emerging trends and promising strategies. *J. Appl. Microbiol.* **2018**, *126*, 2–13.

Soberón-Chavez, G.; Lépine, F.; Déziel, E. Production of rhamnolipids by *Pseudomonas aeruginosa*. *Appl. Microbiol. Biotechnol.* **2005**, *68*, 718–725.

Usman, M.M.; Dadrasnia, A.; Lim, K.T.; Mahmud, A.F.; Ismail, S. Application of biosurfactants in environmental biotechnology; remediation of oil and heavy metal. *AIMS Bioeng.* **2016**, *3*(3), 289–304. doi: 10.3934/bioeng.2016.3.289.

Zhang, G.L.; Wu, Y.T.; Qian, X.P.; Meng, Q. Biodegradation of crude oil by *Pseudomonasaeruginosa* in the presence of *rhamnolipids*. *J. Zhejiang Univ. Sci. B.* **2005**, *6*(8), 725–730.

CHAPTER 14

Polymeric Nanocomposite for Nanoremediation: Laboratory to Land Approach

DIVYA CHAUHAN,[1] NEETU TALREJA,[2] MOHAMMAD ASHFAQ,[2, 3*] ADRIANA C MERA,[2] and CARLOS A. RODRÍGUEZ[2]

[1]Department of Chemical and Biomedical Engineering, University of South Florida, Tampa, FL, USA

[2]Multidisciplinary Research Institute for Science and Technology, IIMCT, University of La Serena, Benavente, La Serena, Chile

[3]School of Life Sciences, BS Abdur Rahman Crescent Institute of Science and Technology, Chennai, India

*Corresponding author. E-mail: mohdashfaqbiotech@gmail.com; ashfaqm.sls@crescent.education

ABSTRACT

The continuously increasing of contaminants such as heavy metal, pesticides, antibiotics, and biological contaminants is a major concern nowadays that affect water bodies, plants, animal, and life of human being. Nanotechnologies have a potential ability to remove contaminants from water bodies, as nanomaterials are an excellent class of adsorbents. The incorporation of polymers within the nanomaterials might enhance the removal efficiency of contaminants. Moreover, polymeric composite is easily tuned with the functional group that significantly enhances the removal ability of the polymeric nanocomposite materials and also selectivity of contaminants. In this chapter, we discuss nanomaterials, polymeric composite, and the removal process of contaminations, nanomaterials for the removal of contaminations, carbon-based materials for the removal of contaminations, and polymeric composite for the removal of contaminations (heavy metal ions, pharmaceuticals, dyes, and pesticides). We also discuss about the toxicological aspects of nanoma-terials/polymeric nanocomposite. Finally, we conclude and future prospects of nanomaterials/polymeric composite materials for environmental remedia-tion application.

14.1 INTRODUCTION

The environmental contamination (heavy metal, pesticides, antibiotics, and biological contaminants) is a major concern nowadays because of its effect on water bodies, plants, animal, and life of human being. Significant research has been done on the exposure of contamination like pesticides (Salazar et al., 2018), heavy metals (Mahmud et al., 2017), and antibiotics (Pham et al., 2018) in the environment, as this caused significant damage in ecosystems by poisoning drinking water (Khan et al., 2016), air quality, and nutritional contents of crops eventually over human and animal health. Water contamination mainly occurs by variety of contaminants (like organic and inorganic) wastes such as discharge by industries, increase in consump-tion of chemicals, volcanic activity, vegetation fires. Various industries such as leather, pulp, paper, pharmaceutical, etc., contribute to environmental pollution. The sources of anthropogenic persistent pollutants are uses of chemicals, heavy metal ions, pollutants (organic), and pesticides. The organic contaminants mainly polychlorinated biphenyls, dioxins, pesticides (organo-chlorine), polycyclic aromatic hydrocarbons, and dibenzofurans are the main contaminants that bring about the persistent organic pollutant contamination in the environment (Jones and Voogt, 1999), due to their nonbiodegradable nature (Gaur and Narasimhulu, 2018). Moreover, these compounds are still remaining a major concern due to their mobility, tendency of bioaccumula-tion, and tenacity nature. However, these compounds enter to the food chain and accumulated into the fatty tissue of human, thereby enhancing the risk of antagonistic effects on the environment as well as on human health.

Presently, the treatment of contaminants (water and air contamination) is one of the greatest challenges nowadays. In this context, nanotechnology is supposed to be a newer approach for environmental removal technologies as nanostructured materials are an exceptional class of adsorbent, sensor, and catalyst materials (Savage and Diallo, 2005; Saputra et al., 2013; Faisal et al., 2014; Zou et al., 2016; Das et al., 2017; Afreen et al., 2018; Mishra et al., 2018). Nanomaterials possess excellent properties because of its small size such as higher specific surface area, high surface-to-mass ratio with a controlled pore size distribution, and high reactivity that might improve the adsorption capabilities for a specific pollutant (Khin et al., 2012) due to these excellent properties nanomaterials are applied as an excellent adsorbent for environmental remediation. In spite of several properties, there are some drawbacks that limit their application on an industrial scale as smaller particle-size of nanoparticles (NPs), (1) excessive pressure drops (applied in a fixed bed or dynamic flow condition), (2) hard to separate and re-use, and (3) risk associated with the ecosystem and human health because of the release of NPs within the environment. Therefore, polymeric support might overcome such issues. In this context, hybrid nanocomposite materials have a potential ability and are synthesized by impregnation or coated NPs within the polymers. Moreover, the polymeric matrix might be enhancing the sorption ability of NPs by using surface functionalization that enhances specific interaction with pollutants/surface sorption ability. The polymeric nanocomposite materials are synthesized by polymer–polymer hybrid, and polymer–inorganic hybrid materials (NPs with polymeric support). Another advantage of using polymers in polymeric–inorganic hybrid materials is that they maintain the intrinsic properties of NPs. Moreover, polymers have various advantages such as they provide higher stability and processability. In addition, the encapsulation of NPs within the polymeric nanocomposites leads to various characteristics such as mechanical, electrical, and optical (Zhao et al., 2011, 2018; Khin et al., 2012).

The fabrication of polymer–NPs-composite-based membranes is done by using impregnation of NPs onto porous membrane/blending of NPs with polymers. Polymeric membrane incorporated with NPs provides exceptional efficacy in the separation of pollutants, eco-friendly, and economically viable in comparison with that of other techniques mainly separation and purification processes that required high temperature. However, polymeric-membranes are damaged by biofouling and fouling (organic/inorganic) that reduces the shelf life of the membrane. Usually, membrane fouling is caused by the interaction between the surface of membrane and foulants that contain various types of substances (organic, inorganic, and biological) (Khorshidi et al., 2018; Saraswathi et al., 2019). The fabrication of membranes using polymer matrix with NPs might enhance the permeability as well as the fouling resistance. This chapter focuses on polymer–nanomaterial-based composites for environmental remediation application. The advantage of polymer support on NPs can increase the adsorption properties as well as their applicability on an industrial scale will be discussed for remediation application.

14.2   REMOVAL PROCESS OF CONTAMINATION

The supply of clean or pure drinking water has emerged one of the most severe issues nowadays, globally. The water treatment technologies usually associated with a series of processes that designed to remove single or multiple substances from water on the basis of molecular size and characteristic of the target pollutant.

The availability of clean water has emerged as one of the most serious problems facing the global economy in the 21st century. Water treatment systems typically involve a series of coupled processes, each designed to remove one or more different substances in the source water, with the particular treatment process being based on the molecular size and properties of the target contaminants. There are several processes that can be used for wastewater remediation. These processes are as follows: (1) adsorption, (2) catalytic processes, (3) membrane processes, (4) electrochemical oxidation/reduction, (5) ionizing-radiation approach, and (6) magnetically assisted approach. Figure 14.1 shows the different removal processes of contaminants. In brief, adsorption and membrane filtration are discussed in detailed.

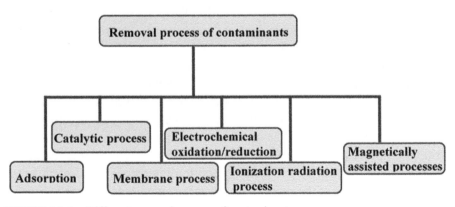

FIGURE 14.1   Different removal process of contaminants.

14.2.1   ADSORPTION PROCESS

Adsorption is the main surface phenomenon, which takes place from the solution to the solid phase. The adsorption process is mainly divided into two categories: (1) physiosorption and (2) chemisorption that cover both adsorption and precipitation reactions. Figure 14.2 shows the types of the adsorption process. Fundamentally, adsorption is a mass transfer process in which a solute molecule is transferred from the liquid phase to the adsorbent surface by various physical or chemical interactions. Adsorption of contaminant from liquid to solid phase involves three major steps: (1) transport of pollutant from the bulk phase to the surface of the adsorbent, (2) pollutant adsorption on the surface of adsorbent with the pores, and (3) transport into the surface of the adsorbent.

Recently, several low-cost adsorbent materials have been developed and reported in the literature such as activated carbon derived from waste of agriculture, by-products from industries, natural materials, modified/functionalized biopolymers, NPs, polymer–composites, etc., have been synthesized that are applied in the removal of various types of contaminants from wastewater. For industrial application, still there is research needed to synthesis cost-effective adsorbent that can be effectively applied to treat wastewater.

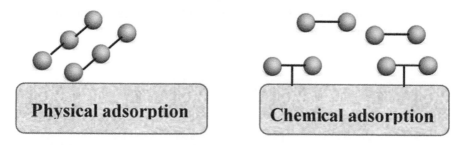

FIGURE 14.2   Schematically representation of types of adsorption process.

14.2.2   MEMBRANE SEPARATION

The membrane-separation technologies are extensively used for the treatment of wastewater as it works without producing any by-products. Polymer composites are mainly used for membrane-filtration process. The membrane-separation works with the principle to apply semipermeable membrane that should be permeable to water and nonpermeable to solute for the filtration of fluids, gases, particles, and solutes. Various pressure-driven membrane processes are available for the rapid separation such as microfiltration, ultrafiltration, nanofiltration (NF), and reverse osmosis (RO) process that applied for water treatment process, globally.

The NF process is mostly applied for the separation of low-molecular-weight contaminants that include salts, sugar-based compounds (glucose and lactose), micropollutants, etc., in polluted water. Recently, cellulose acetate (Ghaee et al., 2015) and polyamide (Saha and Joshi, 2009) are widely used for the synthesis of NF membranes. Moreover, other polymers, for example, polyacrylonitrile (Fritsch et al., 2012), polyvinyl alcohol (PVA; Jegal and Lee, 1999), and inorganic materials (metal and metal-oxides; Lee et al., 2007) might be applied for the synthesis of NF membranes. The advantage of NF membranes technology enables selective rejection of contaminants and also maintains the quality of water, thereby NF membrane is almost similar to the RO process (Petersen, 1993; Esteban-Fernández de Ávila et al., 2015). The membrane technology also works with two modes of sorption process: (1) physical and (2) chemical adsorption. The physical process is a reversible process, whereas chemical is irreversible due to strong chemical bonding (covalent) and polymerization, while reversible process occurs through weak chemical bonds (hydrogen bonds and complexation or both). The NF membranes work in a single step and are able to remove hardness, natural organic matters, particles, and a number of organic and inorganic substances (Figure 14.3).

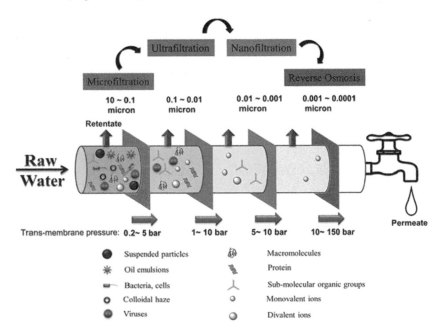

FIGURE 14.3   Membrane process for water remediation technology.

Source: Reprinted with permission from Liao et al., 2018. © Elsevier.

14.3 NANOMATERIALS FOR THE REMOVAL OF CONTAMINANTS

The unique properties of metal oxides and metal nanoparticles are the most widely used materials for environmental remediation applications. Metal oxides and NPs can be synthesized in various shapes and sizes based on the process and reaction time and other process parameters. Common metal NPs used for environmental applications are Fe, Al, Au, Ag, TiO_2, Al_2O_3, Fe_2O_3, etc. Metal oxides and NPs are synthesized by co-precipitation, thermal decomposition, reduction, and hydrothermal process. These methods are extensively used and can be simply industrialized. Fe is the most common metal used for environmental remediation technologies as it can easily form complex with various pollutants such as As, Cr, etc. Fe-loaded adsorbents, such as sulfide, oxy-hydroxides, and aluminosilicate, have been effectively used in the reduction and precipitation of metal contaminants. Several literatures report the advantage of using Fe for water treatment. For example. Niu et al. (2005) synthesized zero-valent Fe-NPs for Cr removal and observed that 100% of Cr(VI) was degraded at a dose of 0.4 g/L. The adsorption is mainly pH dependent that decrease with increasing the pH of the solution. Another study reported the supported Fe-NPs onto cations exchange resin for Cr removal. They observed that the resin-supported material performs better for Cr(IV) adsorption as compared to unsupported Fe-NPs (Ali et al., 2015). Among all Fe species, zero-valent-Fe (Fe^0) was considered as the most effective for wastewater remediation (Xue et al., 2018). With the introduction of nanotechnology, Fe-NPs/nanomaterials swapped the use of bulky Fe/other metals-based systems for water purification technologies. For example, Zhu et al. demonstrate increase adsorption efficiency of biochar-based material for cadmium (Cd) adsorption after adding zero-valent Fe. Adsorption efficiency was increased to 22.37 and 26.43 mg/g using composites for Cd adsorption (Zhu et al., 2019).

14.3.1 CARBON-BASED NANOMATERIALS FOR THE REMOVAL OF CONTAMINANTS

Carbon-based nanomaterials such as activated carbon, charcoal, carbon nanotubes, activated carbon fibers, carbon nanofibers (CNFs), graphene, and fullerenes have been extensively used in the environmental application. Among all of them, CNFs are relatively newer materials that incorporate with metal NPs. The incorporation of metal NPs within the CNFs makes an exceptional candidate for various end applications such as environmental remediation, lithium-ion battery, antibiotic removal, vitamin removal, calorimetry sensor, electrode material, agricultural, wound dressing, antibiotic material, drug delivery system, sensor, and microbial fuel cell (Saraswat et al., 2012; Sharma et al., 2013; Singh et al., 2013; Ashfaq et al., 2014, 2016; Sankararamakrishnan et al., 2016; Ashfaq et al., 2017a, 2017b, 2018; Bhadauriya et al., 2018; Kumar et al., 2018; Omar et al., 2019)

Talreja et al. (2014) synthesized Fe-CNFs on porous carbon beads for the removal of Cr from aqueous solution. The data suggested that the Fe-CNFs on beads efficiently remove Cr from water with ~41 mg/g adsorption capacity. The produced Fe-CNFs on beads also effectively removes Cr in a continuous flow system (Talreja et al., 2014). Another study (Talreja et al., 2016) synthesized the ethylene-diamine functionalized CNFs effectively used for the removal of salicylic acid from the water system. The data suggested that the surface functionalization of CNFs favor adsorption ability. The higher adsorption ability ~682 mg/g was observed. Moreover, electrostatic interaction and hydrogen bonding might be enhancing adsorption ability because of the functionalization of CNFs surface with ethylene-diamine.

Usually, the removal of contaminants using carbon-based nanomaterials/ nanocomposite mainly depends on several factors; water chemistry (pH of the aqueous solution), the presence of functional groups on the surface of materials, contact time, and extent mixing rate.

14.3.2   REMOVAL OF CONTAMINATIONS USING POLYMERIC NANOCOMPOSITES

14.3.2.1   REMOVAL OF HEAVY-METAL IONS

The most common heavy metals pollutants are chromium (Cr), copper (Cu), arsenic (As), and lead (Pb), globally, which are mostly discharged by industries mainly leather, textile, fertilizer, metallurgy, synthesis of pigment and dye, and electroplating process. Several polymeric composites and noncomposites-based materials are used for the removal of heavy metal pollutants from water bodies. It is reported in most of the literature that the polymer nanocomposites are far better in removal of heavy metal as compared to NPs, distinctly. For example, Kumar et al. fabricated a chitosan-magnetite nanocomposite strip for Cr removal. They observe the effect of Fe_2O_3 on chitosan and found that Cr removal efficiency (92.33%) of chitosan-magnetite nanocomposite strip compared to chitosan strip (29.39%), which is far lesser than polymer composite (Sureshkumar et al., 2016). Another study by Ali et al. suggested that the synthesis of chitosan-1,2-cyclohexylenedinitrilotetraacetic acid graphene oxide (Cs/CDTA/GO) nanocomposite for Cr removal with maximum adsorption capacity ~166.98 mg/g of the adsorbent was observed (Ali et al., 2018). Yan et al. synthesis Fe_3O_4, poly(acrylic acid) blended with chitosan by using the co-precipitation process for Cu(II) adsorption, which was further applied for phosphate removal (Yan et al., 2012). Savina et al. (2011) reported iron (Fe) NPs are embedded into the walls of a macroporous polymer for the removal of As(III) with a total capacity of up to 3 mg As/g of nanoparticles with minimal iron leaching at pH range 3–9. Kumar et al. (2011) synthesized bi-metallic (Fe-Al) polymeric micro-nano-sized adsorbent for As and fluoride removal from the wastewater. The produce metal-polymeric beads effectively remove both contaminants (arsenic and flouride) from water with loading efficiency 40 and 100 mg/g, respectively.

Polymeric composites with noble metal-NPs (Au and Pt) were also used for water remediation process, for example, Sanchez et al. used Pt^0 and Pd^0 nanoparticles into a poly(pyrrole-alkyl-ammonium) matrix for As(III) and As(V) removal by electrocatalytic oxidation process (Chauhan et al., 2014, 2017). The detection limit was found to be ~0.17 ppm. The use of water-soluble poly(quaternary-ammonium) salt acted as supporting electrolyte as well as an extracting agent that allows us to efficiently remove the electro-catalytically generated As(V) species by using liquid-phase polymer-assisted retention technique (Sánchez et al., 2010). Another study by Huang et al. report the application of polyethylene-imine/PVA nanofibers based mat embedded with Pd-NPs for Cr removal. The polymeric support provides uniform distribution of metal-NPs with high mechanical stability (Huang et al., 2012). Therefore, these studies suggested that the inclusion of polymer over metal nanoparticles can efficiently increase the removal efficiency as compared to their NPs, improved NPs stability, high sorption ability, and excellent reusability.

14.3.2.2   REMOVAL OF DYES

The dye molecules are mainly discharged from paper, textile, paint, and pigment industry. However, dyes have been renowned as solemn contaminants (Adeyemo et al., 2017). As dye molecules have their conjugated electron structure, thereby stable organic compounds and their removal from water is quite a complex process. The dye molecules remain intact in aqueous solution for longer duration that affects the natural photosynthetic process (Adeyemo et al., 2017). The product formed from the degradation of dye molecules is considered to be toxic and carcinogenic in nature (Sen et al., 2018). Several polymeric composites have been used for the adsorption or removal of dyes from aqueous solution. The recent study focuses on the synthesis of polyethylene glycol and Ag–ZnO grafted polyaniline-based nanocomposites for the removal of brilliant green dye with an adsorption capacity of 94.46 mg/g, attributed high removal efficiency of polymeric nanocomposite at both acidic and basic medium for dye (Zhang et al., 2019). Ghorai et al. (2014) synthesized polyacrylamide grafted onto xanthan gum incorporated with nanosilica by sol–gel process for methylene blue (MB) and methyl violet (MV) adsorption. The removal efficiency of nanocomposite for dye molecules was 99.4% for MB and 99.1% for MV was reorganized on the basis of H-bonding interactions, dipole–dipole interaction, and electrostatic interactions between anionic (adsorbent) and cationic (dye molecules) with exceptional regeneration efficacy. Another study by Pal et al. (2012) suggested that the MB adsorption onto polyacrylamide-grafted-carboxy-methyl-tamarind and SiO_2-NPs with the adsorption loading of 43.859 mg/g is due to high hydrodynamic radius of synthesized nanocomposite. Magnetite NPs grafted with polyacrylic acid hydrogels were also used for MB removal. The maximum efficiency for dye removal was found to be 507.7 mg/g over the 10% weight loading of magnetic hydrogel nanocomposites (Pooresmaeil et al., 2018). Based on these research articles, it can be concluded that polymeric nanocomposites material shows good removal efficiency as compared to their individual matrix.

14.3.2.3 REMOVAL OF PHARMACEUTICALS COMPOUNDS

The pharmaceuticals such as ampicillian, amoxicillian, bezafibrate, salicylic acid, clofibric acid, sulfamethoxazole, iopromide, paracetamol, triclosan, carbamazepine, and diclofenac are the group of compounds that are used to sustain hygiene and health such as medicine, cleaning products, fragrances, and bacterial protection purpose (Ternes et al., 2002). These compounds entered in an aquatic chain via industries, hospitals, and through human excretion as metabolites. Several NPs, hybrid materials, and composites have been synthesized to remove these pharmaceuticals from the environment. Among the polymeric composites can efficiently remove these pharmaceuticals from wastewater. For example, Arya et al. (2016) synthesized magnetic polymer clay composite for atenolol, ciprofloxacin, and gemfibrozil removal having a maximum loading capacity of 15.6, 39.1, and 24.8 mg/g, respectively. The electrostatic interaction was considered as a driving force for the adsorption of these pharmaceutical compounds. The adsorption performance was mainly affected by pH and humic acid addition into solution (Arya et al., 2016). Zhuo et al. (2017) synthesized polymeric beads of sodium alginite and chitosan for the selective removal of benzoic acid, ibuprofen, and ketoprofen. MIL-101(Cr)/SA) and MIL-101(Cr)/chitosan composite beads showed excellent adsorption for the selective pharmaceuticals due to electrostatic and π–π interaction not only with amine group but also with centered Cr atom.

14.3.2.4 REMOVAL OF PESTICIDES

The pesticides can directly affect the crop production and protection, their impact on the environment is quite significant. Pesticides removal using polymer nanocomposites were reported by several literatures (Dehaghi et al., 2014), synthesis of chitosan–ZnO NPs-based nanocomposite beads for the removal of permethrin pesticide from the water system. The results show that the high removal efficiency of 99% was observed by using chitosan–ZnO NPs-based nanocomposite beads (Salazar et al., 2018) synthesized β-cyclodextrin decorated with Fe_3O_4-NPs-based materials for the removal of aromatic chlorinated pesticides from water. Another study (El-Said et al., 2018) synthesized mesoporus-silica-polyaniline-based composite for the removal of chloridazen pesticides from water. The study suggested that the removal efficiency increased with the functionalization of polyaniline onto the silica.

In general, the aforementioned studies suggested that the removal efficiency was effectively increased with the incorporation of polymeric composite within the nanomaterials. Moreover, polymeric composite is easily tuned with the functional group that significantly enhances the removal ability of the polymeric nanocomposite materials.

14.4 TOXICOLOGICAL ASPECTS

The nanomaterials/polymeric composite showed tremendous success in various end applications especially in environmental remediation. Upon exposure of these nanomaterials/polymeric composite there might be leakage to the environment that causes severe detrimental effects of plants, animal, and human being (Voigt et al., 2014; Chervenkov et al., 2007). Various studies suggested that there are no adverse effects of nanomaterials/polymeric composite in the environment. Moreover, a lot of studies show that the uses of nanomaterials/polymeric composite have adverse effects (Han et al., 2018). There is a need of more/deep study of nanomaterials/polymeric composite with different toxicological aspects, as the toxicity of nanomaterials/polymeric composite remains a concern. It is necessary to understand that nanomaterials/polymeric composite materials used for the removal of contaminants are not another contaminants themselves after the uses of materials. Therefore, biodegradable materials have attracted great interest for environmental remediation applications. The applicability of the biodegradable materials or polymeric composite has various advantages: (1) enhance consumer confidence, (2) no waste material, and (3) provide greener and safer substitutes for environmental remediation. Moreover, newer technologies based polymeric composite have the potential ability because it depends on target-specific capture of contaminants (Ashfaq et al., 2013; Mittal, 2014; Christou et al., 2016).

14.5   CONCLUSION AND FUTURE CHALLENGES TO LABORATORY TO LAND SCALE

The application of nanomaterials/polymeric composite for the removal of heavy metal ions, antibiotics, dyes, and pesticide provides effective removal efficiency. The synthesis of nanomaterials/polymeric composite is simple, and cost-effective. There is no exact mechanism for the removal of contaminants from water by using nanomaterials/polymeric composite. The mechanism is different for different nanomaterials/polymeric composite materials and different contaminants, as removal of contaminants using materials mainly rely on several factors:

(1) water chemistry (pH of the aqueous solution),
(2) the presence of functional groups on the material surface,
(3) contact time, and
(4) extent mixing rate.

The studies suggested that the removal efficiency was effectively increased with the incorporation of polymeric composite within the nanomaterials. Moreover, polymeric composite is easily tuned with the functional group that significantly enhances the removal ability of the polymeric nanocomposite materials. Furthermore, there is still a need of improvement in the nanomaterials/polymeric composite materials for selective removing contaminants, resistance to changes in pH, high adsorption/removal efficiency, stability, and re-usability. Despite tremendous success of nanomaterials/polymeric composite materials, the toxicological aspect nanomaterials/polymeric composite still remains a concern. It is necessary to understand that the nanomaterials/polymeric composite materials used for the removal of contaminants are not another contaminants themselves.

ACKNOWLEDGMENT

The authors acknowledge the financial support given by CONICYT through the project Fondecyt number 3190515 (Dr Neetu Talreja) and 3190581 (Dr Mohammad Ashfaq).

KEYWORDS

- adsorption
- environmental remediation
- nanomaterials
- pesticides
- polymeric composite

REFERENCES

Adeyemo, A.A.; Adeoye, I.O.; Bello, O.S. Adsorption of dyes using different types of clay: a review. *Applied Water Science*. **2017**, *7*(2), 543–568.

Afreen, S.; et al. Carbon-based nanostructured materials for energy and environmental remediation applications, in *Approaches in Bioremediation: The New Era of Environmental Microbiology and Nanobiotechnology*. Prasad, R.; Aranda, E. (Eds.). **2018**, Springer International Publishing: Cham, 369–392.

Ali, M.E.A. Synthesis and adsorption properties of chitosan-CDTA-GO nanocomposite for removal of hexavalent chromium from aqueous solutions. *Arabian Journal of Chemistry*. **2018**, *11*(7), 1107–1116.

Ali, S.W.; Mirza, M.L.; Bhatti, T.M. Removal of Cr(VI) using iron nanoparticles supported on porous cation-exchange resin. *Hydrometallurgy*. **2015**, *157*, 82–89.

Arya, V.; Philip, L. Adsorption of pharmaceuticals in water using Fe_3O_4 coated polymer clay composite. *Microporous and Mesoporous Materials*. **2016**, *232*, 273–280.

Ashfaq, M.; et al. Cytotoxic evaluation of the hierarchical web of carbon micronanofibers. *Industrial and Engineering Chemistry Research*. **2013**, *52*(12), 4672–4682.

Ashfaq, M.; Verma, N.; Khan, S. Novel polymeric composite grafted with metal nanoparticle-dispersed CNFs as a chemiresistive non-destructive fruit sensor material. *Materials Chemistry and Physics*. **2018**, *217*, 216–227.

Ashfaq, M.; Verma, N.; Khan, S. Carbon nanofibers as a micronutrient carrier in plants: efficient translocation and controlled release of Cu nanoparticles. *Environmental Science: Nano*. **2017b**, *4*(1),138–148.

Ashfaq, M.; Verma, N.; Khan, S. Copper/zinc bimetal nanoparticles-dispersed carbon nanofibers: a novel potential antibiotic material. *Materials Science and Engineering C: Materials for Biological Applications*. **2016**, *59*, 938–947.

Ashfaq, M.; Verma, N.; Khan, S. Highly effective Cu/Zn-carbon micro/nanofiber-polymer nanocomposite-based wound dressing biomaterial against the *P. aeruginosa* multi- and extensively drug-resistant strains. *Materials Science and Engineering C: Materials for Biological Applications*. **2017a**, *77*, 630–641.

Ashfaq, M.; Verma, N.; Khan, S. Synthesis of PVA-CAP-based biomaterial in situ dispersed with Cu nanoparticles and carbon micro-nanofibers for antibiotic drug delivery applications. *Biochemical Engineering Journal*. **2014**, *90*, 79–89.

Bhadauriya, P.; et al. Synthesis of yeast-immobilized and copper nanoparticle-dispersed carbon nanofiber-based diabetic wound dressing material: simultaneous control of glucose and bacterial infections. *ACS Applied Bio Materials*. **2018**, *1*(2), 246–258.

Chauhan, D.; Dwivedi, J.; Sankararamakrishnan, N. Novel chitosan/PVA/zerovalent iron biopolymeric nanofibers with enhanced arsenic removal applications. *Environmental Science and Pollution Research*. **2014**, *21*(15), 9430–9442.

Chauhan, D.; et al. Synthesis, characterization and application of zinc augmented aminated PAN nanofibers towards decontamination of chemical and biological contaminants. *Journal of Industrial and Engineering Chemistry*. **2017**, *55*, 50–64.

Chervenkov, T.; et al. Toxicity of Polymeric Nanoparticles with Respect to their Application as Drug Carriers. **2007**. Springer: Dordrecht, Netherlands.

Christou, A.; et al. A review of exposure and toxicological aspects of carbon nanotubes, and as additives to fire retardants in polymers. *Critical Reviews in Toxicology*. **2016**, *46*(1), 74–95.

Das, R.; et al. Recent advances in nanomaterials for water protection and monitoring. *Chemical Society Reviews*. **2017**, *46*(22), 6946–7020.

Dehaghi, D.S.; et al. Removal of permethrin pesticide from water by chitosan–zinc oxide nanoparticles composite as an adsorbent. *Journal of Saudi Chemical Society*. **2014**, *18*(4), 348–355.

El-Said, W.A.; et al. Synthesis of mesoporous silica-polymer composite for the chloridazon pesticide removal from aqueous media. *Journal of Environmental Chemical Engineering*. **2018**, *6*(2), 2214–2221.

Esteban-Fernández de Ávila, B.; et al. Dual functional graphene derivative-based electrochemical platforms for detection of the TP53 gene with single nucleotide polymorphism selectivity in biological samples. *Analytical Chemistry*. **2015**, *87*(4), 2290–2298.

Faisal, M.; et al. Development of efficient chemi-sensor and photo-catalyst based on wet-chemically prepared ZnO nanorods for environmental remediation. *Journal of the Taiwan Institute of Chemical Engineers.* **2014**, *45*(5), 2733–2741.

Fritsch, D.; et al. High performance organic solvent nanofiltration membranes: development and thorough testing of thin film composite membranes made of polymers of intrinsic microporosity (PIMs). *Journal of Membrane Science.* **2012**, *401–402*, 222–231.

Gaur, N.; Narasimhulu, K.P.Y. Recent advances in the bio-remediation of persistent organic pollutants and its effect on environment. *Journal of Cleaner Production.* **2018**, *198*, 1602–1631.

Ghaee, A.; et al. Preparation of chitosan/cellulose acetate composite nanofiltration membrane for wastewater treatment. *Desalination and Water Treatment.* **2015**, *57*, 1–8.

Ghorai, S.; et al. Enhanced removal of methylene blue and methyl violet dyes from aqueous solution using a nanocomposite of hydrolyzed polyacrylamide grafted xanthan gum and incorporated nanosilica. *ACS Applied Materials and Interfaces.* **2014**, *6*(7), 4766–4777.

Han, J.; et al. Polymer-based nanomaterials and applications for vaccines and drugs. *Polymers.* **2018**, *10*(1), 31.

Huang, Y.; et al. Efficient catalytic reduction of hexavalent chromium using palladium nanoparticle-immobilized electrospun polymer nanofibers. *ACS Applied Materials and Interfaces.* **2012**, *4*(6), 3054–3061.

Jegal, J.; Lee, K.H. Nanofiltration membranes based on poly(vinyl alcohol) and ionic polymers. *Journal of Applied Polymer Science* **1999**, *72*(13), 1755–1762.

Jones, K.C.; deVoogt, P. Persistent organic pollutants (POPs): state of the science. *Environmental Pollution.* **1999**, *100*(1), 209–221.

Khan, S.A.; et al. Recent development of chitosan nanocomposites for environmental applications. *Recent Patents on Nanotechnology.* **2016** *10*(3), 181–188.

Khin, M.M.; et al. A review on nanomaterials for environmental remediation. *Energy and Environmental Science.* **2012**, *5*(8), 8075–8109.

Khorshidi, B.; et al. Robust fabrication of thin film polyamide-TiO_2 nanocomposite membranes with enhanced thermal stability and anti-biofouling propensity. *Scientific Reports.* **2018**, *8*(1), 784.

Kumar, R.; Ashfaq, M.; Verma, N. Synthesis of novel PVA–starch formulation-supported Cu–Zn nanoparticle carrying carbon nanofibers as a nanofertilizer: controlled release of micronutrients. *Journal of Materials Science.* **2018**, *53*(10), 7150–7164.

Kumar, V.; et al. Development of bi-metal doped micro- and nano multi-functional polymeric adsorbents for the removal of fluoride and arsenic(V) from wastewater. *Desalination.* **2011**, *282*, 27–38.

Lee, S.Y.; et al. Silver nanoparticles immobilized on thin film composite polyamide membrane: characterization, nanofiltration, antifouling properties. *Polymers for Advanced Technologies.* **2007**, *18*(7), 562–568.

Liao, Y.; Loh, C-H; Tian, M.; Wang, R.; Fane, A. G. Progress in electrospun polymeric nanofibrous membranes for water treatment: Fabrication, modification and applications, Progress in Polymer Science, **2018**, *77*, 69–94.

Mahmud, H.N.M.E.; Huq, A.K.O.; Yahya, R. Polymer-based adsorbent for heavy metals removal from aqueous solution. IOP Conference Series: *Materials Science and Engineering.* **2017**, *206*, 012100.

Mishra, P.K.; Kumar, R.; Rai, P.K. Surfactant-free one-pot synthesis of CeO_2, TiO_2 and Ti@ Ce oxide nanoparticles for the ultrafast removal of Cr(VI) from aqueous media. *Nanoscale.* **2018**, *10*(15), 7257–7269.

Mittal, V. Functional polymer nanocomposites with graphene: a review. *Macromolecular Materials and Engineering.* **2014**, *299*(8), 906–931.

Niu, S.F.; et al. Removal of hexavalent chromium from aqueous solution by iron nanoparticles. *Journal of Zhejiang University Science.* **2005**, *6*(10),1022–1027.

Omar, R.A.; et al. Impact of nanomaterials in plant systems, in *Plant Nanobionics: Volume 1, Advances in the Understanding of Nanomaterials Research and Applications*, R. Prasad, (Ed.). **2019**, Springer International Publishing: Cham, 117–140.

Pal, S.; et al. Carboxymethyl tamarind-g-poly(acrylamide)/silica: a high performance hybrid nanocomposite for adsorption of methylene blue dye. *Industrial and Engineering Chemistry Research.* **2012**, *51*(48), 15546–15556.

Petersen, R.J. Composite reverse osmosis and nanofiltration membranes. *Journal of Membrane Science.* **1993**, *83*(1), 81–150.

Pham, T.D.; et al. Adsorption of polyelectrolyte onto nanosilica synthesized from rice husk: characteristics, mechanisms, and application for antibiotic removal. *Polymers.* **2018**, *10*(2), 220.

Pooresmaeil, M.; et al. Efficient removal of methylene blue by novel magnetic hydrogel nanocomposites of poly(acrylic acid). *Advances in Polymer Technology.* **2018**, *37*(1), 262–274.

Saha, N.K.; Joshi, S.V. Performance evaluation of thin film composite polyamide nanofiltration membrane with variation in monomer type. *Journal of Membrane Science.* **2009**, *342*(1), 60–69.

Salazar, S.; et al. Removal of aromatic chlorinated pesticides from aqueous solution using β-cyclodextrin polymers decorated with Fe_3O_4 nanoparticles. *Polymers* **2018**, *10*(9), 1038.

Sánchez, J.A.; et al. Electrocatalytic oxidation of As(III) to As(V) using noble metal–polymer nanocomposites. *Electrochimica Acta.* **2010**, *55*(17), 4876–4882.

Sankararamakrishnan, N.; Chauhan, D.; Dwivedi, J. Synthesis of functionalized carbon nanotubes by floating catalytic chemical vapor deposition method and their sorption behavior toward arsenic. *Chemical Engineering Journal.* **2016**, *284*, 599–608.

Saputra, E.; et al. Different crystallographic one-dimensional MnO_2 nanomaterials and their superior performance in catalytic phenol degradation. *Environmental Science and Technology.* **2013**, *47*(11), 5882–5887.

Saraswat, R.; et al. Development of novel in situ nickel-doped, phenolic resin-based micro– nano-activated carbon adsorbents for the removal of vitamin B-12. *Chemical Engineering Journal.* **2012**, *197*, 250–260.

Saraswathi, M.S.; Nagendran, A.; Rana, D. Tailored polymer nanocomposite membranes based on carbon, metal oxide and silicon nanomaterials: a review. *Journal of Materials Chemistry.* **2019** *7*(15), 8723–8745.

Savage, N.; Diallo, M.S. Nanomaterials and water purification: opportunities and challenges. *Journal of Nanoparticle Research.* **2005**, *7*(4), 331–342.

Savina, I.N.; et al. High efficiency removal of dissolved As(III) using iron nanoparticle-embedded macroporous polymer composites. *Journal of Hazardous Materials.* **2011**, *192*(3), 1002–1008.

Selatile, M.K.; et al. Recent developments in polymeric electrospun nanofibrous membranes for seawater desalination. *RSC Advances.* **2018**, *8*(66), 37915–37938.

Sen, F.; et al. The dye removal from aqueous solution using polymer composite films. *Applied Water Science*. **2018**, *8*(7), 206.

Sharma, A.K.; et al. Preparation of novel carbon microfiber/carbon nanofiber-dispersed polyvinyl alcohol-based nanocomposite material for lithium-ion electrolyte battery separator. *Materials Science and Engineering C: Materials for Biological Applications*. **2013**, *33*(3), 1702–1709.

Singh, S.; et al. Preparation of surfactant-mediated silver and copper nanoparticles dispersed in hierarchical carbon micro-nanofibers for antibacterial applications. *New Biotechnology*. **2013**, *30*(6), 656–665.

Sureshkumar, V.; et al. Fabrication of chitosan–magnetite nanocomposite strip for chromium removal. *Applied Nanoscience*. **2016**, *6*(2), 277–285.

Talreja, N.; Kumar, D.; Verma, N. Carbon bead-supported ethylene diamine-functionalized carbon nanofibers: an efficient adsorbent for salicylic acid. *CLEAN—Soil, Air, Water*. **2016**, *44*(11), 1461–1470.

Talreja, N.; Kumar, D.; Verma, N. Removal of hexavalent chromium from water using Fe-grown carbon nanofibers containing porous carbon microbeads. *Journal of Water Process Engineering*. **2014**, *3*, 34–45.

Ternes, T.A.; et al. Removal of pharmaceuticals during drinking water treatment. *Environmental Science and Technology*. **2002**, *36*(17), 3855–3863.

Voigt, N.; et al. Toxicity of polymeric nanoparticles in vivo and in vitro. *Journal of Nanoparticle Research*. **2014**, *16*(6), 2379.

Xue, W.; et al. Performance and toxicity assessment of nanoscale zero valent iron particles in the remediation of contaminated soil: a review. *Chemosphere*. **2018**, *210*, 1145–1156.

Yan, H.; et al. Preparation of chitosan/poly(acrylic acid) magnetic composite microspheres and applications in the removal of copper(II) ions from aqueous solutions. *Journal of Hazardous Materials*. **2012**, *229–230*, 371–380.

Zare, E.N.; Motahari, A.; Sillanpää, M. Nanoadsorbents based on conducting polymer nanocomposites with main focus on polyaniline and its derivatives for removal of heavy metal ions/dyes: a review. *Environmental Research*. **2018**, *162*, 173–195.

Zhang, D.F.; et al. Combustion synthesis of highly efficient Bi/BiOBr visible light photocatalyst with synergetic effects of oxygen vacancies and surface plasma resonance. *Separation and Purification Technology*. **2019**, *218*, 1–7.

Zhao, G.; et al. Polymer-based nanocomposites for heavy metal ions removal from aqueous solution: a review. *Polymer Chemistry*. **2018**, *9*(26), 3562–3582.

Zhao, X.; et al. Polymer-supported nanocomposites for environmental application: a review. *Chemical Engineering Journal*. **2011**, *170*(2), 381–394.

Zhu, L.; et al. Coupling interaction between porous biochar and nano zero valent iron/nano α-hydroxyl iron oxide improves the remediation efficiency of cadmium in aqueous solution. *Chemosphere*. **2019**, *219*, 493–503.

Zhuo, N.; et al. Adsorption of three selected pharmaceuticals and personal care products (PPCPs) onto MIL-101(Cr)/natural polymer composite beads. *Separation and Purification Technology*. **2017**, *177*, 272–280.

Zou, Y.; et al. Environmental remediation and application of nanoscale zero-valent iron and its composites for the removal of heavy metal ions: a review. *Environmental Science and Technology*, **2016**, *50*(14), 7290–7304.

Index

Printed and bound by CPI Group (UK) Ltd, Croydon, CR0 4YY

23/10/2024

01777701-0012